ENCYCLOPEDIA OF INORGANIC CHEMISTRY

ENCYCLOPEDIA OF INORGANIC CHEMISTRY

Vol. 1

By

Sananda Chatterjee

DAE, DCA, CMCNet

Scientific Consultant

Institute for Natural Sciences

Kolkata

(West Bengal)

D P H

DISCOVERY PUBLISHING HOUSE PVT. LTD.

NEW DELHI-110 002

Published by:
Tilak Wasan
DISCOVERY PUBLISHING HOUSE PVT. LTD.
4383/4A, Ansari Road, Darya Ganj
New Delhi-110 002 (India)
Phone : +91-11-23279245, 43596064-65
Fax : +91-11-23253475
E-mail : parul.wasan@gmail.com
discoverypublishinghouse@gmail.com
web : www.discoverypublishinggroup.com

***First Edition:* 2012**

ISBN: 978-81-8356-874-6 1.

Encyclopedia of Inorganic Chemistry

Printed at:
Shree Balaji Art Press
Delhi

Preface

Chemistry is limitless subject but full of experiments and surprises, which is interesting indeed, especially for a new student, though some mathematical and theoretical calculations are there, in spite of all these, chemical world is world of lure.

The origin of chemistry is very deep-rooted and comes from a very early time, the time of the mediaeval history rather from the time of the Alchemists.

Egyptians, whom are thought to be the father of the alchemistry, used the word "kimiya". Kimiya moved from Egypt to Arabia and it flourished in whole world, in England it was named as "Alchemistry" (in India it was named as "*Ap-Rasayana*"). From this alchemistry our modern chemistry has been developed.

Medicines, drugs, paints etc. are all chemical products. Some processes such as distillation, fractionation, and purification, which we use widely now in modern chemistry are all have been given by the ancestors of chemistry, the Alchemistry.

The Modern Periodic Table developed by Dimitri Ivanovich Mendeleev is the modernized systematic formed of the elements (among them some were discovered by the Alchemists). The original table consists of only 107 elements but in this book all 119 elements have been discussed.

In this modern world of chemistry when I were engaged in my chemical lab, I were thinking that there should be something from which a student of chemistry will easily get all the information of chemistry.

Also a man who is anxious of chemistry, will start loving this subject, so to make the entrance of this chemical world easy and nice, I thought to write a book on chemistry, providing all the necessary information about chemistry [though the inorganic part].

While writing this book, I must first thank to the Supra-cosmic force for giving me the power to write this book, after this I must write the name of that person (My Lord) BABA

(Swami Arupanandaji), without whom the book would have been remained as an ordinary work.

And my respected Fairy Godmother my beloved Mummy (Mrs. Dianna Robinson) who gave me different ideas and thoughts.

While writing this book, some informations was needed and that is obvious so here comes the name of Dr. Paul Anderson, (my friend) Director Research Coordinator IUPAC, without whom the book would not have been finished.

I got an immense help from Dr. Palash Gangopadhyay School of Chemistry Central University, Hyderabad for the drawings and different applications such as CHEMSketch, Origin 6.0 and MathLab.

Also must thank to Mr. Dipak Dey and Mr Dipankar Dey of D.K. Scientific, (C-51 College Street Kol-7) for supplying the instruments and chemicals to me.

I am greatful to my three students Miss Ria Banerjee, Miss Moumita Banerjee and Miss Subhomita Banerjee for taking this colossal effort to publish this book.

And last but not the least my father, mother (for the computer) grand-mother and all other members of my family to finish this entire project.

At the end I must only appeal to my audience that, if you are interested in this book that will be my actual reward, I think and also do believe that, if any person is interested in the experimental chemistry, then from the guidance of A chemical Analyser's guide he or she will be able to set up his or her own chemistry lab). If you find any thing interesting at all then I will find that golden opportunity of sharing the secret of divine happiness and joy.

SANANDA CHATTERJEE

Contents

CHAPTER 1

Man's Scientific Attitude

Introduction

From the dawn of the civilization man is travelling in search of truth, the ultimate truth of knowledge. Man is a social being and has lot of thirst of knowledge, which is to be quenched by accumulating them. The five sense organ of man can do it is accumulation; they are eye, nose, tongue, ear and skin and above all MIND the sixth organ. Through these sense organs when a man observes the nature he always tries to note down the observations.

We all know the story of Sir Isaac Newton's apple falling from a tree, the incident was simple but Sir Newton discovered gravitation. So how did it come? It came from the systematic, logical and truthful observations which was organized by Newton's "special knowledge".

So we can say "Science is the man's special knowledge". From these special knowledge man started conducting some experiments to justify the facts of nature.

Hypotheses were created which became convinced theory and after that Law, but this is no end because a law is a law till convinces a law. When we see some natural phenomenon we have to distinguish.

Their characteristics whether they are physical or chemical, now what is physical and Chemical? *The activities, which are related to those conditions, which are reversible by any easy means known as physical phenomenon, e.g. ice converted to water, water converted to steam.* The chemical phenomenon is such an activity *that depends upon some chemical substance which boosts them to react and these reactions are not reversible in nature so easily, e.g. burning of sugar.*

> Science is man's special knowledge

So the subject that deals with this type of experiments observations and their nature in respect with chemical phenomenon can be termed as CHEMISTRY.

History of Chemistry

To know a subject very well we should try to realize the origin of the subject that it's past, that is its history. To know its history we will have to go a long time back in the past about 3000 B.C. where in ancient Egypt, which was better known at that time by "Land of Black

Soil", or *chemea.* That was because the soil was blackish in colour. The ancient Egyptians use to make potteries from this soil. They termed this art as *chemea.* We are all familiar with the Egyptian Mummies; the medically treated dead bodies. They used to know the process of treating these dead bodies and they also termed this art as *chemea.* They also produced perfumes and paints from different herbs and plants, they tried to produce also an immortal extract, they used to think that if anybody drinks that the person would not die. They also tried to extract many metals from their respective ores. In the Alexandria there was a library, which was destroyed by a big fire, about 3 lakes of books were lost at that time. Some were left behind. After the declination of the Egypt the Arabians took this art of *chemea* with them, there a group of men started to develop this art and named it as the *Alchemea.* These men used to think that if the common metals can be extracted form their ores why not noble metals like gold silver and others. It was believed that Kings of that period used to pay these people lot of money so that they will able to full fill the deficit of gold whenever needed. The period was between A.D. 800 to A.D. 1200.

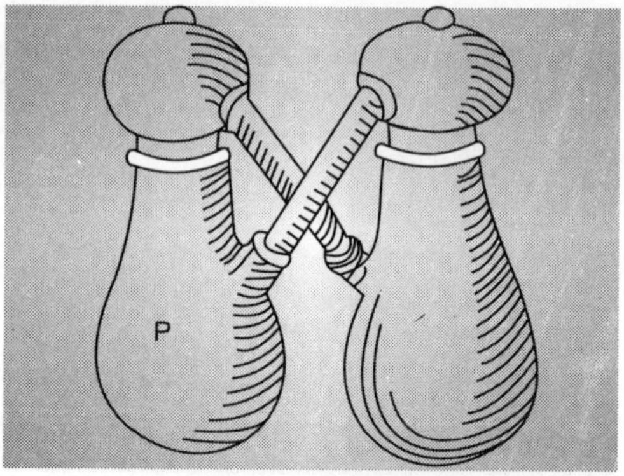

Fig. 1.1. The Alchemical Distillation Apparatus (Aquilae)

JABIR IBAN HYAN (A.D. 800) at that time was a renowned alchemist of Arabia.

Alchemea not only in Arabia it was also started being practised in Europe, in the Spain and in England, where Alchemea was named as Alchemistry. It is believed that our modern chemistry has derived from this ancient alchemistry.

Fig 1.2. A Typical Alchemistry Laboratory

It was not known whether they were successful or not, but they did discovered some useful things now being used in modern chemistry. The one and only solvent called Aqua Regia, Royal water Aqua-forties (strong water) Nitric acid the spirit of Salts hydrochloric acid) and the oil of vitiol (the sulphuric acid.). Alchemistry named as *Ap-Rasayana* when came and practised in India. In Sanskrit *"Rasa"* means the liquid metal mercury; to work with *Rasa* was known as *"rasayan"* and this *Rasayan* is better known as chemistry. Nagarjuna was the well-known Indian alchemist, discovered *"Kadjalli"* a mixture of mercury and sulphur, which became one of the foundation stone in modem medicine, So alchemistry was not a mere subject which we could ignore, due to the immense contribution of Jabir, Roger Beccon, Paracelsus, Nagarjuna and many others.

In 171h century Robert Boyle was the founder of modem chemistry, he said that all the matters in nature could be converted into elements and compounds. After that in 18th century Joseph Priestely (1733 -1804) and Henry Cavendish (1731 -1810) made the subject of modem chemistry a subject of experiments.

Uses of Chemistry

There are hundreds and thousands of uses of chemistry, to discuss about them a separate book necessary. So we will discuss only few of them in this chapter.

Daily life we use various domestic things such as soap, shampoo, and other cosmetics and also medicines, these are all chemical products.

Chemistry has modernized our civilization in many ways. Life saving drugs and other important medicines we use to cure many complicated diseases, the fertilizers we have being used extensively these days to bring about the "green revolution" (to grow more food). The plastics, synthetics, rubber, drugs, fabrics, etc. are also the products of chemistry.

It is being stated that the nation's economy is measured by the yard stick of sulphuric acid that is the development of the nation depends upon the production of sulphuric acid it produces and this sulphuric acid is a chemical product.

Therefore, it is unavoidable that the chemistry has gifted us with useful substances, brought us comfort thus playing an important role in our day to day life (See Fig. 1.3 uses of chemistry). Knowledge of chemistry has helped a lot to mankind to use the raw materials like coal mineral-oil etc. which nature has provided and produced in large number. The useful thing has alloys, insecticides, paints, also fertilizers to boost up the growth of crops. We also use preservatives to keep away anything from the crops, which are nothing but chemical ducts. All these products are strengthening the national economy of our country. Detonators, dynamites, and other explosives, the nuclear explosion in Pokran all are the results of chemical reactions, the uranium oxide used in nuclear bombs can be used as an excellent fertilizer, which makes any type of land a fertile land, so this is the versatile utility of chemistry. Under the next sub-head we shall now learn about the chemical laboratory and the apparatus used in them. We are also going to discuss some of the laboratory rules and regulations; this is necessary because a chemical lab dangerous place, a little careless ness can cause big hazards.

Rules and Regulations, which are to be Observed in the Chemical Lab

To work-in a chemical laboratory one should be very careful; a little sensible manoeuver can gave big hazards. Though there are some dangers which cannot be avoided but can be controlled or minimized by the application of a little care. In this chapter there are some of the common laboratory apparatus have been discussed for those eager students who wants to be a chemist. First of all there are some of the safety rules stated under the common heading called "Always and Never".

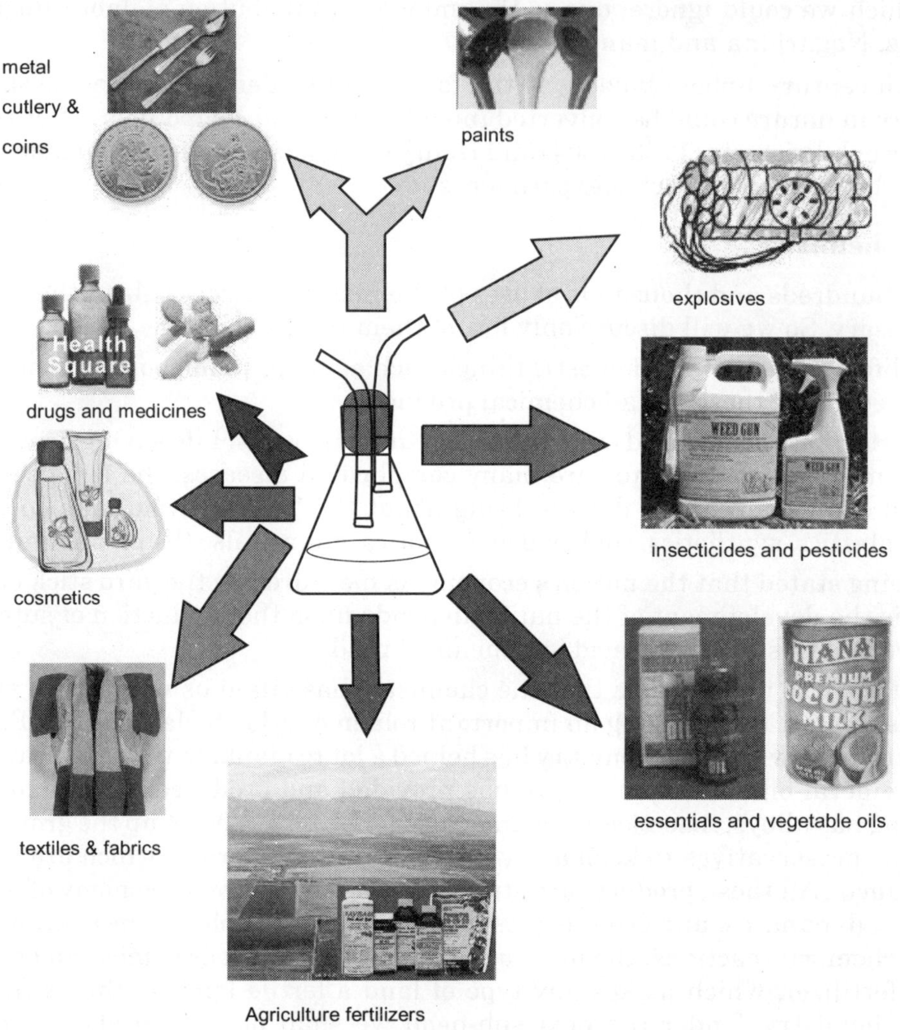

Fig. 1.3. Uses of chemistry

Always

1. Familiarize with laboratory safety procedures.
2. Wear eye protection.
3. Dress sensibly.
4. Wash your hand before leaving the lab.
5. Read the instructions before starting the experiment.
6. Check the experimental set up for correctness before using chemicals.
7. Handle the reagents with utmost care.

8. Keep your working bench clean and tidy.
9. Pour dangerous liquids like bromine and liquid nitrogen using hoods.

Never

1. Eat drink or smoke inside a chemical lab.
2. Directly sniff or inhale the chemicals.
3. Pour water in concentrated acids.
4. Fool around or gossip with other workers or distract them.
5. Run in the lab.
6. Work alone.
7. Carry out unauthorized experiments.
8. Dispose waste here and there.

All the above rules are common sense and needs no further explanation. Indeed if the question is asked what is the common factor that can save the working person from the danger? The answer is "Common Sense".

Common Laboratory Apparatus

We shall now study some of the common laboratory apparatus and their uses, the diagram of some of the apparatus used in chemical lab are shown in the Fig 1.4.

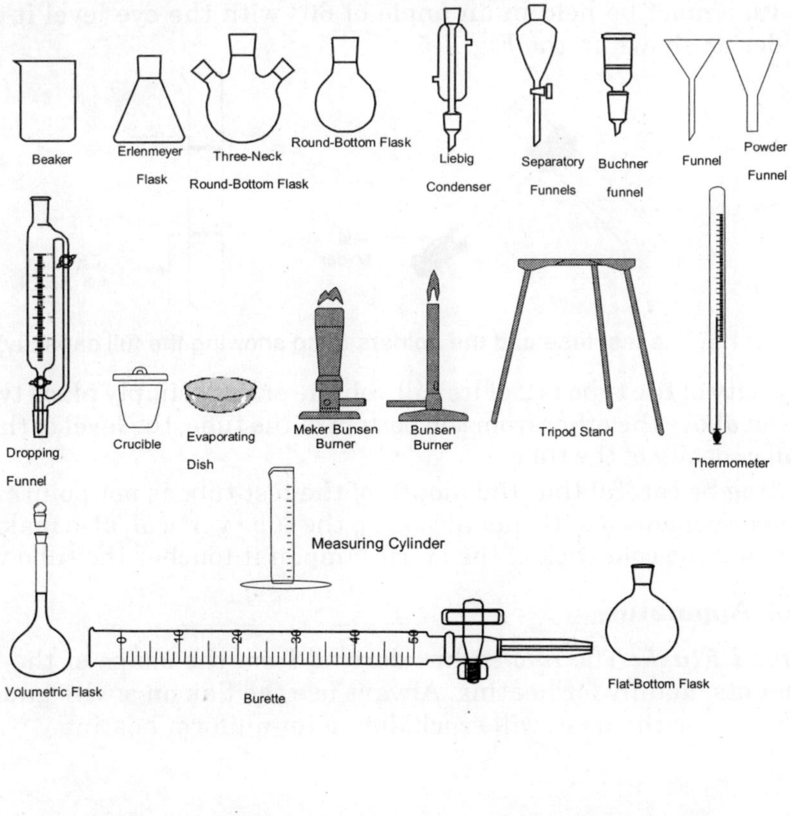

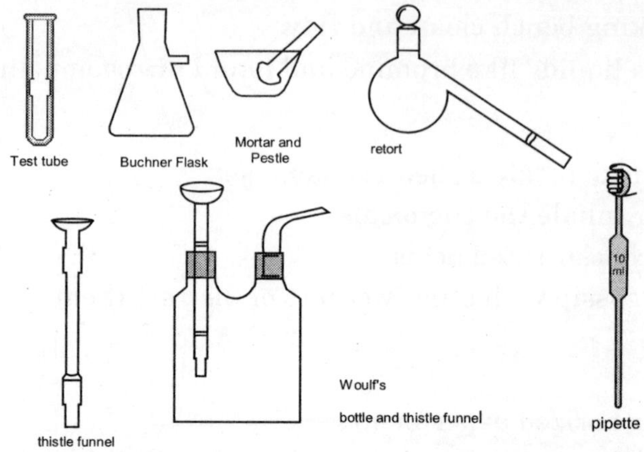

Fig. 1.4 : Some Common Laboratory Apparatus

We are no going to discuss about the uses and their process of handling the apparatus.

Test Tube : They are used for analysis, found in different diameters, e.g $5'' \times \frac{5''}{8}$, $6'' \times \frac{3''}{4}$, $4'' \times \frac{1''}{2}$, $5'' \times \frac{1''}{2}$ etc. They are used for chemical reactions, some of the reactions require heat, for that a special purpose test tubes called boiling tubes are used they are made up of borax glass or Pyrex glass. For heating a test tube must be held in an angle of 60° with the eye level it can be held by hand or by holder as shown in the Fig. 1.5.

> Pyrex is now obsolete, Borosil glass is now available.

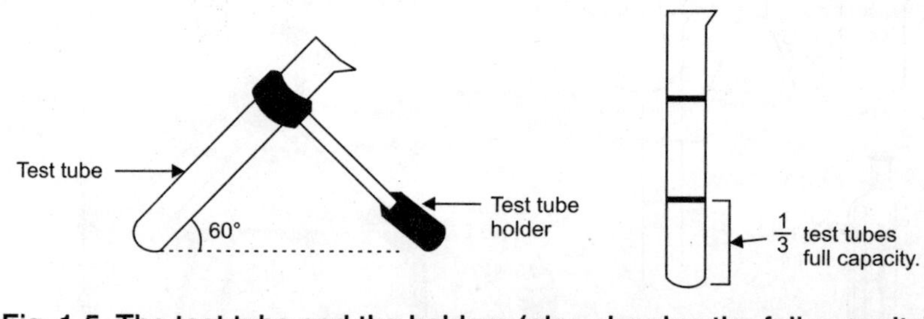

Fig. 1.5. The test tube and the holders (also showing the full capacity)

The full capacity of the tube is 2 of its full volume or very simply place two fingers (fore and the middle on above the other from the bottom of the tube, the level of the finger height will give the full capacity of the tube.

During heating be careful that the mouth of the test tube is not pointing towards you or neighbour or rack chemicals. Do not also keep the tube vertical, also make sure that the tube is not touching the cold wick of the spirit lamp, if it touches the tube will crack.

Description of Apparatus

Round Bottomed Flask: The named has derived from the shape of the flask, used for certain experiments, mainly for heating. Always use the flak on a wire gauge and hold by clamp stand, otherwise the flask will crack due to un-uniform heating.

Conical Flask: This flask has a conical shape, in modem chemical labs the flask is now been frequently used, the use and the process is same as the round bottom flask but it does not require any clamp stand.

Beaker: The name is such because *the vessel has a beak;* the vessel has a projecting lipped out let to pour the liquids safely.

Measuring Cylinder: This is generally a graduated cylinder used for measuring an exact amount of liquid, found in different volumes, 25 ml, 50 ml, 100 ml, 500 ml and 1000 ml.

Funnel: These are generally made up of glass, use for transferring liquids and also solids from one vessel to another, which have, narrow mouths.

Pipettes: These are also measuring apparatus, but can measure only a fixed amount of liquid, the pipette is marked with a white ring near the sucking end according to its capacity. The liquid to be filled is sucked through the other end. The different capacity pipettes are 10 ml, 20 ml 25 ml 50 ml and 100 ml.

Retort: It is a round glass vessel with a long stout outlet fixed with it. It is used for special chemical reactions and experiments.

Clamp Stands: There are various types of stands found with different kinds of rings and clamps for fixing various glass apparatus, always use correct stand for correct apparatus and also use a piece of cotton or cloth for fixing, this minimizes the chance of breaking.

Liebig Condenser: This is a special apparatus named after the inventor Justus Von Liebig (1803-1873). It is used for cooling gases; this apparatus consists an inner capillary tube and an outer glass jacket having two holes one for outlet and another for inlet.

Bunsen Burner: The most important apparatus of the chemical laboratory, invented by Robert Wilhelm Bunsen (1811-1899) we shall be discussing the burner in details.

Spirit Lamp: This is another heating apparatus works on spirit. We shall be discussing later.

Fixing of Stands

There are many clamps and stands used for fixing and holding apparatus for experiments. The mishandling of them can cause severe breakage or destruction; the correct process is shown in the Fig. 1.6.

We shall now see some of the essential apparatus including stands clamps. The diagrams of them are shown below. (See Fig. 1.6a)

Heating

It is often necessary to accelerate a slow reaction by heating or distil a solution and for this purpose heating apparatus are necessary. So various types of burners and spirit lamps are used. We shall now discuss about them.

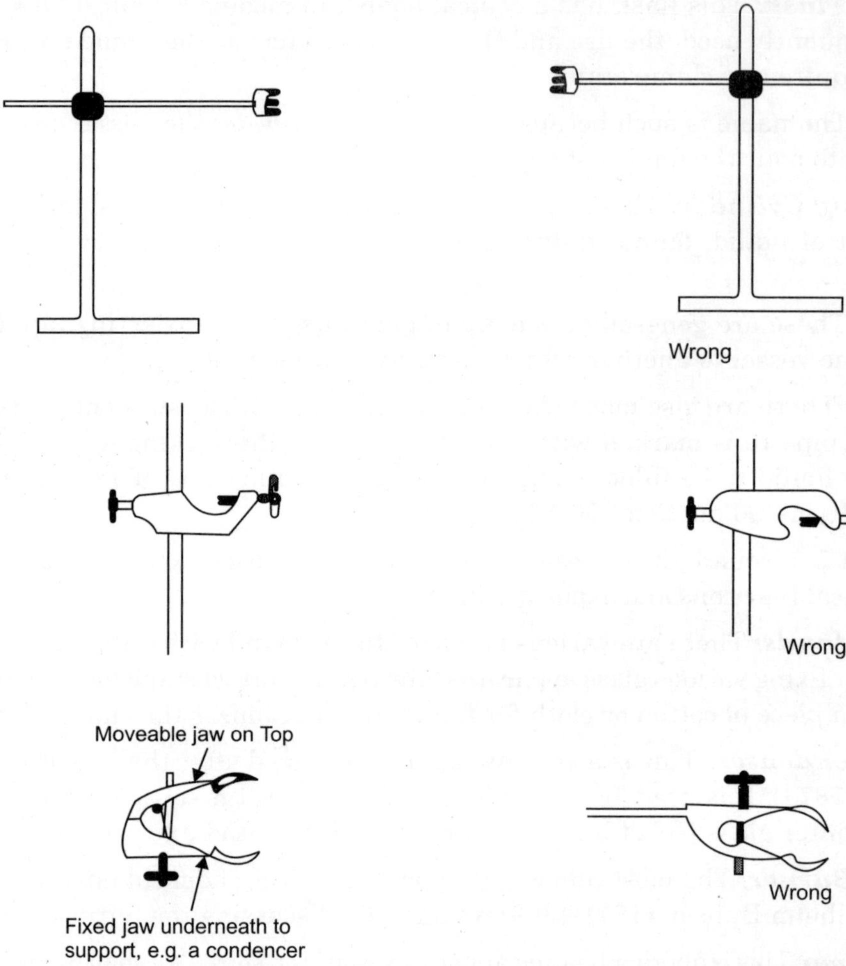

Fig. 1.6. Showing the right and wrong ways of using the stands and clamps

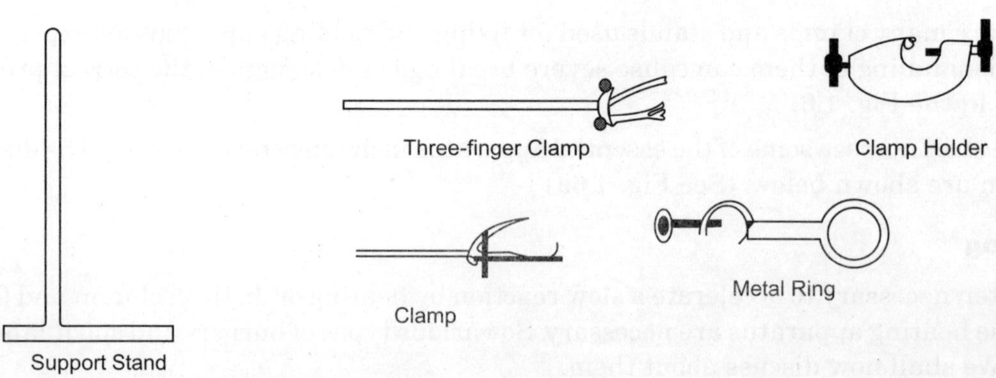

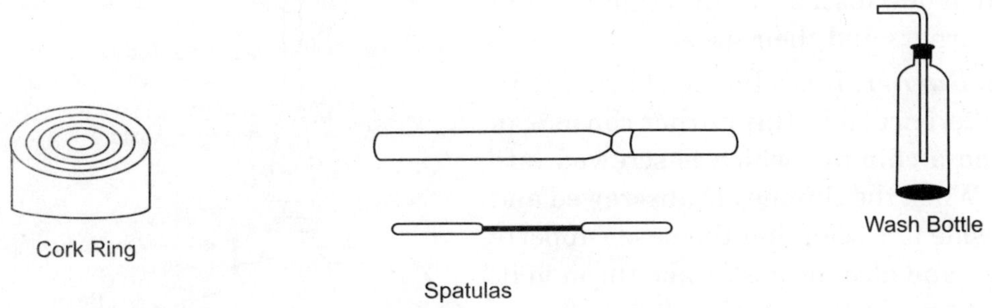

Fig. 1.6(*a*) Some Essential Apparatus

Burners : Robert Wilhelm Bunsen (1811 -1899) invented it. So these are named after him. Bunsen burner works in L.P.G., the household cooking gas or coal gas. Now we shall discuss about the different parts and the use of the Bunsen burner. (See Fig. 1.7 and 1.7a)

(1) *Base :* The burner rests on a cast iron base. The base has a side tube connected to the gas. At the centre of the base there is a small nozzle having a pinhole, through this pinhole the combustible gas enters into the tube when the gas tap is opened.

(2) *The Burner Tube:* It is a long metal tube with two holes at the lower end of air intake. At the top of the tube the gas burns when the gas tap is turned on.

(3) *The Air Regulator:* It is a metal collar having holes corresponding to the holes in the tube. By rotating the ring we can control the air supply sucked into the Bunsen tube. When the air hole is completely closed the burner gives a sooty yellow flame it is known as "luminous" and the air hole when opened partially the flame becomes nearly invisible and the height is also short with a pale blue colour known as "non-luminous" flame, this flame is generally used for work. When the air collar is opened fully the flame gives off a hissing sound called "roaring flame". This can melt the glass rods and tubes, in-fact this flame is used for melting glass rods and tubes.

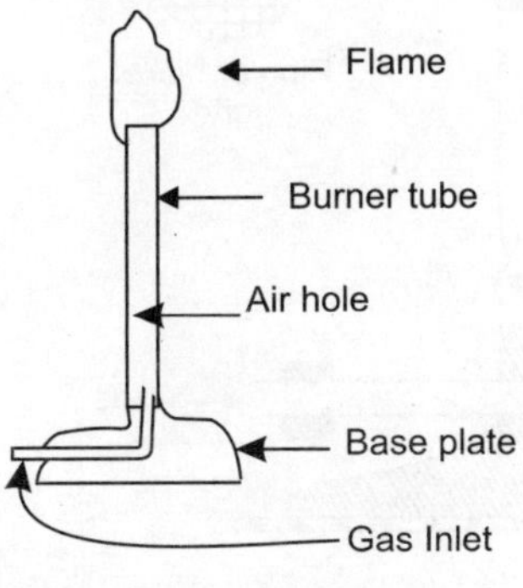

Fig. 1.7. The different parts of a Bunsen Burner

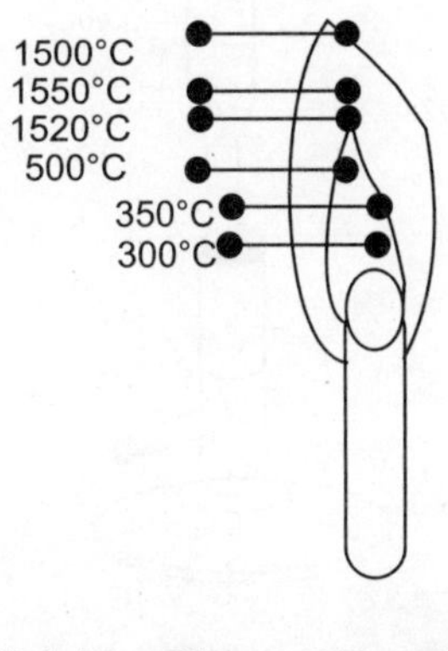

Fig. 1.7(*a*). The Different Temperature Zones of a Bunsen Flame

We shall now discuss the different types of Bunsen burners and their uses.

Teclu Burner: It is a Bunsen burner with a little difference in it, this burner consists of a base and a chimney which is screwed into the base. When the chimney is unscrewed and looked inside it is seen that the base supports the burner and also the gas intake through it. Instead of an air-regulating collar the base consists of a ring with corresponding holes with which the air intake is controlled. The diagram of the Teclu Burner is shown in fig. 1.8.

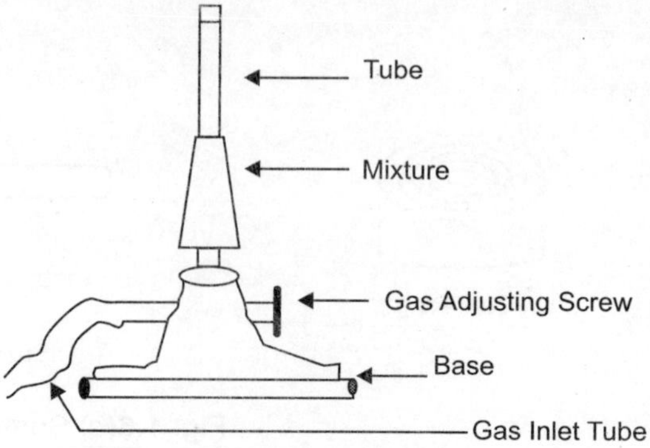

Fig. 1.8. Teclu Burner

Fish-Tail Burner or Bat-wing Burner: This is another kind of Bunsen burner having long tube. Its air hole is completely open and the position is fixed. The mouth of the tube where the burner is lighted has a slit; thus the flame takes a shape life the fish tail. Fig. 1.9 shows the Fish tail burner. These burners are used for making bent tubes and pipes.

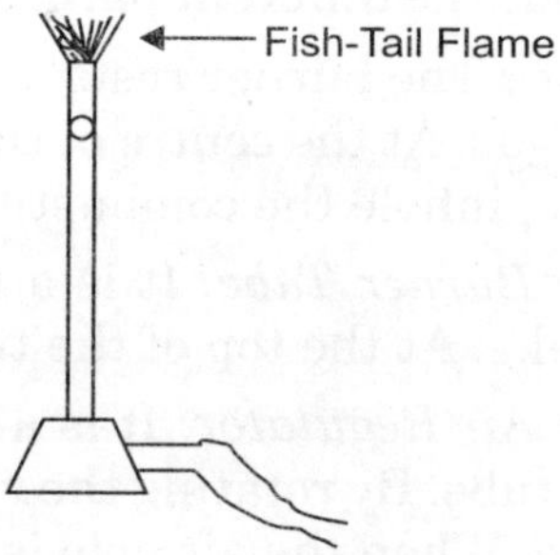

Fig. 1.9. Fish-Tail Burner

Meeker Burner: It is the best type of Bunsen Burner, and can be made within long temperature ranges, in modern chemistry labs this burner is used. Fig. 1.10 shows the Meeker, burner.

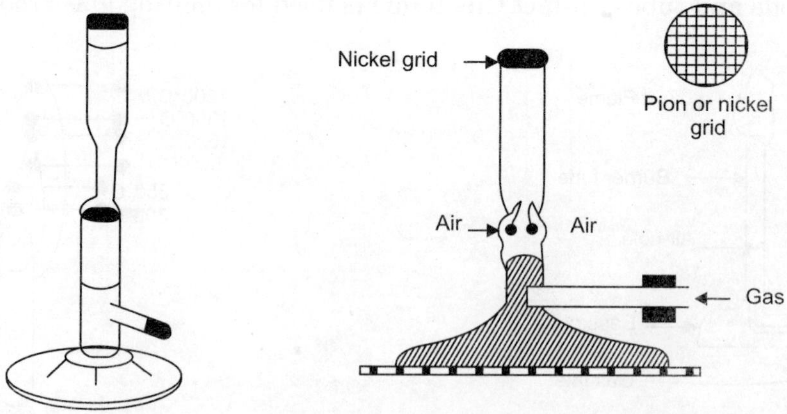

Fig. 1.10. Meeker Burner

Ring Burner: It is made up of a series of small Bunsen burners placed in a circular ring with a cast iron setting. These are widely used in domestic kitchen for cooking.

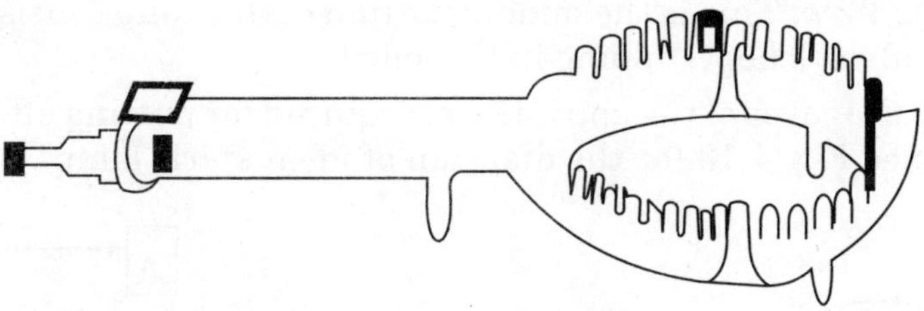

Fig. 1.11. Ring Burner

At first take a burner and connect it to the gas supply, before opening the gas tap, remove all the naked flames, inflammable oils, explosives etc. Now carefully open the gas tap and apply some soap solution on the pipe ends. If the foam produces bubbles *there is a leakage.* Immediately close the gas tap and check for leakage, if the pipe is leaky change the pipe, if it is loose, tight the pipe. Now light the burner, if it produces a luminous flame adjust the air collar and make the flame non-luminous. The burner is now ready for working. But some safety rules must be followed, that is when the burner is not in use it should be put off because the non-luminous flame is hardly visible can cause unnecessary fire hazards and burns. The burner must not be used continuously for long time; it may cause strike back. Sometimes it is found that a burner flame is drawn back within the tube of the burner and burn in the burner near the air intake hole and a loud hissing sound is heard.

This condition is called striking back. The striking back must be avoided at any circumstances.

This happens firstly due to insufficient gas supply and over sufficient supply of air. This results the combustion of the air gas mixture much faster than the supply of gas which causes more gas intake and the upward thrust of the gas causes the flame to come downwards, thus striking back the flame. This also sometimes does happen when the burner is over heated, in this case the burner must be allowed to cool for a period of time, or it can be carefully cooled under the tap water.

Burner must be kept aside for cooling

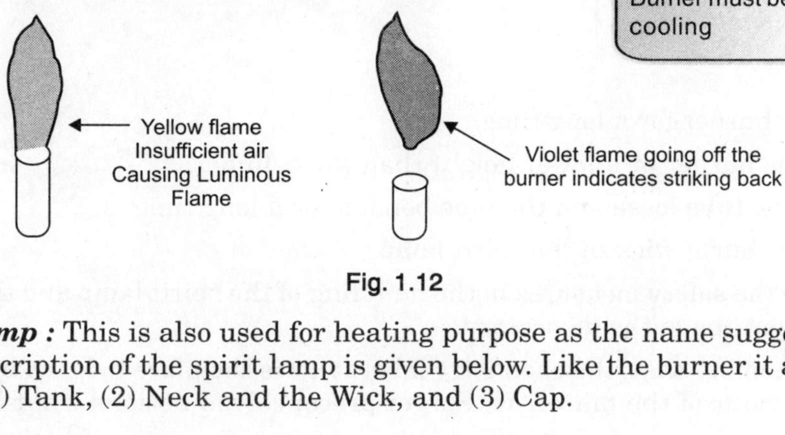

Fig. 1.12

Spirit Lamp : This is also used for heating purpose as the name suggests it work on spirit. The description of the spirit lamp is given below. Like the burner it also consists of three parts. (1) Tank, (2) Neck and the Wick, and (3) Cap.

(1) *Tank:* The tank is the lower portion of the spirit lamp, which is filled with spirit or alcohol. It is made up of glass, brass or aluminium.

(2) *Neck and the Wick:* This is the middle portion of the lamp consisting a cotton wick, this wick is always kept dipping in the spirit.

(3) *Cap:* This is the cover of the spirit lamp, required for putting off and put it on when no in use. (See Fig. 1.13 for the diagram of ideal spirit lamp).

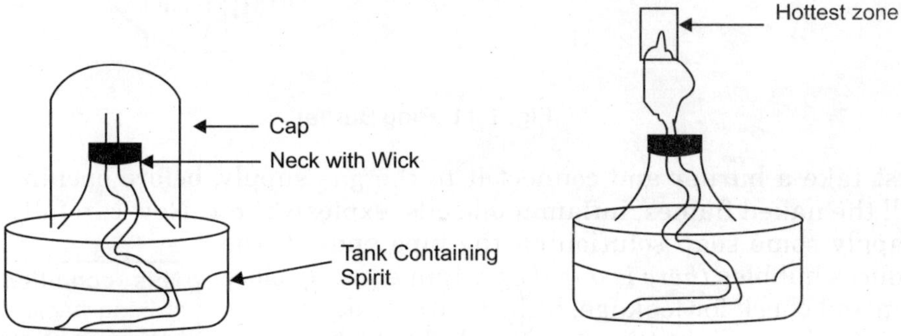

Fig. 1.13. The spirit lamp

Use of Spirit Lamps: A spirit lamp is at first seen that whether it is filled with spirit or not. Then the cap is removed and the lamp is lighted. The wick can be pulled up or down to increase or decrease the flame. To put off the flame, do not blow but replace the cap. Some of the do and don'ts are stated below about the spirit lamp. Some common do and don'ts of lamp and the burner both shown below:

Do

✓ Always check the gas tap supply tube for leak.

✓ Always keep the burner or lamp in one corner of the bench, (always adjust the air collar carefully to avoid striking back.

✓ Always try to use a glass spirit lamp because due to transparency the level of the spirit will be visible.

Don'ts

× Use the burner for a long time.

× Keep the burner at a lower height than the cylinder.

× Leave the tube loose and the pipe bended for a long time.

× Keep the burnt wick in the spirit lamp.

These were the safety measures in the handling of the spirit lamp and the burner, and also all the apparatus in the laboratory.

Always remember that, the chemistry laboratory is the place to learn by observations what the behaviour of the matter is. Forget preconceived notions what is supposed to happen. Follow the directions carefully; and see that actually what "does happen" be meticulous in reading the true observations carefully even though you 'know' something, else should happen. Do not hesitate to seek instructions from your teacher. Recall the ancient dictum. *He who asks a question remains ignored for once, but he who does not remains forever.*

Process of Study Chemistry

Chemistry is not a subject to be 'learned by heart' or "committing in the memory and vomiting in the paper". Chemistry is the subject of realization of truths. Chemistry should be realized by heart, the most important aspects of the chemistry are the chemical equations and formulae. This subject is full of experiments and practical which must be clearly understood by the student. During description of an experiment an appropriate diagram of the experimental set up must be given following it a very short description of the experiment with the detail chemical equation and the precaution measure must be stated at the end of the experiment prominently.

Branches of Chemistry

Chemistry is the vast subject and numerous chemical reactions occur in the various systems, continuous chemical changes are also occurring in the biological systems. In order to study the entire chemical changes reactions and transformation; various branches of chemistry have been developed. They are:

1. Physical Chemistry
2. Organic Chemistry
3. Inorganic Chemistry
4. Analytical Chemistry
5. Industrial Chemistry
6. Biochemistry

Physical Chemistry : Physical chemistry deals with the fundamental principles upon which all the other phases of chemistry are based. Thus we can say that the physical chemistry deals with the investigation of laws and theories of the science of chemistry. The laws and theories of the science then serve to explain the facts that we observe.

Organic Chemistry : Mainly organic chemistry studies the structure of the reactions of carbon compound and mainly deals with the natural products. (The name organic comes from the word organisms). An important part of this chemical science is the process of forming complex compounds from simple materials.

Inorganic Chemistry : This branch of chemistry mainly deals with the elements other than carbon and their compounds, study of mineral ore, radioactivity metals and alloys acids bases and salts.

Analytical Chemistry : As the name suggests it deals with the separation, identification of compounds and also analyzing them (qualitative and quantitative). This is the most important branch of chemistry and no other branch of chemistry can be under stood without this branch.

***Industrial Chemistry*:** It is concentrated with the application of chemistry to industry and deals with the development of chemical processing for the production of useful articles on large scale. The manufacture of soaps, cement, paper, paints and varnishes etc. In addition to these branches there are several other branches, which are of special types such as agriculture chemistry, hydro geochemistry, and radiation chemistry.

***Biochemistry*:** This branch acts as the interpreter between chemistry and biology. It deals with the chemical reactions taking place inside the human body; therefore it is the science concerned with the chemical behaviour of the living matter.

***In one sentence*:** Chemistry is the international language linking the physical and the biological sciences the atmospheric and the earth sciences the medical and the agricultural sciences chemical language is reach and fascinating and creates images of great aesthetic beauty.

CHAPTER

2

Concept of Chemistry

Introduction

To understand the basic concepts of chemistry, we should first try to understand the concept of the matter because our universe is full of matters, which are changing frequently. The matter can be defined as the substance consisting of a definite proportion. Our universe is consisting of matters of different kinds and of different state, the classification of the matter in universe is shown below in the form of a chart :

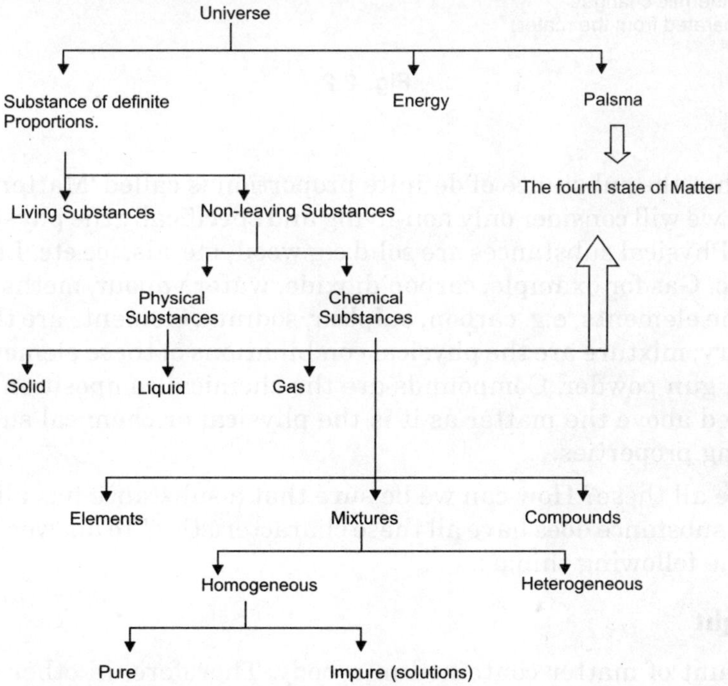

Fig. 2.1

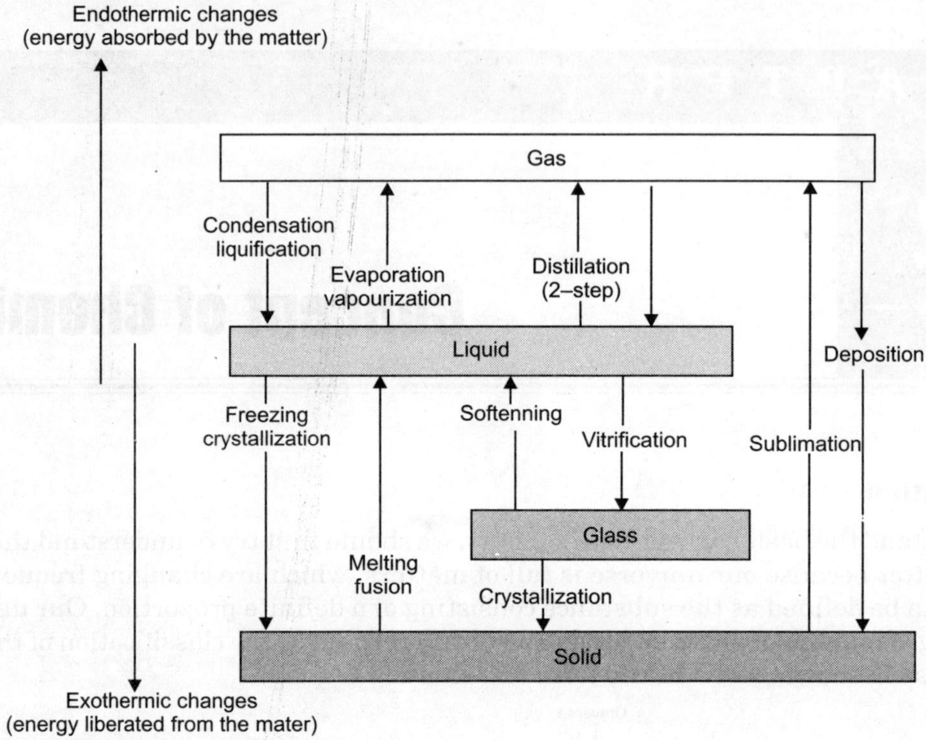

Fig. 2.2

Matter

As we all know that the substance of definite proportion is called 'Matter', it can be living or dead. But here we will consider only non-living and specifically the physical and chemical substances only. Physical substances are solid e.g wood, metals, ice etc. Liquids, e.g. water alcohol, spirit, etc. Gas for example, carbon dioxide, water vapour, methane etc. Chemical substances are the elements, e.g. carbon, sulphur, sodium; elements are the main building blocks of chemistry; mixture are the physical combinations of these elements. Example the air, salt solution, gun powder. Compounds are the chemical composition of the elements. Now as it is stated above the matter as it is the physical or chemical substances it must have few following properties.

Now what are all these? How can we be sure that a substance has all these? How can we be sure that a substance does have all these characteristics? To answer the first question we must know the following things:

Mass and Weight

Mass is the amount of matter contained in a body. Therefore, in other words mass of a metal block is the amount of the metal the block is consisting with. Generally speaking, we do not differentiate mass and weight. We say that the weight of the body is 30 kg and also the mass of the same body is 30 kg, but this is not true, because the mass and the weight of the body are the two entirely different characteristics.

It is no doubt that every thing has its certain weight; we feel the weight when we place the body on out palm, and some bodies are extremely heavy.

From the above information, we can deduct the definition of weight. The weight of the body is the attractive force of the earth exerted upon the body; thus higher the weight stronger the attraction.

Difference Between Mass and Weight

Now we shall study the differences of the mass and the weight, which is shown in the table below:

Mass	Weight
1. Mass is the amount of matter contained in a body.	1. Weight of the body is the force of attraction forwards the centre of earth.
2. Mass of the body is measured in a beam balance.	2. Weight of the body is measured in the spring balance.
3. The mass of the body is always fixed.	3. Weight of the body always varies from place to place.
4. The mass is a scalar quantity, because it has only magnitude but no direction.	4. Weight of the body is vector quantity, because it has both the direction and the magnitude.
5. Motion, position, or the temperature of the body does not influence Mass in the body. So the mass is the intrinstic property of the body.	5. Changing place, motion, temperature of the body can change weight, so the weight is not the intrinsic property of the body.
6. Mass is expressed in grams and pounds.	6. Weight is expressed in dynes and pounds.

Conservation of Mass

It says that the total amount of the quantity of matter in the universe is fixed and it cannot be changed (that is increased or decreased) by any means. When we prepare any substance by either physical or chemical reaction, we see there is only a transfer of mass (matter) no change in quantity of the matter takes place. For example, when a candle is lighted some gas is left to air, the wax melts and finally some ash is left over. In an ordinary look we may see that here the principle of conservation of mass does not come to use. If we look carefully and if we measure the gas released to air and also the amount of the molten wax together with the small amount of the residue left behind we will find that they are equal to the mass of the candle.

When the magnesium wire burns in air a brilliant white ash is produced which is the oxide of the magnesium, if we find the ash heavier than magnesium the conservation of mass comes here like this. The atmospheric oxygen here reacts with the magnesium ribbon thereby increasing the mass of the magnesium ribbon, but the atmosphere at the same time looses the same amount of oxygen that the magnesium has consumed. So no new mass is created or destroyed, this phenomenon will become clearer when we will perform some interesting experiments.

Experiments on Principle on Conservation on Mass

Exp: 1 Landolt's Experiment: In 1893 and subsequent years Landolt's carried out a very accurate sets of experiments in order to test the validity of the principle of the conservation of mass (Fig. 2.3). A solution of lead nitrate and potassium iodide are taken in the two limbs of the 'H' tube, also known as Landolt's tube, as shown in Fig. 2.3.

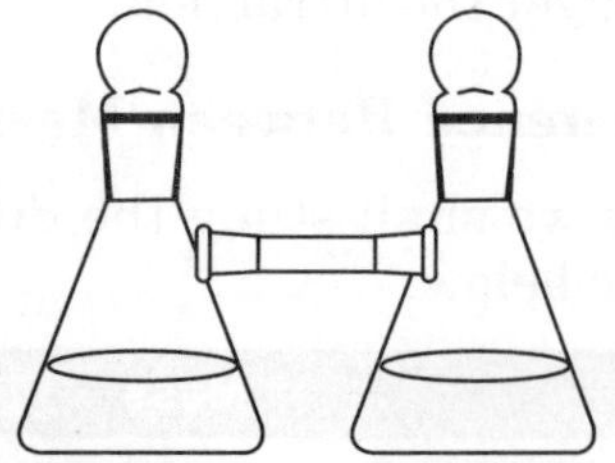

Fig. 2.3. The experimental set up for the Landolt's experiment

The whole apparatus is then sealed off and carefully weighted, then the tube is tilted so that the two liquids mix with each other, a react to form a new product the Lead iodide. After this reaction the apparatus is cooled and weighted carefully for the second time. Weight shows no difference, which means no new mass has been created or destroyed - thus proving the validity of the principle of the conservation of mass.

Experiment

Antoine Laurent Lavoisier (1734-1794) a French chemist performed an experiment to prove the validity of the principle of the conservation of mass. He used a platinum retort in which he introduced tin and he completely sealed that retort. Then he weighted that retort and heated over a furnace after the entire tin has been converted to oxide, the retort was cooled and again carefully weighted, the result showed that there was no change in weight, this proved the validity of the conservation of mass principle. (Fig. 2.4 shows the diagram of the experimental set up).

Fig. 2.4. The experimental set up for the Landolt's experiment

Rusting of Iron

Any student or any common person who are interested in chemistry can perform this experiment. It consists of a glass test tube containing some tap water in which some nails have been introduced, the tube should be sealed by a tight rubber cork, and carefully weighted then the whole set up should be kept in a proper aerated place for few days. When all the nails will be rusted the test tube should be weighted again, the weight are to be

Fig. 2.5. The experiment of the iron nails

compared. There will be no change in the initial and the final weight, which will prove the validity of this principle. The experimental set up is shown in Fig. 2.5.

States of Matter

As we all know that solid liquid and gas are the three states of matter, these are regarded as the three physical States of matter. Solid has a definite shape, size, weight and volume, e.g. wood, ice, etc.

Solid State of Matter

The matter is regarded as the solid when the particles within the matter are very much congested and the gap between them is least. Solid has a definite shape, size weight and volume, e.g. wood, ice.

Liquid State of Matter

The matter is said to be liquid when the particles of the matter is not so much congested like the solid and the gap between them is wider than least. A liquid does not have shape and size but weight and volume does exist. The liquid can adopt the shape and size of the vessel in which it is stored, e.g. water, glass (glass is not a solid it is a super cooled liquid).

Gaseous State of Matter

The matter is said to be gas when the particles of matter are less congested and the gaps between the particles are widest among the three. Gas does not have a definite shape and size, but does have weight and volume, e.g. nitrogen, oxygen, carbon dioxide, etc.

Plasma State of Matter

Solid liquid and gas are regarded as the only three states of matter. But, comparatively recently, the properties of matter in a fourth and exceptionally unique state - "plasma" - have been subject to intensive studies. Two American physicists, Langmuir and Tonks first used the term, plasma, in 1923.

Difference Between Vapour and Gas

The vapour is the air dispersion of the particles of liquid; the process by which it is done is called vapourization. It should be noted that gas and vapour are very much different in nature. The temperature above which the liquifecation by physical means is not possible is called *critical temperature.* The liquid's air dispersion having the temperature below this critical value is called *vapour* and the air dispersion having the temperature above this critical value is called *gas.*

Changing States of Matter

As we all know that matter exists in three states solid, liquid and gas. When the alteration of the state takes place between them, physically or chemically the phenomenon is called *changing the state.* For example when ice changes to water and the water-to-water vapour. It indicates the change from solid to liquid and the liquid to vapour.

Melting, Freezing and Boiling Point

When a solid substance is transformed to liquid at a certain temperature, that particular temperature is called the *melting point* of the solid.

When a liquid substance is transformed to solid at a certain temperature, that particular temperature is called the *freezing point* of the liquid.

When a liquid substance is transformed to vapour at a certain temperature, that particular temperature is called the *boiling point* of the liquid.

For example the melting point of ice is 0°C and the freezing point of the water is also 0°C. The boiling point of the water is 100°C, when the water changes to vapour.

Effect of Pressure on Melting Point

The melting point of a substance depends very certainly on pressure way how it depends is of two types: (1) It increases and (2) It decreases.

(1) Substances like iron and other metals, whose volume *decreases on melting,* have their melting points lowered due to an increase of pressure. Ice also have low melting point about 0.007°C because increase of atmospheric pressure helps the contraction of volume more and therefore and lowers down the melting point.

(2) Substances like wax etc. whose volume *increases on melting* having their melting points raised due to increase of pressure. In case of wax, the melting point is found to be raised by 0.04°C due to an increase of one atmospheric pressure. Here also, the simple reason is that increased pressure causes hindrance to expansion of volume more, and therefore, raises the melting point.

The Phenomenon of Regilation

This is a very interesting phenomenon of physical chemistry. It can be defined as *the melting of a substance with the application of pressure and resolidification.* There is an interesting experiment on this.

Definition of Regilation

Phenomenon, it is as follows:

An ice block is kept on a stand, a heavy weight is kept hanging from the ice block, as shown in the Fig. 2.6. After some time the iron weight slides down cutting through the ice-block, but the ice block remains same one piece. The explanation is like this, the wire cuts through the block of ice because due to the application of pressure which reduces the melting point of the ice. On the other hand the melted ice (i.e. water) increases in volume and rises above the height of the wire position where the pressure is minimum. Thus resolidifies the ice causing

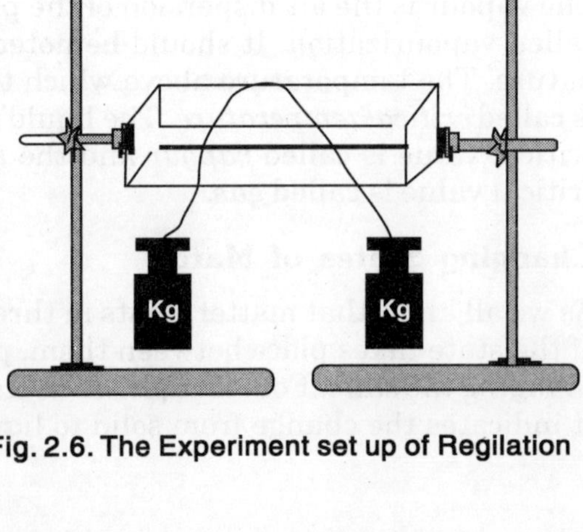

Fig. 2.6. The Experiment set up of Regilation

the block of ice remaining intact. One thing should be kept in mind that the wire should be very thin and the wire should be made up of any metal because the non-metal is not a good conductor of heat.

Filtration

The process of filtration includes the mechanical or physical separation of the suspending or semi-dissolved solid particles. The process includes the filter paper placed in the funnel in which the solution is poured slowly against the glass rod and the clear filtrate is dropped in the reservoir and the residue is left in the filter paper (See Fig. 2.7)

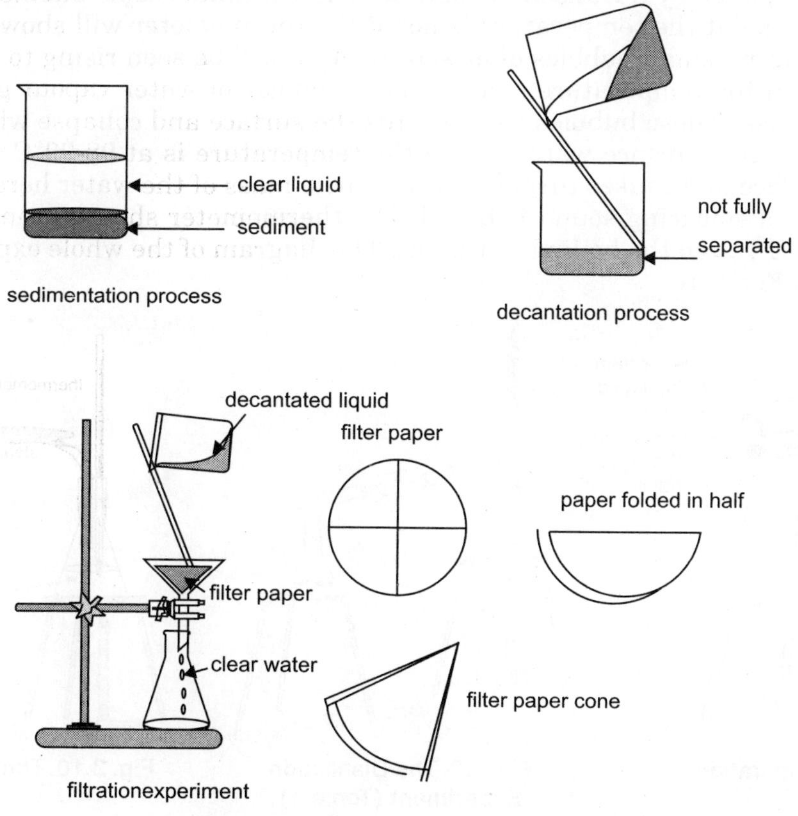

Fig. 2.7. The filtration experiment

BOILING AND EVAPORATION

Evaporation

Evaporation is the process by which the liquid vapourizes, in air from its surface, so evaporation occurs in the surface. So evaporation does not need any temperature, but only surface area

Boiling

The process by which a liquid substance starts vapourizing it is known as boiling and the temperature at which this boiling start is called boiling point. The boiling point of the water is 100°C, which can be proved by the following experiment. Boiling takes place though out the whole mass of the liquid.

Experiment

Some water is taken in a conical flask which can be closed by a rubber cork having two holes through these a delivery tube and a thermometer is passed. The flask is heated on a tripod and wire gauge by a Bunsen burner. After few minutes some bubbles and smokes are seen at this point the temperature is noted the thermometer will show a continuous rise in temperature, small bubbles of dissolved gases will be seen rising to the surface of the water. When the temperature is at 75-80°C bubbles of water vapour gathers at the bottom of the flask. These bubbles rise towards the surface and collapse while coming in contact with the cold surface water. When the temperature is at 98-99°C these bubbles rush to the surface and causes turmoil[ã] to the whole mass of the water here at this stage dense smoke and smearing sound is heard. The thermometer shows a constant reading between 100-101°C and the boiling continues. The diagram of the whole experimental set up is shown in Fig. 2.10.

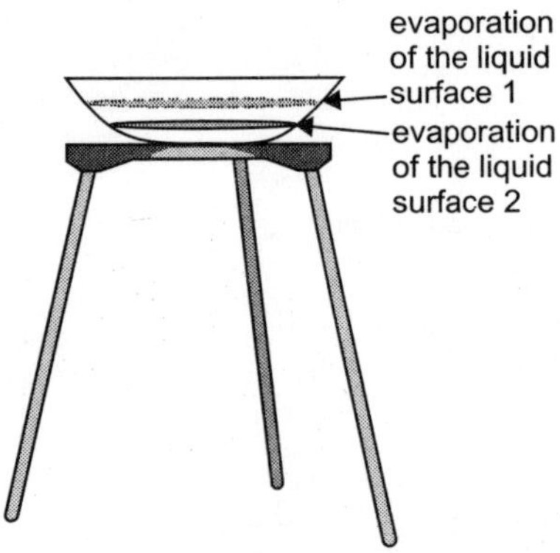

Fig. 2.8. Evaporation

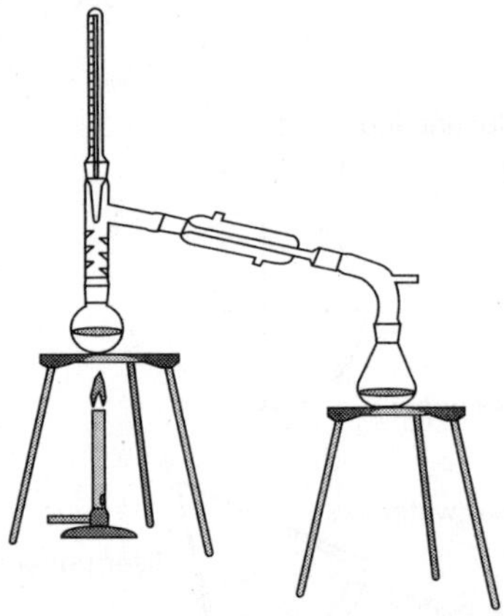

Fig. 2.9. The Distillation Experiment (Tome 1)

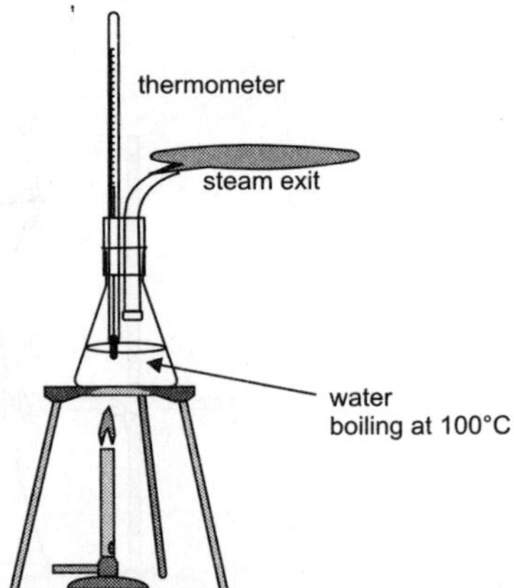

Fig. 2.10. The boiling of water

Factors Influencing the Evaporation

1. *Types of liquid* : That is if the liquids are volatile (like alcohol) the rate of the evaporation will be faster and if the liquid is viscous (like oil) the process of evaporation will be slower.
2. *Presence of Moisture in the Atmosphere* : This greatly affects the process of evaporation. If the atmosphere is moist then the evaporation process becomes slow and if the atmosphere is dry the process is quick. So the wet cloths become dry quickly during summer and very slowly during rainy season.

3. *Area of the Exposed Surface :* This means that more the surface of the liquid is exposed more quickly the process of evaporation will take place. So the cloths are always hanged over the ropes in order to expose the maximum surface area of the cloths. A cup of tea cools much faster when it is poured in a dish.

Condensation

The opposite reaction of the process of evaporation is called condensation. That means *the retransformation of the vapour to the liquid is called condensation.* We can perform an experiment on this phenomenon. Some water is kept boiling in a retort when the water is seen boiling and some vapour is escaping through the nozzle. A flask containing some cold water is held near the nozzle after sometimes it will be seen that a good amount of water has gathered around the outer wall of the flask, which is nothing but water. Now we can prove that the evaporation and the condensation are the two back-to-back processes.

Distillation

This process is the process of boiling combined with the process of condensation the liquid water is boiled in the distillation flask which is mixed with potassium permanganate and the steam is collected and cooled by passing it though the Liebig's condenser, the liquid is collected in the reservoir flask which is safe and purified water.

Difference between Boiling Evaporation and Condensation

Evaporation	Condensation	Boiling
1. This is a slow process of conversion of liquid to vapour.	1. This is the reverse process of evaporation in which the vapour comes back to liquid.	1. This is the fastest process of conversion of liquid to vapour.
2. Evaporation occurs on the surface of theliquid.	2. This process transfers the whole mass of the vapour into liquid.	2. Boiling occur in the whole mass of the liquid.
3. Evaporation require no particular temperature or pressure.	3. The condensation reduces the temperature of the vapour and then the vapour is converted into liquid.	3. Boiling occurs at a particular temperature.
4. The evaporation depends upon the exposed surface.	4. The condensation depends upon the amount of vapour evaporated.	4. Boiling also depends upon the mass of the liquid.
5. The evaporation depends upon the atmospheric moisture.	5. Condensation occurs when the vapour comes in contact with the moist surface.	5. Boiling does not depend upon the moisture of the liquid.
6. The evaporation does not require any superincumbent pressure.	6. Condensation also does not require any superincumbent pressure.	6. Boiling requires a superincumbent pressure.

Factors Influencing the Point

The following factors influence the boiling point of the liquid :

1. The Superincumbent Pressure of the Liquid : The boiling point of a liquid depends upon superincumbent pressure under which a liquid boils. Increase of pressure raises the boiling point while the decrease of pressure lowers it. It has been found that the normal boiling point of water (100°C) is raised or lowered by 1°C due to an increase or decrease of every 27mrn of atmospheric pressure respectively.

When the air pressure above a liquid increases, the molecule of the liquid comes nearer to each other. This opposes the molecules from coming out of the liquid and the liquid molecules then require more heat to make them free from each other and to get out of the liquid as vapour. So the boiling point rises. When the air pressure decreases, the velocity of the liquid molecule increases and the molecules can easily leave the liquid. It makes the liquid cold and hence the boiling point falls.

2. Dissolved Impurities in the Liquid : The liquids having impurity boils at a high temperature than normal pure liquids. For example, the boiling point of the pure water is 100°C and that of the water containing the impurity like common salt and other substances have a boiling point above 100°C (nearly 109°C). For this reason the thermometer is not kept dipped in the water during checking the boiling point of any liquid.

3. Nature of the Liquid : The boiling point of different liquid is not same as because of their nature and chemical composition. For example, the boiling point of water is 100°C and the alcohol is 78.3°C and ether is having a boiling temperature of 35°C. Thus, we can say that boiling point depends upon the nature of the liquid.

Effects of Pressure on Boiling Points

The boiling point of a liquid as stated earlier depends upon the pressure in which the liquid is allowed to boil. If the superincumbent pressure is increased the liquid boils at much higher temperature, but if the pressure is decreased the liquid boils at a lower temperature its usual boiling point. Now we shall see the experiment on the above-mentioned topic.

Lowering of the Boiling Point Due to Reduction of Boiling Point Franklin's Experiment

Some water is taken in a round-bottomed flask and it is set to boil. After some times when the seam has driven out all the air from the flask the flask is tightly closed and a thermometer is inserted, then the flask is inverted and then cooled under water. See Fig. 2.11. It is seen that the water starts boiling again and the thermometer indicates a lowered boiling point.

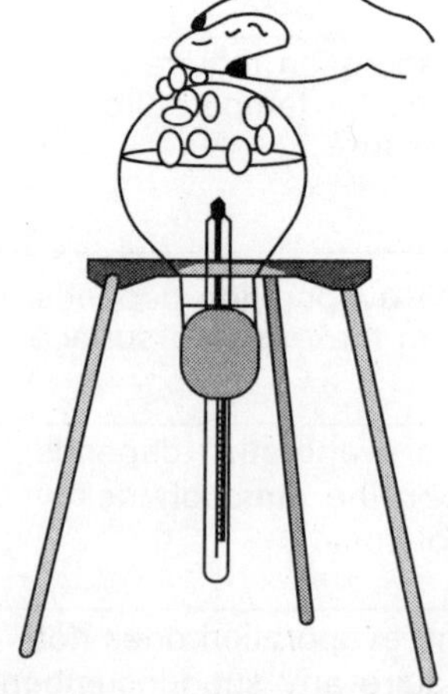

Fig. 2.11. The Franklin's Experiment

The above phenomenon is possible because when the cold water is poured some water vapour in the flask condenses as the liquid, which reduces the pressure around it. The reduced pressure reduces the usual boiling point of the water. When the temperature of the liquid is equal to the reduced boiling point the liquid the water starts boiling.

Effects of Altitude on Boiling Point

Atmospheric pressure gradually decreases as the altitude increases *and the rate 85 mm (of mercury for every kilometre of rise*. At a considerable high altitude the pressure decrease is not linear, now we know that the boiling point of a liquid depends upon the superincumbent pressure of the liquid surface. As the air pressure at the top of a mountain is low, the boiling point of water will also become low, that is, the water will boil at much lower temperature. It has been estimated that water boils at a temperature of 70°C only, at the top of Everest has an altitude of 4884m. The boiling point of water at Darjeeling (altitude about 7000ft) nearly 90°C.

We will now read a practical application of tile phenomenon about effect of pressure on boiling point.

Pressure Cooker

It is a unique scientific wonder is described below with a diagram (Fig. 2.12). It is an airtight seal pot made up of steel. This produces steam heat to cook food as quick as possible. It first appeared in the year 1679 as "PAPINS DIGESTER" after the name of its inventor Denis Pepins. The cooker heats water to produce very hot steam, which forces the pressure inside the cooker to rise as high as 266°F (120°C) which is much higher than the maximum heat possible in an ordinary sauce-pan. The higher temperature of a pressure cooker penetrates food quality reducing the cooking time without diminishing the food value, that is the vitamins and minerals.

Pressure cookers are extremely helpful at the higher altitudes, where there is a common problem of lower temperature boiling.

Modern innovations in pressure cooker design include safety valve, pressure regulators portable cookers, and low-pressure fryers. The diagram of the pressure cooker is shown below in fig. 2.11.

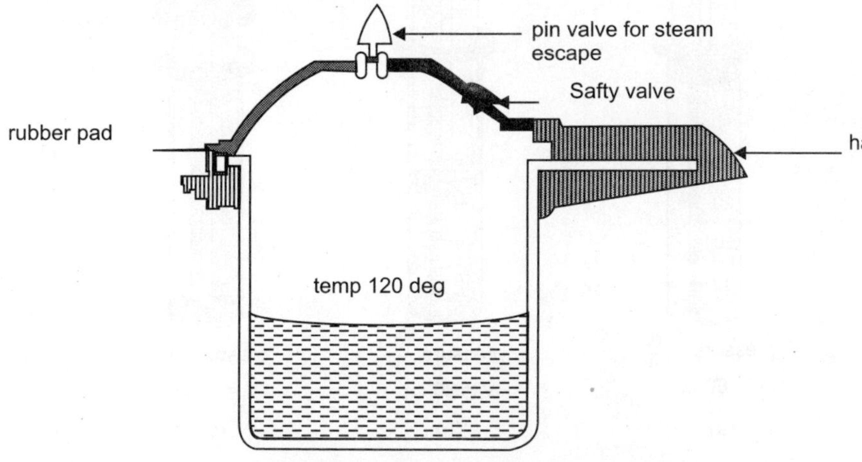

Fig. 2.12. The diagram of a pressure cooker

GASEOUS STATE OF MATTER

In an earlier subhead of this chapter we have learnt about the vapours and gases. Now we shall learn about the different properties of the: gases and the laws they obey.

Boyle's Law

The peculiar properties of gases are that a very small quantity of gas can occupy an infinity large volume, if it is allowed. Which suggests that the volume of the gas can be easily changed. This was actually proposed by altering the pressure exerted on the gas. This phenomenon was studied by a British scientist *J Robert Boyle (1627-1691)*. He observed that *the volume of the gas decreases with the increasing of pressure and the volume increases with the decrease of pressure.* The increasing and decreasing of volume with the application of pressure was established by Robert Boyle was called Boyle's Law.

Which states that : Temperature remaining constant the volume of a certain mass of gas is inversely proportional to the pressure applied to it.

Let V be the volume of a certain mass of gas at certain temperature and P be the pressure applied to it then according to Boyle's Law we get:

$$V \alpha \frac{1}{P} \qquad \text{where temperature is constant.}$$

Or, $$VP = k \qquad \text{(constant)}$$

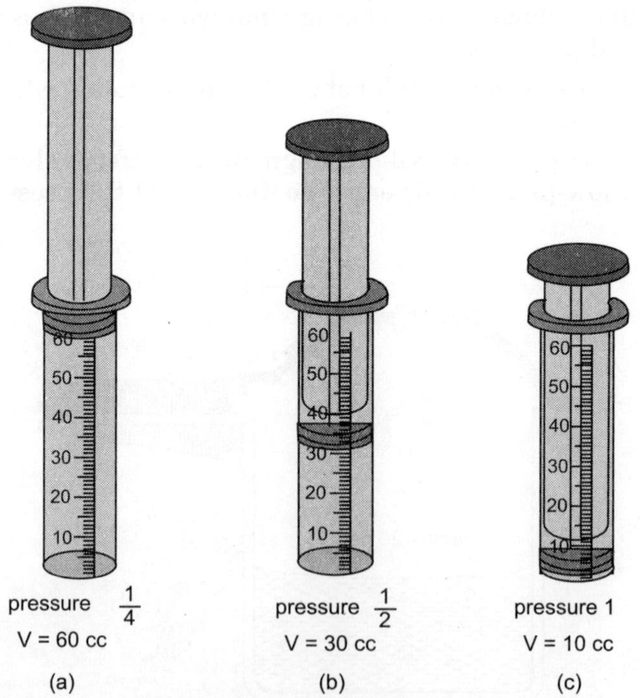

Fig. 2.13. Experiment of Compression of Gas

So, if the volumes of the gases having a true mass is V_1, V_2, V_3 etc., and the pressures are P_1, P_2, P_3 respectively then $V_1 P_1 = V_2 P_2$ and $V_2 P_2 = V_3 P_3$.

Suppose we take 100cc of gas in a vessel, the pressure exerted on the gas by a piston is equal to 1 atmospheric pressure and its temperature is 30°C (as shown in Fig. 2.13a). Now increasing the weight to double, keeping the temperature of the gas constant increases the pressure of the gas to double. According to Boyle's law the volume of the gas will be 100/2 = 50. See Fig. 2.13b. Now the pressure of the gas is increased four times, by increasing the weights on the piston four times, then the volume of the gas becomes 100/4 = 25 (Fig. 2.13c). Now if the pressure is made exactly half that is decreased by two times the volume becomes double, which is equal to 25 × 2 = 50cc. Similarly if the pressure is reduced four times the volume also increases by four times which is also equals to 25 × 4 = 100cc.

Verification of Boyle's Law

Boyle's law can be verified by an experiment, which consists of an apparatus called Boyle's Law apparatus. This apparatus consists of two stout glass tubing T and fixed to a vertical stand. A scale S running parallel to the tubes is also attached to the stand. AB is a long narrow tube, closed at the upper end and CD is the sort tube having a wide upper end and can slide up and down on the scale. The tubing and the both of the tubes contain mercury. The upper portion of the tube contains dry air or the gas to be tested upon.

To start with let the mercury level be at F in CD vessel lower than in the vessel AB, the vessel CD is lowered up or down to adjust the mercury level. Taking the reading on the scale at levels A G and F. As the closed tube AB is uniform in cross-section, the volume of the air enclosed in it is proportional to the length L (difference between A and a) of the tube it occupies. Hence, for volume we consider the length only. Consider a point E, in the same horizontal plane as F. Then pressure at E equal to that due to air (say P) in the tube plus the vertical distance between G and E, while the pressure at F is the atmospheric pressure (say H), which could be read on a barometer. Points E and F being at same level, $p + h = H$, i.e., the net pressure is $p = H - h$. The product $(H - h_1)$ is calculated. Now raise the open tube CD, the levels F and G again, say it is h; then net pressure will be $(H - h_1)$ or $H + h_1$ according as level of mercury in the open tube is lower or higher than that of mercury in the closed tube. New volume will be as before. The product $(H \pm h_1)$ is found to be same as before. If several such readings are taken for pressure and volume it will be noticed that their product is constant at the same temperature. If on a graph paper P is plotted against V the curve so obtained is a rectangular hyperbola (Fig. 2.13a), the figure is shown as isothermal as the temperature between P and V is constant. It is necessary to point out that the above law is not rigorously obeyed over a wide range of pressure; PV first decreases as p (or pressure) as it increases, reaches a minimum value and then begins to show a increase for large pressure. The gases, which obey the Boyle's Law, are called Ideal Gas or the Perfect Gas, for example air, hydrogen, nitrogen are nearly perfect gases.

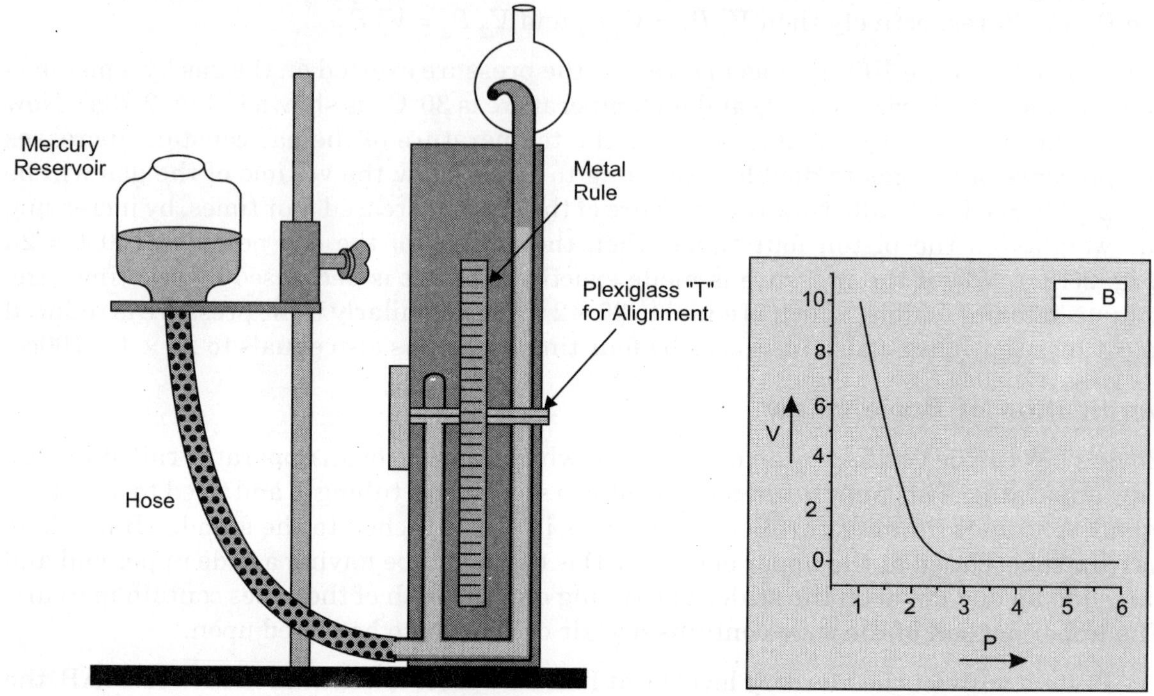

Fig. 2.14. The Boyle's Law apparatus

Charle's Law

Pressure remaining constant, the volume of mass of gas (given) increases or decreases by a common fraction $\left(=\frac{1}{273}\right)$ of its volume at 0°C for each degree Celsius rise or fall of temperature.

If V_0 represents the volume of a given mass of gas at 0°C, then from the law, easily we can have:

These are the mathematical expressions of Charl's Law

Volume at 1°C = $V_0 + V_0 \bullet \frac{1}{273}$

Volume at 2°C = $V_0 + V_0 \bullet \frac{1}{273}$

Volume at t°C = $V_0 + V_0 \bullet \frac{1}{273}$

If V_1 represents the volume at t°C then

$$V_t = V_0 - V_0 \bullet \frac{1}{273} = V_0\left(1 - \frac{t}{273}\right)$$

or, $V_t = V_0\left(1 - \frac{1}{273} \times 273\right) = 0$

Verification of Charles's Law

The gas therefore, at –273°C would occupy no volume. This temperature 273°C where all the gases have zero volume is termed as absolute zero.

A thermonic scale is derived, with this temperature as the starting point, the scale is known as perfect scale or absolute scale. The temperature measures on this scale are commonly goes by the name of absolute zero. In this scale each degree is equal to that of centigrade scale and the only difference between the two lines in the starting points, the absolute scale starting at -273°C of the centigrade scale. Hence, adding 273 to the former does the conversion of degrees centigrade into degree absolute (or degree Kelvin) simply. There are other scales of temperature in wide use. All these and their mutual relations are shown in the following steps.

Hence the equation for volumes we have the substitution of the value of $T = 273 + t$

$$V = V_0\left(1 + \frac{1}{273t}\right) = V_0\frac{273+t}{273} = \frac{V \bullet T}{273}$$

The verification of Charles law is total mathematical because no chemical experiment is possible at absolute zero temperature

But 273°C absolute is the same as 0°C and is generally by T_0 hence

$$V = V_0 \frac{T}{T_0}$$

or $$\frac{V}{T} = \frac{V_0}{T_0}$$

In similar way if the volume of a gas is kept constant and the temperature altered the pressure of the gas has been found to change. It was observed that with increase of temperature at

$$P = P_0 273\left(1 + \frac{1}{273}t\right)$$

$$= P_0 273\frac{t}{273}$$

$$= \frac{P_0 T}{T_0} \text{ or, } \frac{P}{T} = \frac{P_0}{T_0}$$

Combination of Charles's Law and Boyle's Law

From Boyle's Law, we get a relation between the volume (V) of a certain mass of gas and T pressure (P) when the temperature of the gas is kept constant. On the other hand, Charle's gives us a relation between the volume (V) of the given mass of gas and its absolute. Temperature and pressure change, the consequent change in volume of the given mass of gas may be obtained by combining the above two laws.

From Boyle's Law we see $V \propto \frac{1}{P}$ when temperature remains constant. Again, from Charle's Law, we have $V \propto T$ when pressure (P) is constant. So when both the temperature and pressure of a certain mass of gas at any time be V, T and prespectively and if the quantities change to V_2 T_2 and P_2 subsequently then :

Combined Boyle's Law and Charles's Law

$$\frac{P_1V_1}{T_1} = \frac{P_2V_2}{T_2} \text{ etc. ... = constant.}$$

It is to be noted that the above constant becomes same for all gases, if l gm-mole of the gas is considered; it is termed as R, the universal gas constant. The combine laws give us the equation $PV = RT$. This equation is known as perfect gas equation. The gases, which follow these equations, are called perfect gases.

Evaluation of '*R*'

R has the dimension of ergs/tem/mole as shown below :

P = Pressure = $\frac{\text{Force}}{\text{Area}} = \frac{\text{Force}}{L^2}$, where L is the dimension of length,

$V = L^3$

Hence, $R = \frac{PV}{T} = \frac{\text{Force}}{L^2} \times \frac{L^3}{\text{deg}} = \text{Force} \times \frac{L}{\text{Deg}}$

But, force × L = work + erg

Hence, R = Erg/deg (temp)/mole

(*i*) If P is expressed in atmospheres and V in litres

$$R = P \bullet \frac{V}{T} = \frac{1 \times 22.414}{273} = 0.082 \text{ litres atmosphere.}$$

(Since a gram of any gas occupies 22.414 litres of volume at N.T.P. that is when the pressure is 1 atmospheric and the temperature is 0°C by Avogadro's Hypothesis.

(*ii*) If P is in atmosphere and V in ml then, $R = P \bullet \frac{V}{T} = 1 \times \frac{22414}{273} = 82.06$ ml atmos

(*iii*) If P is in dynes and V is an ml (that is both in absolute units)

$$R = P \bullet \frac{V}{T} = 1.013 \times 10^6 \text{ dynes (approx).*}$$

Discussion on Theory of Matter

John Dalton (1766-1844) to explain the formation of compounds first introduced the concept of atoms from smaller substances known as elements. He told that substances are composed of tiny particles, which are invisible. He called them atoms. He further stated that the

*1 megadyne = 10^6 dynes) V = 22414 ml.

atoms of anyone particular element are identical in all properties as well as in weights. This is in brief Dalton's atomic theory.

Many of the elements and compounds are available in gaseous state. It was found that the volumes of the elements and the compounds involved in a chemical reaction, in which both initial and final constituents were gases, which were connected by a simple relationship, this was first noticed by a French Chemist Joseph Louis Gay-Lussac, (1778-1850). He found experimentally that hydrogen and oxygen related chemically in gaseous state to form water vapour, which was also a gas. He further stated that at the same temperature and pressure *two volumes* of hydrogen and *one volume* of oxygen combined to form *two volumes* of water vapour. Gay Lussac was surprised to see that a simple ratio 2: 1 : 2 existing between the volume hydrogen oxygen and water vapour. He asked himself, "*Do all other gases combine in such simple ratios to form new compounds?*" He then, prepared hydrogen chloride gas by combinill hydrogen and chlorine gas and also he found the same simple ratio is existing between the volumes of the reacting gases and the product gas. In this way, experimenting with other gases he found the same rule to exist. Then he formulated a law, known as Gay Lussac's Law. This law states that when different gases react with each other chemically to produce gaseous substances, then under the same condition of temperature and pressure, the volume of reacting gases and the product gases bear a simple ratio among one another. In trying to explain the above law by Dalton's atomic theory, a Swedish scientist name J.J. Berzzelius (1779-1848) came to a very interesting conclusion, which was, under same condition of the temperature and pressure equal volumes of all gases contain equal number of atoms. This was known as Berzzelius Hypothesis.

Avogadro's Hypothesis

But a serious difficulty arises in the application of the hypothesis in the chemical reaction of gases. It was found that for a satisfactory explanation of the chemical behaviour of gases, atoms needed to be subdivided which was forbidden according to Dalton's theory. To overcome this difficulty an Italian scientist named Amedeo Avogadro (1776-1856) took a bold step. He visualized two kinds of ultimate particles called atoms and molecules. The following properties were attributed to the molecules and the atoms :

(*i*) Molecules remain in the properties of the elements or of the compounds to which, they belong and have free existence. For example, molecules of water, molecules of hydrogen chloride gas, etc.

(*ii*) Atoms do not exist in free State but take part in chemical reaction.

(*iii*) Molecules are divisible but atoms are not. Two or more atoms combining together make a molecule for example by spilling one water molecule; we get two hydrogen and one oxygen atom. The hydrogen atom or the oxygen atom cannot be further split up. Similarly if we split one hydrogen chloride molecule we will get one hydrogen and one oxygen atom. With the idea of atoms and molecules, Avogadro modified Berzzelius hypothesis in the following way:

Under the same condition of temperature and pressure, equal volume of all gases will contain equal number of molecules. This is Avogadro's Hypothesis. This hypothesis therefore means that at a certain temperature and pressure, if a Vessel is filled up with various gaseous elements or compounds. Then in each case the number of gas molecules in the vessel will be Equal, for example 1 cc of oxygen or 1 cc of hydrogen will contain. As many molecules as 1 cc of chlorine or 1 cc of nitrogen or any other elementary or compound gas contains under the same condition of temperature and pressure. If 1 cc of hydrogen contains 1000 millions of molecules at a certain temperature and pressure then 1 cc of any other gas at the same temperature and pressure will contain the same number of molecules.

Relation between Vapour Density and Avogadro's Hypothesis

We can prove that, by application of Avogadro's hypothesis that the vapour density of a gas is half its molecular weight.

Now vapour density of a gas means *the ratio of the weight of a certain volume of gas to the weight of an equal volume of hydrogen taken under the same condition of pressure and temperature.* Thus if D were the vapour density then:

$$D = \frac{\text{Weight of the V volume of gas}}{\text{Weight of V volume of hydrogen gas at same temperature and pressure}}$$

We know that according to Avogadro's hypothesis, that under the same condition of temperature and pressure equal volume of all gases contain equal number of molecules, hence:

$$D = \frac{\text{Weight of the one molecule of the gas}}{\text{Weight of the one molecule of the hydrogen}}$$

$$D = \frac{\text{Weight of the one molecule of the gas}}{\text{Weight of the 2 atoms of the gas}}$$

$$D = \frac{\text{Weight of the one molecule of the gas}}{\text{Weight of the one atom of hydrogen} \times 2}$$

$D = \frac{M}{2}$ where M is the molecular weight of the gas.

Thus the molecular weight of carbon-dioxide (CO_2) M= molecular weight is 44. Then the vapour density is 44/2 = 22.

Alternatively determining the vapour density of a gas in the laboratory, we can fix up its molecular weight.

Molecular Weight and Avogadro's Number

The molecules of an element or a compound are so small that direct determination of their weight is not possible like atomic weight, molecular weight of a substance is also measured will the composition with the substance, e.g. oxygen atom whose weight is taken as 16. It may be defined as the number of times of a molecule of a substance is heavier than one sixteenth of the weight of an atom of oxygen so :

$$\text{The molecular weight} = \frac{\text{Weight of a molecule of a substance}}{\frac{1^{th}}{16}\text{ of the weight of oxygen atom}}$$

In order to calculate the molecular weight of an element or compound at first, the number of constituent atoms are to be ascertained from its particular formula because *the molecular weight is the sum total of the atomic weights of the atoms which constitute the substance that an element or compound.*

Avogadro's Number and Its Determination

To determine the Avogadro's Number we should understand what is called gram molecular weight? The gram molecular weight is *the quantity of a substance whose mass in gram is numerically equal to its molecular weight,* for example the molecular weight of oxygen (O_2) therefore, the gram molecular weight of oxygen is also 32. The molecular weight of water 18.01, therefore, the gram molecular weight of water is also 18.01.

The volume occupied by the mole of any gas is called gram molecular volume. According to Avogadro's hypothesis the gram molecular volume of a certain mass of gas at N.T.P. (natural temperature and pressure) is 22.4 litres, and the every mole of the substance contains equal number of molecules, this number is known as Avogadro's number, which is 6.06×10^{23} N.T.P.

NUMERICAL CHEMISTRY

In this section we shall now discuss various problems of the gas laws.

Example 1 : *The volume of a certain mass of gas at 60 mm of pressure is 450 ml. What will be the volume of the gas if the pressure is changed to 17mm? Temperature is constant.*

Ans. : Since the temperature is constant we can apply Boyle's Law in this problem,

$$P_1V_1 = P_2V_2$$

where $P_1 = 60$ mm

$V_1 = 450$ ml

$V_2 = ?$

$P_2 = 17$ mm

$$\therefore V_2 = \frac{60 \times 450}{17} = 1588 \text{ ml}$$

Thus the volume of the gas after reducing the pressure in 1588 ml

Example 2 : *The volume of a certain mass of a gas at 86mm of pressure and 0°C temperature is 130ml. At what pressure (assuming the temperature is constant), will the volume be (i) 2 times, and (ii) 1 times of the initial volume?*

Ans. : As the temperature is constant, Boyle's Law is applied in this case also,

(*i*) Here, $P_1 = 86$ mm $V_1 = 130$ ml

$\therefore V_2 = 130 \times 2 = 260$ ml {since Volume is doubled)

Now we have $P_1V_1 = P_2V_2$

or $86 \times 130 = 260 \times P_2$

or $P_2 = 86 \times 130/260 = 43$mm.

(*ii*) Here $P_1 = 86$ mm $V_1 = 130$ ml

Now $V_2 = 130 \times 1 = 65$ ml $P_2 = ?$ Now applying the above formula $86 \times 130/65 = 172$ mm.

Example 3 : *A certain mass of gas at 10°C is so heated that both its pressure and volume are doubled. What will be the final temperature of the gas?*

Ans. : Let the volume of the gas at 10°C is V_0 and the pressure P_0 suppose at first the pressure of the gas is raised to T_1 so that its volume is V_1 at constant pressure P_0. As the pressure is constant we can use Charle's Law.

$$\frac{V_1}{T_1} = \frac{V_0}{T_0} = \frac{V_0}{273+10} \quad [T_i = 10°C = 273°K]$$

$$\therefore \quad V_1 = V_0 \times \frac{T_1}{273+10}.$$

Now keeping the temperature constant at T_1 the pressure is doubled so that the volume of the gas becomes $2V_0$ then applying the Boyle's Law:

$$2P_0 \times 2V_0 = P_0 \times V_0 \times \frac{T_1}{273=10}$$

$$\therefore \quad T_1 = 4 \times 273 + 10 = 1093°K$$

$$\text{or} \quad (1093 - 273) = 820°C$$

PRACTICE PROBLEM

Prob. 1. The volume of a certain mass of gas is 240ml, at absolute temperature, 400°K keeping the pressure constant if the temperature is reduced to 100°K, find out the volume of the gas? (**Ans.** = 60mm)

Prob. 2. The volume of a certain mass of gas at 20°C and 60cm pressure is 120cc Calculate the volume at 100mm pressure assume that temperature is kept constant. (**Ans.** = 72ml)

Prob. 3. The volume of a gas changes from 1500ml to 3000ml at a constant temperature of 12°C, If the initial pressure is 70mm of Hg what will be the fin pressure? (**Ans.** = 35mm)

Prob. 4. The pressure of a gas is 80 mm, the volume of the gas at that pressure 600ml. If the temperature remains unchanged and the pressure is reduced to 40mm, then calculation the volume occupied by the gas at that pressure? (**Ans.** = 1200ml)

Prob. 5 A certain amount of a gas occupies a volume of 600ml at a pressure of 760mm Hg at 0°C temperature. What will be the volume of the gas if the pressure increase to 800mm of Hg at 0°C? (**Ans.** = 570ml)

CHAPTER

3

Architecture of Atom

Introduction

As we have discussed in the previous chapter about the matter and energy the matter, that can be of any kind consists of tiny particles called *molecules* these molecules can be divided further into most fundamental particle the atom; which is the main building block of the substance.

Now-days the experiments have proved that atoms are no more the fundamental particles of matter, they are subdivided into three fundamental particles the *electron, proton* and *neutron.* Now even the electron can be divided and the most finite particle is derived called *Tao-leptons.*

Discovery of Electron

Before speaking about the discovery of electrons, we must know something about cathode rays, which was discovered by Julius Pluker in 1859 and later studied fully be J.J Thomson in 1887. Then they knew it that under excessively low pressure in order of 0.01mm of mercury produced a stream of rays from the cathode, which enlightened the fluorescent screen as they are emitted from cathode, which gave the name as Cathode Rays.

Some of the properties of cathode rays are stated below :

(*i*) The Cathode rays travel in straight and cast shadows of the object in their paths.

(*ii*) When they fall on certain substances, they produce fluorescence, depending upon the colour of the substance.

(*iii*) They are emitted right-angled from the cathode substance.

(*iv*) Cathode Rays are permeable through the thin sheets of mater, without puncturing it and can pass through the thick volume of air.

(*v*) Cathode rays produce a rise of temperature when targeted on a substance.

(*vi*) They ionize the gases through which they pass.

(*vii*) They travel with the velocity ranging 113 to 1/10 of that of light.

(*viii*) When they fall on heavy solid substance (like platinum, tungsten, they produce a highly penetrating rays called X-Rays, or Roentgen Rays.

(*ix*) The cathode rays cause mechanical effort in a small pedal wheel placed in their path. This shows that cathode rays consists of electrons, which produce mechanical pressure

(*x*) These rays are deflected in an electrical or magnetic field from their normal path in a direction indicating that they are not only gets deflected but also negatively charged particles.

The cathode rays consisting of the negatively charged particles called electrons, which are formed by disintegration of atom of the gas under high electrical tension or from the material of the cathode. The electrons were found to possess the same properties, irrespective of the material of the electrodes or the nature of the gas in the discharge tube.

Mass of Electron

The charge to ratio of an electron (*e: a)* of and a hydrogen atom was measured to be 1.759 x 107 e.m.u 1 g. Hence

$$\frac{\text{Mass of an Electron}}{\text{Mass of an H atom}} = \frac{m}{m_H} = \frac{.9577 \times 10^4}{1.759 \times 10} = \frac{1}{1840}.$$

The is the mass of an electron is smaller than that of hydrogen atom and the ratio is 1:1840.

Discovery of Protons the Positive Rays

Goldstein observed in the year 1886 that there is a presence of some another kind of rays in the discharge tube. These rays are known as Positive Rays or Canal rays. The positive rays are found to have the following properties :

(*i*) These rays follow the direction of the route from anode towards cathode.

(*ii*) They are found to get deflect in an electric or magnetic field in the direction showing that they are positively charged material particles.

Experiments have shown that the charge to mass ratio *(elm)* of such a particle depends upon the nature of the gas in the discharge tube. The value of this ratio is found to be minimum for hydrogen i.e. 105 coulombs 1 g. Such particles in case of hydrogen are called Protons or positively charged electrons.

Origin of Positive Rays

On the application of a high voltage through the discharge tube, electrons are ejected from the cathode. These electrons move towards the anode and on their way they collide with the gaseous molecule. The electrons being very energetic cause the removal of electrons from molecules of gas.

$$H_2 \rightarrow H + H$$
$$H \rightarrow H^+ + e$$
$$H \rightarrow H_2^+ + e$$
$$O_2 \rightarrow O^+ + e$$
$$O_2 \rightarrow O_2^{++} + e$$

The gaseous ions thus formed travel towards the cathode. The rays consisting positive ions are known as protons.

Neutron

James Chadwick in 1932 obtained a new radiation from the bombardment of $_4Be^9$

$$_4Be^9 + _2He \longrightarrow _6C^{12} + _0n^1.$$

This new radiation was found to consist of particles having no charge. The mass of the particle was almost equal to the mass of proton. This particle was named as neutron, as it is electrically neutral. These are the three main subatomic particles of the matter.

A comparative study of the three particles is shown in the table below :

Particle	Mass			Charge
Electron	0.0054 a.m.u	9.183×10^{-28} g	–1	4.80216×10^{-10} e.s.u or 1.60206×10^{-19}C
Proton	1.0073 a.m.u	1.6735×10^{-24} g	+1	4.80216×10^{-10} or 1.6020×10^{-19} C
Neitron	1.0086 a.m.u	1.674×10^{-24} g	0	0

About the structure of the atom in 1910, the renowned British physicist, Lord Rutherford did an experiment on the scattering of alpha particles, which were left to pass through a gold foil and fall on a cylindrical fluorescent screen, which shows the scattering spectrum. The whole apparatus was enclosed in an evacuated chamber to prevent the extraneous* deflections by air molecules. A slit was there at one side through that alpha rays could enter and hit the gold foil. The course of the alpha particles could be deduced from the position of the flash of the light. Whenever alpha particles collided with the screen. By such studies it was established that the bulk of the beam passed through the gold foil without deflection, very few particles were deflected. Later on it was found that a few of the alpha particles did bounced back from the direction, which they came. The number that did was found to be directly proportional to the thickness of the gold foil. It was also found that if gold foil was made

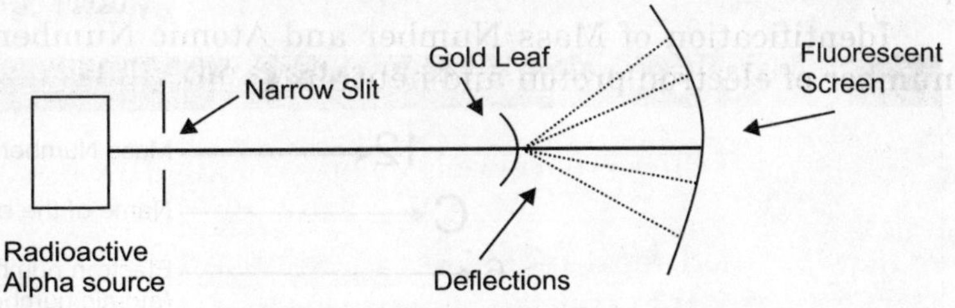

Fig. 3.1. Lord Rutherford's experiment

*Extraneous : Outside external.

1 atom thick then only 108 (see Fig. 3.1. Lord Rutherford's experiment).

Constitution of Nucleus

The atomic nuclei constitute proton and neutron bound together and the radius of the nucleus is very thin about 2×10^{-13} as compared to 10^{-8} cm of the atom (hydrogen atom). As most mass of the atom is concentrated in the nucleus and its density is enormously high approximately 1013 g cm^3. The densities of different nuclei are almost same.

Each nucleus consists of a definite number of protons and the charge on the nucleus is determined by this number, for example the nucleus of hydrogen atom contains one proton. Hence, the nuclear charge is + 1, similarly the nucleus of lithium consists of 3 protons and the charge of the nucleus is +3.

Atomic Number

The number of molecular charge is known as atomic number. It is also known as that an atom is neutral so the nuclear charge must have an equal amount of opposite charge. Here the electron~ come into account, that is the number of protons in an atomic nuclei always equals to the number of electrons in an atom, which means that: -

> Atomic Number is considered as the most fundamental property of atom.

Atomic Number = Number of protons = No. of Nuclear charge = Number electrons

Atoms with the same atomic number can have different atomic masses but will have almost identical properties.

Mass Number

From the total number of protons and neutrons determines the mass number of an atom.

Mass number = Number of Protons +Number of Neutrons

If the atomic number and the mass number are given then the number of electrons protons and neutrons and also vise- versa can easily be determined.

Identification of Mass Number and Atomic Number and also determination of the number of electron proton and neutron.

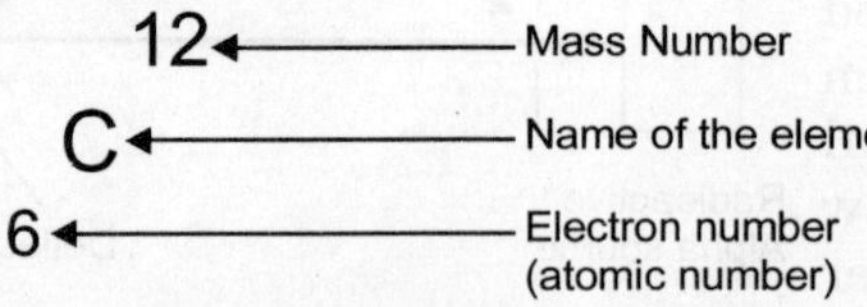

∴ We know that the electron number is the number of electrons which is 6. The mass number the total number of the protons and neutrons which is 12 and we also know that the number of protons is always equal to the number of electrons which is also 6.

∴ The number of Neutrons are (12 – 6) = 6. and the name of the element is carbon.

Nature and the Forces in the Nucleus

Y1U1bwa in the year 1935 said that pi mesons act as a commenting* force binding the proton- proton, proton-neutron and neutron-neutron together. This pi meson may have a positive charge (π^+), a negative charge (π^-) or no charge (π^∞) These pi mesons are continuously exchanged as follows (Fig. 3.2):

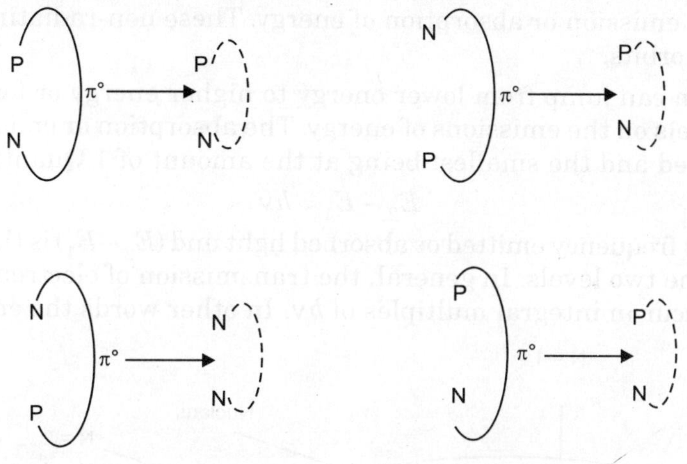

Fig. 3.2. The configuration of the nucleus the pi-mesons

But all the mesons are unstable outside the nucleus.

An Atom is Neutral, Why?

An electron has a charge but opposite in sign to that of a proton. As the number of protons in an atom is equal to the number of electron the total positive charge in the nucleus is equal to the total negative charge of all the electrons in the atom. Hence an ordinary atom is electrically neutral.

Considering the atom of oxygen, it has 8 protons in the nucleus and 8 extra nuclear electrons :

∴ Total positive charge = +8 and total negative charge is –8.

∴ The net charge is +8 +(–8) = 0.

Deffects of Rutherford's Model

1. According to Maxwell's theory, a charged particle moving under the influence of an attractive force must emit electromagnetic radiation continuously. In giving out radiation an electron, which is negatively charged particle should continuously loose energy. Its orbit should therefore; become steadily smaller, and in a very short time it should collide with the nucleus. But we know that atomic nucleus does not behave in this way.
2. Also Rutherford's model does not explain the existence of line spectra.

*Commenting : Interrupting

Bohr's Theory

Due to their defects of the model explained by Lord Rutherford Prof. Niels Bohr (1936) proposed some amendments:

1. An atom possesses several stable orbits, which are circular in nature. In these orbits electrons revolve, but the number of electrons in a specified orbit is fixed, so there is no emission or absorption of energy. These non-radiating orbits are called stationary orbits.
2. An electron can jump from lower energy to higher energy or form higher to lower energy levels on the emissions of energy. The absorption or emission takes place in a fixed rated and the smallest being at the amount of 1 Quantum. Hence.

$$E_2 - E_1 = h\nu$$

Where, v is frequency emitted or absorbed light and $(E_2 - E_1)$ is the energy difference between the two levels. In general, the transmission of electrons from one orbit to another occur in integral multiples of $h\nu$. In other words the energy is quantised.

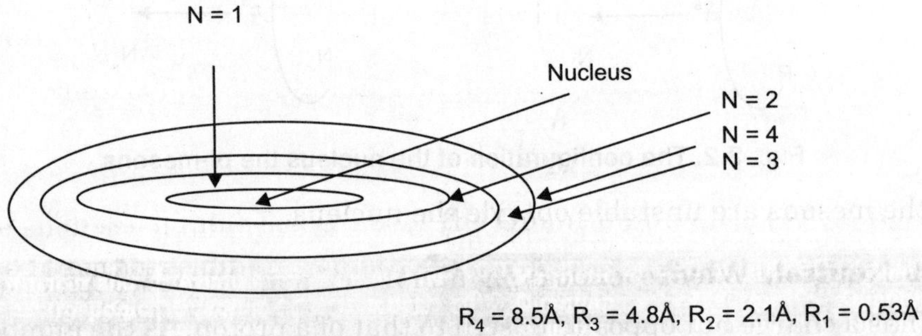

Fig. 3.3. The diagrammatic representation of an atomic nucleus

Quantisation mean that a quantity does not change vigorously.

3. The angular momentum of an electron moving in an orbit is an electromagnetic multiple of h/21t. That is $mvr = nhl2x$, where n = 1, 2, 3, 4, 5, 6 for the number of orbits and h is the Plank's constant, 6.624×10^{-27} ergs-sec.

Radius and Energy of Orbit

Let us consider a circular orbit of radius π and the linear velocity of an electron v, the mass of the electron being m; for an orbit to be stable, which has been shown in Fig. 3.4.

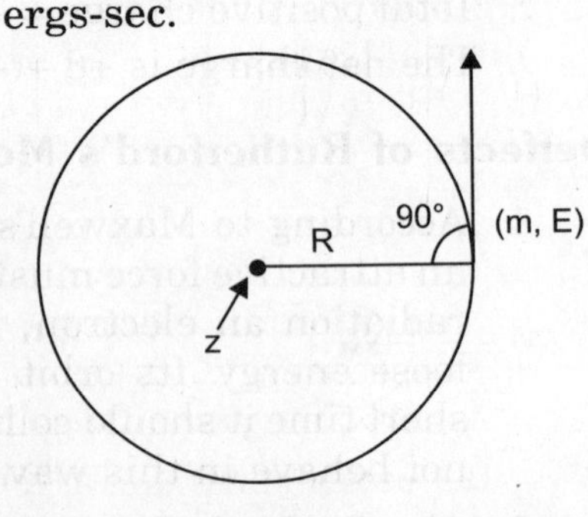

Fig. 3.4. The Radius and the energy level of the orbit

The centrifugal force produced by the moving electron must be equal to the attractive force: between the nucleus and the electron.

$$\text{Centrifugal Force} = \frac{mv}{r}$$

Electronic Force Attraction $\frac{E^2}{R^2}$ ...(3.1)

$$\therefore \quad \frac{mv}{R} = \frac{E^2}{mR} \text{ or, } v_2 = \frac{E^2}{mR}$$

Now according to the third proposal, which is being stated below :

$$\text{mvr} = \frac{Nk}{2\pi}$$

$$v^2 = \frac{N^2h^2}{4\lambda^2m^2N^2}$$

$$\text{or,} \quad R = \frac{N^2h^2}{4\lambda^2e^2m}$$

For a particular atom all quantities on the R.H.S. except N is constant. Hence, $R \propto N^2$.

In other words the radii of successive orbits are proportional to the squares of the integers 1, 2, 3, 4 ...etc. These integers are called *Principle Quantum Numbers*.

To Calculate the Total Energy of the Electron

$$\text{Total energy} = \text{K.E.} + \text{P.E.} = \frac{1}{2}mv^2 = V = \frac{E^2}{2R}$$

$$\text{Substituting the value of R, E} = \frac{E^2}{2} \times \frac{4\pi^2mE^2}{N^2h^2} = -\frac{2\pi^2mE^2}{N^2h^2}$$

That is $E \propto \frac{1}{N^2}$

Thus for a given orbit the value of energy will remain constant.

Merits and Demerits of Bohr's Theory

(1) Bohr's theory explains successfully the spectra of the simple one-electron system, such as H, He^+, Li^{++}

(2) Electronic transitions as predicted by Bohr's theory closely agree with those obtained experimentally by Rydberg

(3) The radius of the first orbit in hydrogen atom calculated from Bohr's theory is found to be in good agreement with the most probable radius of the ground state predicted by Scrodinger *Wave model*. But the theory fails to explain the spectra of more complicated multi-electron atoms.

Bohr's Burry scheme of the arrangement of Electrons in the Valence Orbits of an atom

The atomic orbits are arranged circularly around the nucleus. The orbit nearest to the nucleus is called first orbit (N=1) or K-shell, the next orbit is called second orbit (N=2) or

L-shell. Similarly the, M-shell (N=3) and N-shell (N=4). The number of electrons in these orbits is ascertained by Bohr's Burry scheme. According to this scheme: -

(1) The maximum number of electrons in an orbit is determined by the formula $2n^2$ where the n is the number of the orbit. Thus,

The maximum number of electrons in the first orbit is $2n^2 = 2 \times 1 \times 1 = 2$.

The maximum number of electrons in the second orbit is $2n^2 = 2 \times 2 \times 2 = 8$.

The maximum number of electrons in the third orbit is $2n^2 = 2 \times 3 \times 3 = 18$ and so on.

(2) Now the electron number is given the orbit number can be calculated by

$$2n^2 = E$$

or, $$E = 2n^2$$

or, $$n^2 = \frac{E}{2}$$

or, $$n = \sqrt{\frac{E}{2}},$$

where n is the number of orbit and E is the number of electrons.

Heisenberg's Uncertainty Principle

According to this principle, the simultaneous determination of the position and momentum of an electron is impossible. The product of uncertainty is equal to or greater than Plank's constant 'h'.

> This principle is regarded as the fundamental principle of nature.

$$\Delta x \bullet \Delta y \geq \frac{h}{4\pi}$$ where Δx is the uncertainity indetermination of the position and Δy is the determination of momenutum

Wave Mechanical Concept of Atom

Louis de Broglie (1923) introduced a new concept that electron possesses corpuscular and wave character.

$E = h\nu$, $E = mc^2$ (where c is the velocity of light).

$\therefore h\nu = mc^2$ or hc/A. (since A.c = A.v) or $hlA. = mc$.

For an electron, let the velocity be y, then $hl1.. = mv = p$, where p is the momentum of the electron. $\therefore 1 = hlp$.

This is known as De Broglie equation showing the particle as well as wave character of electron.

In wave mechanics the complete definition of an electron is given by *a mathematical function called the wave function and the wave function of an electron is called orbital.*

Quantum Number

Heisenberg, Schrodinger and Dirac (1925) developed a theory, which governs the mechanics of a small particle like electron. This theory is called the Quantum Mechanics or the wave mechanics.

According to the quantum mechanical theory of the atoms total of three quantum *n*, 1 and *m* is needed to specify the physical state of electron in a hydrogen atom. The rate of the 4th quantum that is spin quantum (s) to specify the orientation of the axis of spin of the electron is also to be considered.

(a) Principal Quantum Number : This gives an idea about the number of main energy levels in which the electrons reside. It is denoted by *n*. The energy of transmission of the electron from one circular orbit to another depending upon the value of *n*.

The value of *n* gives the information about the energy of the electron and the size. Large *n* means large size, the *n* can have any value of *n*, which should be an integer and lie between O and infinity.

As *n* is increased the, energy of the electron increases. The energy of the electron is expressed as :

$$E = \frac{2\pi^2 me^4 z^2}{h^2 n^2}$$

Where *n* is the principal quantum number

When $n = 1, L = 0$, i.e. one sub-shell

When $n = 2, L = 0, 1$ i.e. two sub-shell

When $n = 3, L = 0, 1, 2$ i.e. three sub-shells

When $n = 4, L = 0, 1, 2, 3$ i.e. four sub-shells.

Thus sub-shells corresponding to $L = 0, 1, 2$ and 3 are called *s*, *p*, *d* and *f* sub-shells respectively. The energy content of the sub-shell of a given shell increases in the order of $S<p<d<f$.

(c) Magnetic Quantum Number : The sub-shells defined by definite *n* and *L* values in the presence of a magnetic field are further subdivided. The further sub divisions are denoted by a third quantum number, *m*, which is known as *Magnetic Quantum Number.* This quantum number actually *defines the orientation of sub-shells in space, m* can have any integral value from $-L$ to $+L$ including 0. Thus, the total possible values of *m* will be as follows:

If L = 0 m = 0,

"L = lm = –1, 0, +1

II L = 2, m = –2, –1, 0, +1, +2

II L = 3, m = –3, –2, –1, 0, +1, +2, +3

All the possible orientations corresponding to $L=3$ are shown in the Fig. 3.5.

(d) Spin Quantum Number : It has been assumed that each electron spins around its own a - like a top, this behaves like a small D18gnet. A spinning electron can have only two possibilities The electron can either spin clock-wise or anti-clock-wise this is shown in the figure 3.5a below:

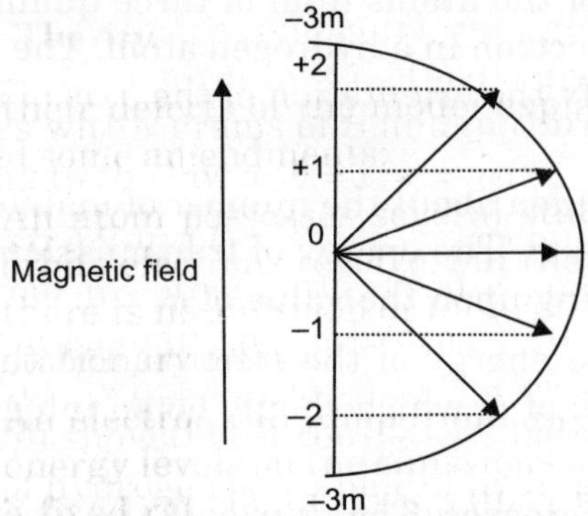

Fig. 3.5. All the possible orientations of L = 3

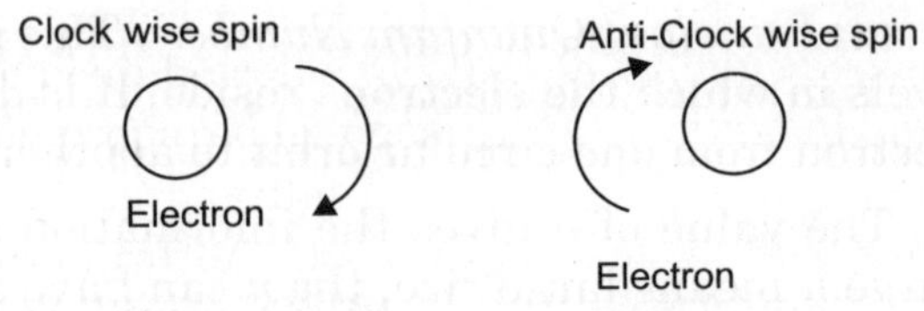

Fig. 3.5(*a*): The spin of the electron

The two directions of spin are usually represented by arrows pointing upwards (*t*) or down wards (.). In order to account for the energy originated out of spinning of the electron, a fourth quantum number is independent of the other three quantum numbers, *n*, *I*, and *s* can have two possible values +1 and –1 depending on the direction of the spin. The spin angular momentum is given by :

$$\sqrt{s(s+1)}\frac{h}{2\pi}$$

Now we shall see a table below about the Quantum number of electrons.

Principal Quantum Number(n)	Number and Spectroscopic Sign.	Azimuth Quantum Number(L)	Number and Symbol	Magnetic Quantum Number	Total Electron Number	Spin Quantum Number	Number of electron in subshell
1	K	0	s	0	2	+1/2, –1/2	2
2	L	0, 1	s, p	0	8	+1/2, –1/2	2
3	M	0, 1, 2,	s, p, d	+1, 0, –1	18	+1/2, –1/2	6
4	N	0, 1, 2, 3	s, p, d, f	0,	32		
				+1, 0, –1			10
				+2, +1, 0, –2, –1		$\begin{Bmatrix}+1/2\\-1/2\end{Bmatrix}$	14

Atomic Structure

The path of an electron around the nucleus cannot be described in terms of orbits. An orbit means a definite path, which is not correct according to the Uncertainty Principle. The motion of electrons around the nucleus is descri-

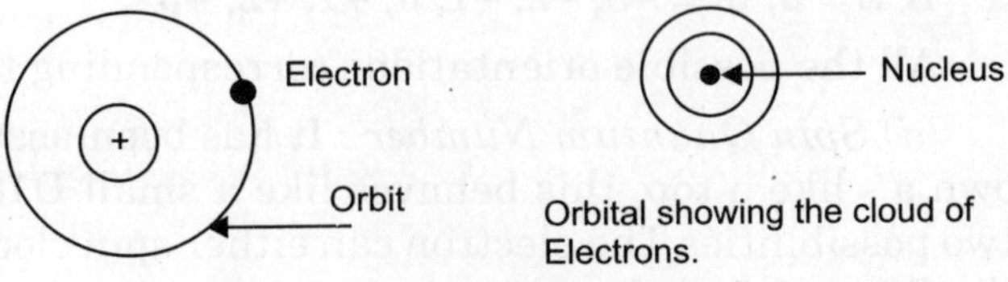

Fig. 3.6. The Orbit and Orbital.

bed in terms of a set of mathematical expressions, which allow for the dual behaviour of electrons. These expressions are called Orbital Wave Functions, or simply Orbitals, denoted by *Y*. Figure 3.6 shows the structure of orbital and orbit.

In this physical sense, an *Orbital is a region in space around the nucleus in which there is the maximum probability of the electrons being located This region is a cloud of negative charge*. However, it is very difficult to depict an original picture of an orbital.

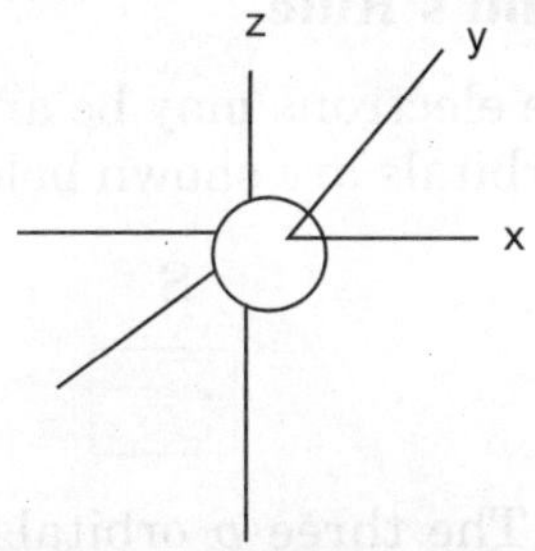

Fig. 3.7. The diagrammatic example of s-subshell

Sub-shells and Orbitals

The sub-shell (i.e when $L = 0$) has only one possible. This means that s-subshell has only one orbital and nucleus. The size of the s-orbital depends upon the of s-subshell is shown in Fig. 3.7.

Orientation (i.e. *m*, has only one value) in space. What is s-orbital? It is specially symetrical to value of the principal quantum number. A diagram of the size of the s-orbital depends upon the s-subshell is *shown* in Fig. 3.7.

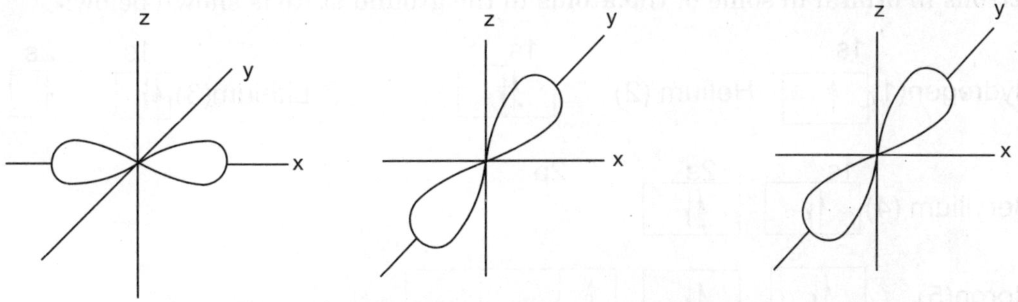

Fig. 3.8. The diagram of three orbitals of p-subshell

For p-subshell (i.e., when $L = 1$), *m* has three values (+1, 0, –1). Therefore, p-subshell has three orbitals. These orbitals are designed as *Px Py* and *Pz*, the diagram of them are *shown* in Fig. 3.8.

The d-subshell (i.e $L = 2$), m has five values (+2, +1, 0, –1, –2). Therefore, d-subshell has five orbital, being designed as dxy, dyz, $dx^2 - y^2$.

The f-subshell (i.e. $L = 3$), has seven values for *m* (+3, +2, +1, 0, –3, –2, –1), therefore, *f*-subshell will have seven possible orientations, this means that f-subshell has seven orbitals.

Pauli's Excleution Principle

Now two electrons in an atom have all the four quantum numbers identical.

The principle has been found to be very useful in determining the maximum number of electron in each shell. Let the azimuth quantum number is $(2L + 1)$. Now L can have values ranging from O to $(n - 1)$. Hence, total number of electrons in a given shell is

$$\sum_{L=0}^{L=n-1} 2(L+1) = 2n^2$$

Hund's Rule

The electrons may be arranged into boxes. Each box represents an orbital. Thus s, p and d orbitals are shown below:

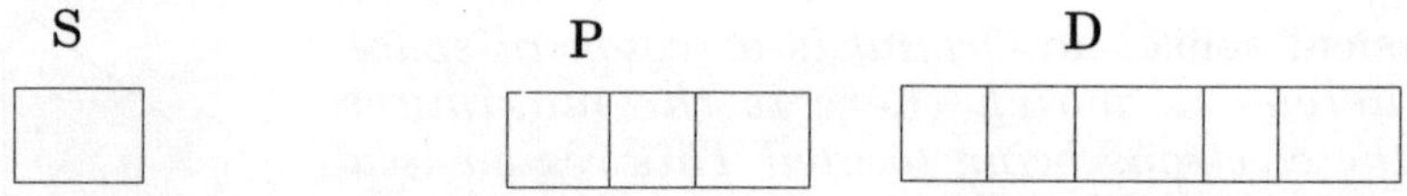

The three *p* orbitals are of identical energy; similarly, the five d are of equivalent energy, such orbits are called degenerate orbitals. According to Hund's rule, one electron is placed in all orbitals of equal energy (degenerate orbitals) before two electrons are placed in anyone of the degenerate set and the electrons in the singly occupied orbitals have spins. If two electrons are to added to the p-orbital, these are two possible ways:

The first arrangement is not true according to Hund's rule. The second arrangement of electrons in orbital in some of the atoms in the ground state is shown below:

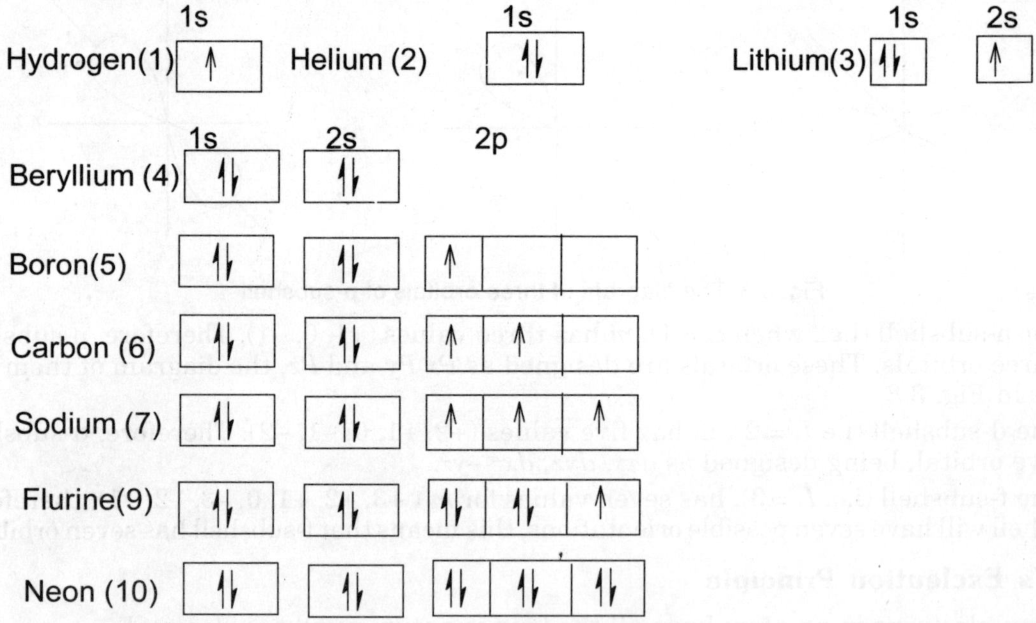

The maximum number of electrons corresponding to $n = 3$

$$[2(L+1) \quad + \quad [2(2L+1) \quad + \quad [2(2L+1)$$

$$L = 0 \qquad\qquad L = 1 \qquad\qquad L = 2$$

$$= 2 + 6 + 10 = 18$$

The configuration of He (2) can be shown by any of the following two ways as in the figure below :

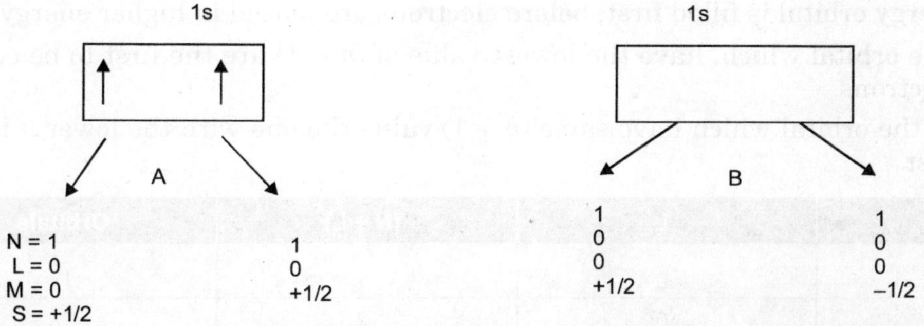

The configuration as shown in (A) is incorrect to Pauli's Exclusion principle because the electrons having same values. (+1, –1).

In configuration,(B) both the electrons are having different values so this configuration is correct.

As consequence, of this principle two is the maximum number of electrons permitted in the. s-orbital, six in the the p-orbitals and 10 in the five d-orbitals. This can be shown in the chart given below :

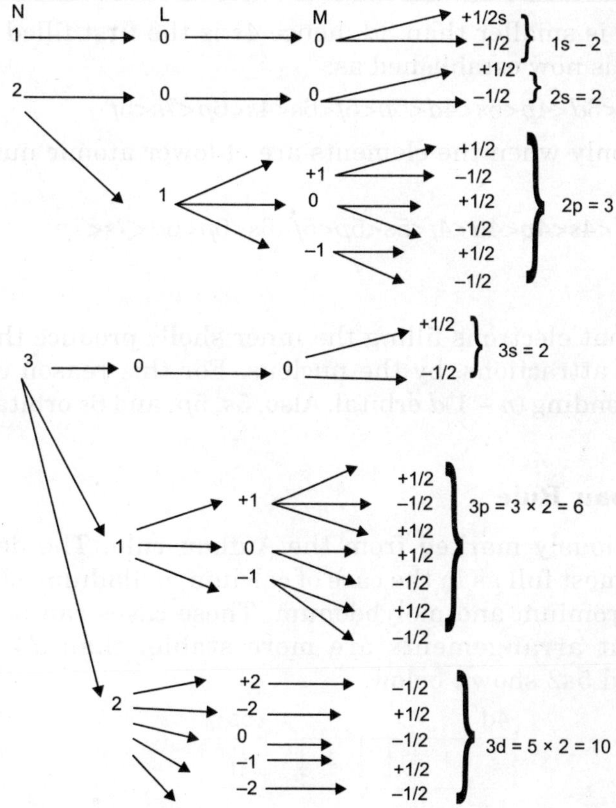

Aufbau Principle or Building *p* Principle

According to this principle electrons are put into order of the energy of the orbitals. The lowest energy orbital is filled first, before electrons are placed in higher energy orbitals.

1. The orbital which, have the lowest value of $(n + 1)$ are the first to be occupied by electrons.
2. In the orbital which have same $(n + 1)$ value the one with the lower n is occupied first.

N	L	(N + L)	Orbitals
1	0	1	1*s*
2	0	2	2*s*
2	1	3	2*p*
3	0	3	3*s*
3	1	4	3*p*
3	2	5	3*d*
4	0	4	4*s*
4	1	5	4*p*

$(N + L)$ values of 4*s* is smaller than 3*d*, hence 4*s* is the first filled up before 3*d*. The order of orbitals energy is now established as:

$1s<2s<2p<3s<3p<4s<3d<4p<5s<4d<5p<5f<6s<4s<6p<7s<5f$

This order is valid only when the elements are of lower atomic numbers for element higher atomic number.

$1s<2s<2p<3s<3p<3d<4s<4p<4d<4f<5s<5p<5f<6s<6p<6d<7s<7p$

Screening Rule

Screening rule states that electrons filling the inner shells produce the screening of the outer electrons against attractions, by the nucleus. For this reason every orbital filled earlier than the corresponding $(n - 1)d$ orbital. Also, 5*s*, 5*p*, and 6*s* orbitals are earlier than 4*f* orbital.

Deviations from Aufbau Rule

Some departure is obviously marked from the Aufbau rule. The departure comes in d-orbitals are earlier almost full as in the case of calcium, palladium, silver, and platinum gold. Or half full as chromium and molybdenum. These cases can be explained by fact that completely vacant arrangements are more stable, than *d*4 and *d*9, consider configuration of 4d4 and 5s2 shown below.

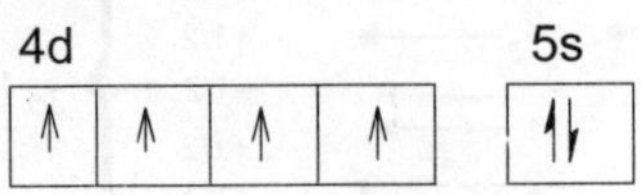

If one electron moves from 5d to 4d the arrangement becomes somewhat like the arrangement as shown in the figure below:

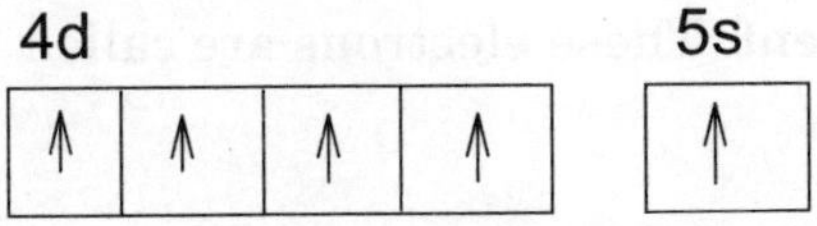

In this arrangement the 4d is complete half-filled hence this arrangement is preferred.

Electronic Structure

The chemical behaviour of an atom is mainly controlled by electronic structure. The term electronic structure means actually three things they are:

I. Distribution in space around the molecules of these electrons and

II. The relative energies of these electrons.

III. The number of electrons.

Ground State Electronic Configurations of Elements

Atom in it's lowest energy state is state to be in the normal or ground state. Below there are some atoms of chemical elements showing the configuration at ground state.

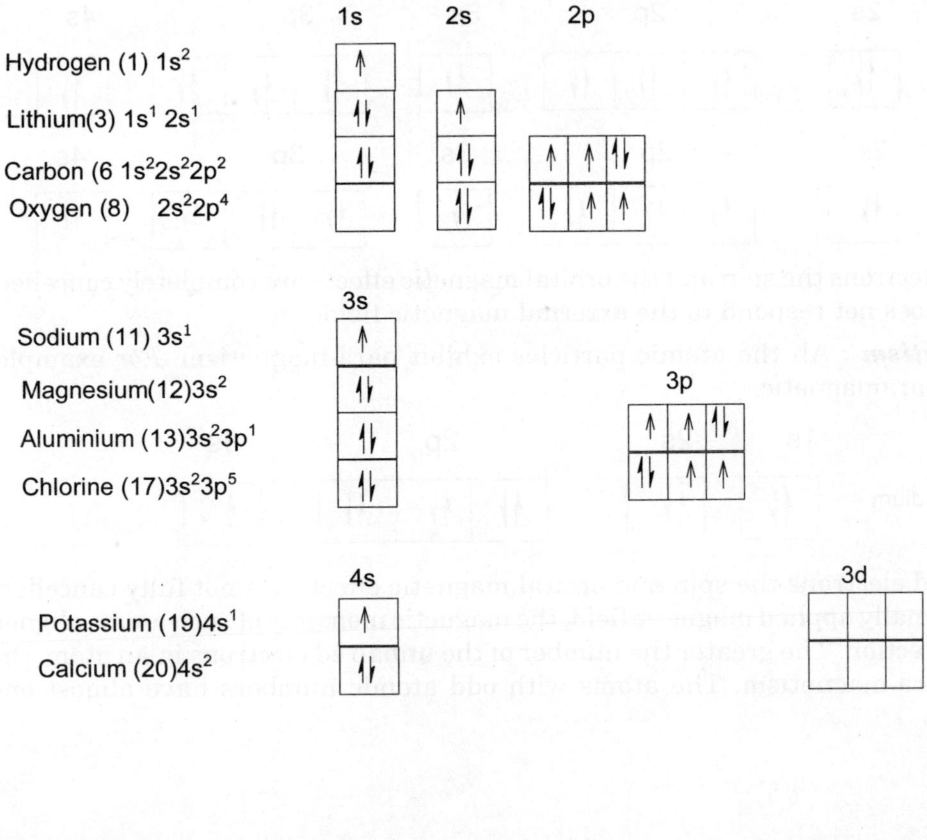

Valence Electrons

The number of electrons in the outer most shell of an atom determines the maximum bondable electrons of the element. These electrons are called *Valence Electrons.*

Core Electrons

The electrons occupying the inner shells preceding the outermost are called Core Electrons, for example, in the atom there are only one valence electron Sodium (II) $Is^2\ 2s^2\ 2p^6\ 3s^1$ Valence electron = 1, Core Electron = 0. Valence electrons take part in chemical reaction and bonding.

Kernel

The nucleus and the low frequency electrons, except the electron in the outermost shell that are never affected by a chemical change are called *Kernel* of the atom.

Magnetic Properties if An Atom

Atoms exhibit magnetic properties, there are two kinds of magnetism: (1) Dia-magnetism and (II) Para-magnetism.

Dia-magnetism : Atoms, ions or molecules containing only paired electrons or completely filled orbitals show dia-magnetism, because all the electrons are paired.

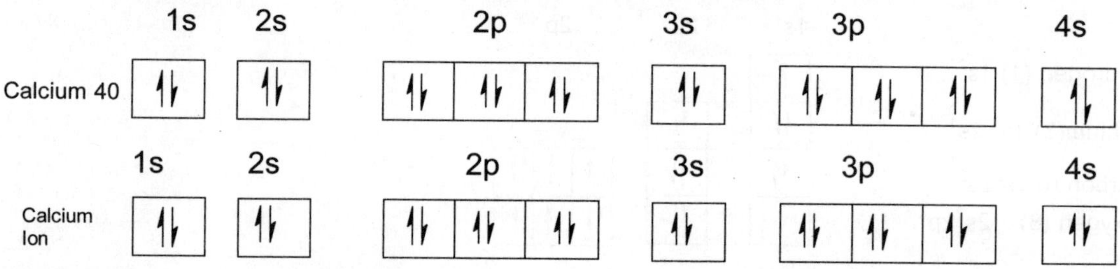

With paired electrons the spin and the orbital magnetic effects are completely cancelled. Hence, the pair does not respond to the external magnetic field.

Para-magnetism : All the atomic particles exhibit para-magnetism. *For* example, Sodium atom is paramagnetic.

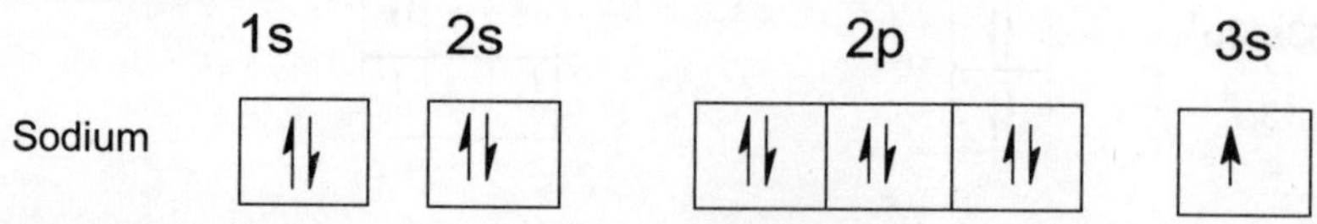

With unpaired electrons the spin and orbital magnetic effects are not fully cancelled. Hence, in an externally applied magnetic field, the magnetic moments of such atoms aligned in a particular direction. The greater the number of the unpaired electrons in an atom, the greater is the para-magnetism. The atoms with odd atomic numbers have almost one unpaired electron.

Determination of Quantum Member for an Electron

In order to completely identify an electron all the four quantum numbers, (*n, l, m* and *s*) are needed. The following guidelines should be followed to know the values of *n, l, m* and *s* for an electron:

(*i*) Write sown the electronic configuration of the atom and get the value of *n* and *l* directly as an example

Oxygen (8) $2s^2\ 2p^4$

Suppose we have to know the value of *n* and *l* for the last electron, eventually the last electron lives in the p-subshell. Thus, for this electron, $n = 2$ and $l = 1$

(*ii*) The values of the magnetic quantum number is determined from the value of the electron.

This means that both the electrons in the s-subshell have, when $l = 1$ (i.e. p-subshell) $m = +4, 0, 1$.

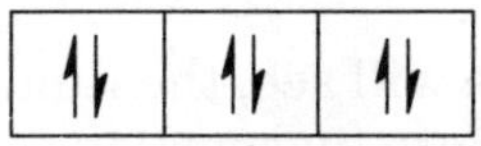

L =1

M = +1 0 -1

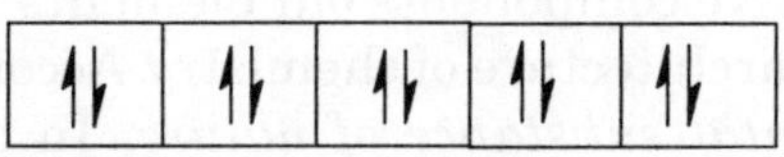

L = 2

M = +2 +1 0 -1 -2

The values of m for the all 10 electrons are given below:

Electron Number	m
1, 6	+ 2
2, 7	+ 1
3, 8	0
4, 9	–1
5, 10	–2

The evaluation of *m* for electrons in the *f*-subshell can be done similarly.

(*iii*) We know that an orbital can accommodate a maximum of two electrons having opposite spin. Hence, the value of the spin quantum number(s) for the electron spinning clockwise and that spinning anti-clockwise are given as +1 and –1

CHAPTER 4

Periodic Table of Chemical Substance

Elements

What are elements? If we consider a dictionary we will see, the fundamental substance of nature". In *Rigveda,* the nature has been divided into five portions (*i.e.* pancha bhutas) or five elements. These are space *(aakaash)* air *(Vaayu),* five *(agni)*, water *(apah)* and earth *(prithvi)*. It was also believed that a person takes birth from these five components of nature. It was also believed that from these five components our elements were born. Now let us turn our attention from Rigveda to the architecture of chemistry. According to modern chemistry *elements are the most fundamental substance of nature.* In other words *an element is such an elementary substance, which is not possibly divisible to any other elements.* As per as the examples are concerned, Sulphur, Oxygen, Nitrogen, Carbon are the suitable ones. In our nature the number of elements are innumerable, only 109 are known to us and these are tabulated in the systematic way by Dimitri Ivanovich Mendeleev (1834 -1907) a Russian scientist, and his work is known as *Mendeleev periodic table*. But this work was done later, after John Dalton (1766-1884).

Dalton's Atomic Theory

John Dalton profounded most fascinating theory about chemical elements and their properties, which are summarized below :

(*a*) The union of elementary atoms in simple proportion forms compounds atoms.

(*b*) Elements are made up of atoms, which are indestructible. All the atoms of the same element have the same weight and are similar in size and shape.

Modified Dalton's Atomic Theory

Elements have different *atomic weight,* but the number of the weight is finite. Each element consists of same kind of atoms at a definite proportion only. Actually the atoms move with a speed very much less than that of light. Dalton did not investigate the speed of the atoms but he investigated the velocities of molecules of gases.

The theory also explains the various laws of chemical combination, which we are going to discuss later. The fact that the compounds always contain constituent elements, which are combined, in fixed proportions. This happens because combinations can only take place between atoms and also these have definite weights. If for examples, two elements A and B combine to form more than one compound, must be made up of one atom of $A + 1$ atom of B or 1 atom of $A + 2$ atoms of B or 1 atom of $A + 3$ atoms of B or 1 atoms of $A + 1$ atoms of B etc. The compounds must obey the Law of Multiple proportions. But when it faced Gay Lussac's law Dalton's theory faced difficulties.

It was known for example that one volume of hydrogen gas combined with one volume of chlorine to form two volumes hydrogen chloride. Dalton's theory stated that combinations took place between simply related numbers of atoms. Berzzelius indeed believed that, as a result of combining of Dalton's theory with Gay Lussac's law, that equal volume of all gases under the same condition of temperature and pressure contained equal number of atoms. On applying this hypothesis to the data that we have before us, we see that one atom of hydrogen combines with one atom of chlorine to form two atoms of hydrogen chloride. The conception of an atom necessitated the possibility of obtaining a single atom of hydrogen chloride and this single atom would contain half an atom of hydrogen and half an atom of chlorine. This means that atoms are also divisible into two parts (but this is not possible according to Dalton's atomic theory). Instead of looking for this error; however; in the assumption that equal volumes of gases contains some number of atoms, Dalton made the statement that gases do not combine exactly in simple proportions and that when ever the figures are obtained which agreed with that result there were experiment errors which was the most remarkable statement to make in the face of all the evidence.

The atomic weight of an element is defined *as the ratio of weight of one atom of element to that of an atom of hydrogen.* (Which is always I) but the above definition was amended as "The ratio of the weight of an atom of an element to the I/16th of the weight of oxygen atom" or in the other way.

$$\text{Atomic Weight} = \frac{\text{Weight of an Atom of a substance or element}}{\frac{1}{16}\text{th of the weight of an oxygen atom}}$$

At one time the atomic weight of an element was regarded as the most important constants of the elements and now after the introduction of the periodic law this constant has the factor of determining the properties of the element.

Periodic Table

Elements have been systematically listed in this table, near about 109 elements have been yet discovered, 10 elements are gases, 2 are liquid and 97 are solid. Elements from 92 to 109 are radioactive. The periodic table also consists of atomic numbers of the elements and they have placed serially in the numeric order, from 1 to 119.

But before presenting the periodic table we shall see the name and the origin of the name of the chemical elements. We shall have to learn the symbols because the symbols are used for the construction of chemical formula of different chemical compounds, which we will read in this chapter.

Name of the Different Chemical Compounds and their Symbols

Name of the Element	Source of the Name	Symbol
Hydrogen	Hydrous = water, gen. = genous, producer Named by Henry Cavendish	H
Helium	Hellos in Greek means something related to sun, as because it was originated from sun Lithos in Greek earth's crust	He Li
Lithium	Lithos in Greek means earth's crust	
Beryllium	Beryl which is the name of its major ore	Be
Boron	•••	B
Carbon	•••	C
Nitrogen	•••	N
Oxygen	Oxous means acid producer, Lavoisier named it	O
Fluorine	•••	F
Neon	Neo = new in Latin	Ne
Sodium	Nattrum, Latin	Na
Magnesium	•••	Mg
Aluminium	•••	Al
Silicon	•••	Si
Phosphorous	•••	P
Sulphur	•••	S
Chlorine	•••	Cl
	Chlorous means green, In 1810 Davy named it.	
Argon	In Greek argon means lazy	Ar
Potossium	Latin Kalium	K
Calcium	•••	Ca
Scandium	Discovered in Scandinavia thus named has derived	Sc
Titanium	Came from the name of the natural satellite of Saturn the titan.	Ti
Chromium	•••	Cr
Manganese	•••	Mn
Manganese	•••	Co
Nickel	"Old Nick"	Ni

Name of the Element	Source of the Name	Symbol
Copper	Cuprum	Cu
Zinc	Zincum	Zn
Gallium	Gala, in Latin it means France	Ga
Arsenic	•••	As
Scelenium	Selene means moon	Se
Bromine	Bromos, disagreeble odour	Br
Krypton	In Greek it means hidden	Kr
Rubidium	•••	Rb
Strontium	•••	Sr
Yittrium	•••	Y
Zirconium	•••	Zr
Niobium		Nb
Molybdenum	•••	Mo
Technetium	•••	Tc
Ruthenium	•••	Ru
Rhodium	•••	Rh
Palladium	•••	Pa
Silver	Argentium	Ag
Cadmium	•••	Cd
Indium	•••	In
Tin	Stannum	Sn
Antinony	Stibamine	Sb
Telluaium	Tellus, in Latin means earth	Te
Iodine	Iodos, in Greek means violet Stranger	I
Xenon	Stranger	Xe
Cesium	•••	Cs
Barium	•••	Bs
Cerium	Came from the name of the country Ceria	Ce
Osmium	•••	Os
Iridium	•••	Ir
Gold	Aurum	Au
Mercury	Hydrogyrum, (Latin, means Liquid silver)	Hg
Talium	•••	Ti
Lead	Plumbum	Pb

Name of the Element	Source of the Name	Symbol
Bismuth	•••	Bi
Polonium	Named by Madam Curie after her motherland Poland	Po
Astatine	•••	At
Radon	•••	Rn
Francium	After the name of France	
Radium	Came from the word Radiant this was more radiant than polonium, which was discovered first. Named by Madan Curie.	Ra
Actinium	•••	At
Thorium	After the name of Greek God of thunder, Thor	Th
Protactinium	•••	Pa
Uranium	Named after the seventh planet Uranus	U
Neptunium	Named after the eight planet Neptune	Np
Plutonium	From the name of the ninth planet pluto	Pl
Americum	Named after America	Am
Curium	After the name of Curies	Cm
Berkelium	This compound was discovered in the Berkeley campus of California University	Bk
Californium	Named after the place of invention, California	Cf
Einsteinium	After the name of the Famous scientist Albert Einstein	Es
Fermium	After the name of Enrio Fermi	Fm
Mendelevium	After the name of D.I Mendeleef	Md
Nobelium	After the name of Sir Alfred Nobel	No
Lawrencium	After the name of the Inventor Ernest O Lawrence	Lw
Hahnium	•••	Ha
Hafnium	•••	Hf
Tungsten	Wothemium	W
Ytterbium	•••	Yb
Actinium	•••	Ac
Unnihexium	This is the 106th element	Unh
Unnilseptium	This is the 107th element	Uns
Unniloctium	This is the 108th element	Uno
Unnilennium	This is the 109th element	Une

We shall now discuss about the Periodic table after Mendeleev. In this table the elements have been placed in a systematic way with the increasing of the atomic number. This table is consisting of nine vertical columns known as groups and seven horizontal periods. From this Periodic table many characteristics of the elements are also reviled.

Description of Periodic Table

The Periodic table is consisting of periods and groups excepting eight and zero (the ninth group is known as zero) are divided into subgroups a and b. The elements in group 1a consists of Lighter metals or Alkali metals, e.g. Lithium, sodium, potassium (Li, Na, K). Those of group 1b are copper, silver, gold (Cu, Ag, Au) they are regarded as coinage metals or heavy metals since they are heavier than the metals of group 1a. The elements of group seven b are halogens they are chlorine, fluorine, iodine bromine and astatine (CI, F , Br, I, At). The elements of group zero are regarded as the inert gases. This group also acts as a bridge between the electropositive metals and the electronegative halogens; therefore they are also called bridge elements, for example, Helium (at number 2) is placed between hydrogen at number 1 and lithium at number 3. The further subdivision of groups into subgroups shows that at least they have some significance between them.

The first period consisting only hydrogen and helium is regarded as the very short period. The second and the third period consists of eight elements is termed as short period. The elements of third period have a very close similarity with the elements in the second period for example, sodium (Na) is the metal of same characteristics with lithium (Li). Also fluorine and chlorine are both halogens and resembles their properties. It can be easily stated that, Starting from any element in this table the ninth element from the first will bear the similar property of the first.

The elements of the second and third short period are known as typical elements. All total we can see there are eighteen in number. In each of the fourth and fifth period, due to the capacity these periods are called long periods. These periods have been again further subdivided into two series known as *Evens* and *Odds*.

The sixth period contains 32 elements thus regarded as very long period. But this theory could not prove its worth in all the cases. As we all know that after each seven elements the eighth element has similar valence electron, atomic nature, etc. It was Newland in the year, 1840 proposed this phenomenon as Law of Octaves.

But above all the periodic table as described by Mendeleef is the best of all; here mainly the elements are selected according to the ascending order of their atomic numbers, which is the best way to understand the properties of elements. For example if we look at the halogen group the way of the arrangement shows that fluorine has least number of electrons than chlorine, bromine and iodine its orbits are more tightly wrapped, therefore it is more negative than chlorine. Similarly chlorine comes after fluorine then comes Bromine and lastly iodine or more precise, the arrangement is shown below :

*F< Cl < Br < I

---

*This is question is being frequently in I.I.T. exams.

Periodic Law

Before Mendeleef the periodic table law was based on weights and was defined as the periodic properties of elements is a periodic function of their atomic weights. But due to further investigations it has been realized that the properties of elements mainly depends upon the atomic numbers and not on the atomic weights. So the definition was again redefined, as the elements are a periodic functions of their atomic numbers.

Formation of Compounds

Compounds are formed from elements. Two or more elements react with each other to form compounds. The some of the examples of compounds are sugar common salt, iron sulphate etc. A compound does not possess any similarity (physical) or properties of combining elements, e.g. hydrogen and chlorine combine with each other, hydrogen chloride is formed in the presence of sunlight, this hydrogen chloride entirely differs in properties than hydrogen and chlorine.

Physical Chemical Change

The changes in the behaviour and the properties of elements during formation of compounds can be of two types the *physical* and the *chemical* changes. When two or more elements combine physically, that is they just mix with each of the like sugar and water or iron fillings in sulphur powder or salt in water. These can be separate by easy physical means. In the last cases the solution will be allowed to boil where the water will be evaporated and the salt and the sugar will remain as residue. In the second case if a magnet is moved over the mixture the iron fillings will be attracted and thus will be separated.

But when the same iron filling is heated with sulphur in a test tube the mixture of the sulphur and the iron fillings gets converted into a solid mass of dark brown colour and this substance is known as iron sulphate, which consists of neither the property of sulphur nor the property of iron, it is totally a new substance called compound. The change, which made the compound, is obviously a chemical change. There are lots of differences between the physical and chemical changes they are listed below:

Compounds

The compounds as said earlier are formed when two or more than two elements combine chemically. There are lots of compounds. The elements are the main sources of compounds. Now we shall study the differences between the elements and the compounds.

So a compound can be defined as a chemical substance which as derived as a result of the chemical conversion of two or elements in a definite proportion with the absorption or evolution of energy. Compounds are formed due to the combination of element, which is called reaction of the elements and those elements, which react to form the compounds, are called reactants.

Physical Change	Chemical Change
1. Here two or more substances mix physically and can be separated also by easy physical means.	1. Here also two or more substances combine to from entirely different substances than element, which cannot be separated by easy physical means.
2. The properties of elements do not change in physical alteration.	2. Here the properties of the elements are completely changed.
3. Physical change gives rise to mixtures.	3. Chemical change gives rise to compounds.
4. Physical change does not require any special conditions like temperature, pressure etc.	4. The chemical change requires all these special conditions.
5. When sulphur is mixed with iron fillings, before heating it is a physical mixture.	5. After the heating of the mixture of iron fillings and the sulphur mixture it is a chemcial conversion.
6. Physical change requires no definite proportion of the ingredients.	6. Compunds are formed when the elements react with each other in a definite proportion.

Elements	Compounds
1. Elements is fundamental chemical substances, which are not further divisible into more elements.	1. Compounds are formed from the elements, thus when further divided gets separated into elements.
2. Elements are simple substances.	2. Compounds are complex substances.
3. As elements cannot be divisible into further elements they are considered as fundamental substances.	3. Compounds derive from the elements they are regarded as derived substances.
4. Elements are known as reactants.	4. Compound is regarded as products.

Valency

Let us consider some elements with chemical symbols for example oxygen O_2 (exact symbol of oxygen). Similarly nitrogen N_2, hydrogen H and chlorine Cl, why is difference? The reason is as follows:

H means one molecules of hydrogen consisting one atom of hydrogen

O_2 means one molecules of oxygen consisting one atom of oxygen

Cl means one molecules of chlorine consisting one atom of chlorine

We can say from the above explanation in one word that, the numbers are valences of their respective elements. Therefore, valency is defined *as the number of atoms of the element which can combine with the other elements.*

Valency is also known as the maximum reaction capacity of the element. Thus in O_2 the maximum number of oxygen atom available is 2 to react. The valency helps to from different compounds and from the valency of elements when they have reacted to form compounds the status of the element can be visualized.

The valences if different elements are tabulated below :

Name of Elements	Valency of Elements
Hydrogen	1
Chlorine	1
Iodine	1
Sodium	1
Potassium/manganese	1
Silver	1
Oxygen	2
Magnesium	2
Calcium	2
Zinc	2
Aluminium	2
Copper	1, 2
Mercury	1, 2
Iron	2, 3
Phosphorous	3, 5
Sulphur	2, 5, 6, 43

The elements having only one valance atom are regarded as monovalent, e.g. Hydrogen, Chlorine. The elements having two valence atoms are regarded as divalent. For example silver, oxygen, nitrogen. The elements having more than two valence atoms or different valance atoms are regarded as elements of variable valency, e.g. sulphur, phosphorous.

Chemical Formula and Equation Framing

This is the most important portion of the architecture of chemistry, because without the chemical formula and equation has no reaction stands. So an ideal reaction must have ideal equation. Here, in this portion we shall study about the chemical formula and equation formation, also balancing them.

We shall start with the creation of formula of certain chemical compounds. To create a formula, symbol of elements of the compound is written first then respective valences are selected and written on the top of the elements and they are interchanged and written below which makes the formula of the compound, for example,

To Write the Formula of Water

Symbols: - H O

H^1 O^2 Since the valency of Hydrogen and Oxygen is 1 and 2 respectively

H_2O So the formula of water is H_2O

To write the formula of water

Symbols: - K Mn O

K^1 Mn^2 O^2

K_2Mn_1

K_2Mn_1 + O^2

$KMnO_4$

To Write the Formula of Water

Some of the formulae of the important compounds are listed below :

Aluminium Oxide	Al_2O_2
Barium Hydroxide	$Ba(OH)_2$
Calcium Hydroxide	$Ca(OH)_2$
Hydrochloric Acid	HCl
Hydrogen peroxide	H_2O_2
Manganese Dioxide	MnO_2
Potassium Hydroxide	KOH
Potassium Permanganate	$KMnO_4$
Sulphuric Acid	H_2SO_4

Chemical Equation

Chemical equation is formed when number of elements or compounds react with each other in some definite proportions. The mass of the combined reactants and the mass of the combined product always remain equal. When the reaction is completed the formula of the compound or the element, which has taken part in the reaction and also has produced some more compounds, is noted down. Always the reactants are written in the left hand side of the equal to(=) sign and the resultant products on the right side.

$$\underset{\text{Reactants}}{[H_2SO_4 + Zn]} = \underset{\text{Products}}{[ZnSO_4 \downarrow + H_2 \uparrow]}$$

From the above reaction we see that sulphuric acid is reacting with zinc to produce zinc sulphate and hydrogen gas. The arrow downwards indicates precipitation and the upwards indicate an evolution of gas.

As said earlier that the mass of the elements in the reaction must be equal, if during reaction is observed that this mass is not equal, then the equation is made mathematically equal, rather the equation is said to be made balanced. If the equation is found to be balanced at the time of the formation that equation is said to be self-balanced equation. Thus, we can say that an equation is said to be balanced when both of the sides will have it mass that is mass reacted is mass-produced.

The above definition will be much clear by the following examples :

Example 1:

$\Downarrow$ precipitation of salt $\quad$ $\Uparrow$ evolution of gas

Unbalanced: $KCl + O_2 = KClO_3$

Balanced: $2KCl + 3O_2 \Uparrow = 2KClO_3 \Downarrow$

Example 2:

Unbalanced: $Cu + HNO_2 = Cu(NO_3)_2 \Downarrow + 2NO_2 \Uparrow + H_2O$

Balanced: $Cu + 4HNO_3 = Cu(NO_3)_2 \Downarrow + 2NO_2 \Uparrow + H_2O$

Techniques of Balancing the Equation

To balance an equation the techniques mentioned below must be followed :

(*i*) At first the reactants and the products all are arranged in their respective sides.

(*ii*) Now one by one the elements are selected and their valence atoms are compared with either side.

(*iii*) After comparison the extra number is written by the side of the compounds or elements. Here the extra number is known as molecules.

(*iv*) These molecules are always multiplied with the respective valency of the same compound the mathematical product of the valences is added to the mathematical product of the other compounds and then the total is obtained these totals on the either sides (that is on the reaction side and the product side must be equal then only we can say that the equation is balanced).

The process will be clearer by observing the following examples: Let us consider the example below :

Here Mg = 1 in both the sides but O = 2 in the product side. So, if we multiply Mg with 2 in both the sides we get, 2MgO = 2Mg + O_2. Thus, we got both Mg and O equals to 2 in both the sides.

Example 3 : Cu + HNO_3 = Cu $(NO_3)_2$ + NO_2 + H_2O

Here we have: Cu = 1 in both sides, H = 2 in the P side and 1 in the R side.

N = 1 in the P side, O = 3 in the R side and 9 in the P side (3 × 2 + 2 + 1 = 9). So, at first O is balanced.

∴ Cu + $3HNO_3$ = $Cu(NO_3)_2$ + NO_2 + H_2O

Now to balance H

Cu + $4HNO_3$ = Cu $(NO_3)_2$ + $2NO_2$ + $2H_2O$, Now the equation is balanced.

Balance the equations: (with the indication of precipitation and evolution)

1. CaO + H_2O = $Ca(OH)_2$
2. $Ca(OH)_2$ + HCl = $CaC1_2$ + H_2O
3. HCl + Zn = $ZnC1_2$ + H_2O
4. H + I = HI
5. H_2 + CO = C_2H_6O + H_2O
6. Fe + H_2O = Fe_3O_4 + H_2

Laws of Chemical Reaction

We shall now discuss about some laws which governs the chemical reactions, they are as follows :

Law of Definite or Constant Proportions : This Law states that every pure compound is made up of some definite elements combined in together in fixed proportions by weight so that its composition is independent of the methods by which it is prepared.

Explanation : The same compound may be obtained from different sources or can be prepared by application of different methods but it is an established fact that the composition of the particular compound its pure form is always the same and does not vary with the methods of its preparation or the source from which it is collected. The law further implies that in each and every compound, the constituent elements are present in an invariable weight to weight ratio which is characteristic of the compound.

Let a particular compound composed of two elements A and B be prepared by three different methods, in first method a grams of A and b grams of B, in second x grams of A combines with y grams of B and in third m grams of A combines with n grams of B. Then according to the law $a/b = x/y = m/n$.

Illustration (1) : Water is a compound which can be collected from a well or a river, sea or from any other natural source, it can also be prepared by passing an electric current through a gas mixture of hydrogen and oxygen or by passing hydrogen over hot cupric oxide. Now, whatever may be the method of preparation, the composition of the pure water will show that it consists of hydrogen and oxygen in the ratio 2 : 1 by volume, that 2 parts of hydrogen by volume of hydrogen has combined with 1 part of oxygen by volume.

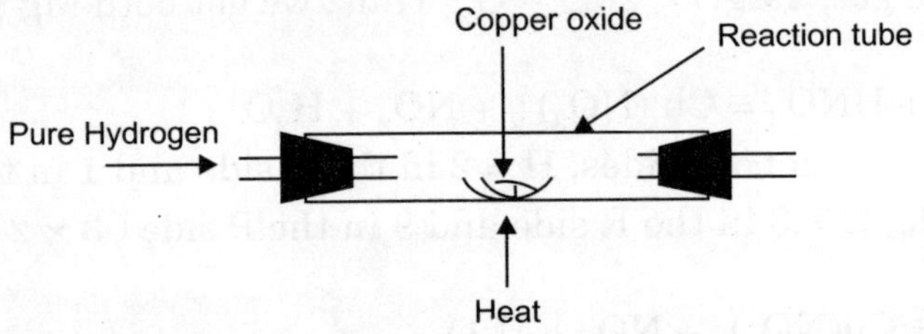

Fig. 4.1. The exp setup to prove the law of constant proportion

Experimental Verification of the Law of Definite Proportion

In this experiment we shall prove the above law by the separation of copper from cupric oxide prepared from three different methods.

Three samples of cupric oxide are taken separately; one of them is first taken in a porcelain boat (about 19 m). The boat is kept in a hard glass tube, at one end a source of pure hydrogen gas is connected, and at the other a jet tube for the escape of the excess of hydrogen gas is connected. At first the hydrogen gas is passed for some time then the boat is strongly heated, it will be seen that the black copper oxide is changing its colour from black to brown. The hydrogen gas has changed the cupric oxide to metallic copper and it has combined with oxygen to form steam.

Calculation

Let the weight of the porcelain boat is $= W_1$ g

Weight of the boat + the cupric oxide $= W_2$ g

Weight of the boat + copper $= W_3$ g

$\therefore$ Weight of the cupric oxide $= (W_2 - W_1)$g, weight of copper $= (W_3 - W_1)$ g and the weight of the oxygen in the cupric oxide $(W_2 - W_1) - (W_3 - W_1)\text{g} = (W_2 - W_3)$ g

$$\therefore \quad \frac{\text{Weight of Copper}}{\text{Weight of Oxygen}} = \frac{W_2 - W_1}{W_2 - W_3}$$

$$\therefore \quad \% \text{ of copper in cupric oxide} = \frac{100 \times (W_3 - W_1)}{W_2 - W_1}$$

$$\therefore \quad \% \text{ of copper in cupric oxide} = \frac{10 \times (W_3 - W)_1}{W_2 - W_1}$$

Same experiment is repeated with the other samples. It will be found that neglecting the experimental error, which is limited the percentage ratio of the copper and the oxygen are same in each case. Thus the law is confirmed.

Law of Multiple Proportions: The law states that when two elements combine chemically to form two or more compounds, the different weights of one of the elements that combine with fixed weight of the other element bear to each other a simple ratio of small whole number.

Illustration : *(1) Compounds formed by the combination of hydrogen and oxygen:* When combined chemically they form two different compounds that water and hydrogen peroxide. In water 2 part by volume of hydrogen and 1 part of oxygen by volume combines. In hydrogen peroxide, 2 parts of hydrogen by volume and 4 parts of oxygen by volume combine. So it is found that different volumes of oxygen (1 and 4 parts) which unite with a fixed volume by 2 parts of hydrogen. The reaction ratio is simple enough.

(2) Compounds formed by the reaction of carbon and oxygen: Carbon and oxygen combine to form two oxides the carbon dioxide and carbon mono oxide. The ratio of carbon and oxygen in the carbon dioxide is 3: 4 by weight and that in the carbon mono oxide is 3 : 8 by weight.

∴ The weights of oxygen, which combine with the same weight of carbon to form these oxides bear to one another the simple ratio 1 : 2 (3: 4: 8). In the similar manner it is also found that the different weights of carbon that combine with the fixed weight of oxygen (16 parts) are in the ratio 12 : 6 that is 2 : 1.

(3) *Compounds formed by the combination of nitrogen and oxygen:* Nitrogen forms as many as five stable oxides and it is found that the composition of the oxides of nitrogen illustrates the law of multiple proportions.

Compounds of Nitrogen	Ratio of Nitrogen and Oxygen
1. Nitrous Oxide	7 : 2
2. Nitric oxide	7 : 8
3. Nitrogen trioxide	7 : 12
4. Nitrogen dioxde	7 : 16
5. Nitrogen pentoxide	7 : 20

∴ The different weights of oxygen combining with a definite proportion by weight (14 parts) of nitrogen are in the ratio 1 : 2 : 3 : 4 : 5. This is a simple integral ratio in accordance with the law.

(4) *Compounds formed by the combination of carbon and hydrogen:* In the four compounds formed by the chemical combination of carbon and hydrogen (acetylene, ethylene, ethane and methane), 12 parts by weight of carbon combines with 1.008, 2×1.008, 3×1.008, 4×1.008 parts of hydrogen respectively. These figures are in the ratio of 1 : 2 : 3 : 4

Compounds	Ratio by weight of Carbon and hydrogen	Constant weight of carbon	Simple ratio of hydrogen.
Acetylene	24 : 2 × 1.008	12	
Ethylene	24 : 4 × 1.008	12	1 : 2 : 3 : 4
Ethane	24 : 6 × 1.008	12	
Methane	12 : 4 × 1.008	12	

Experimental Verification of the Law of Multiple Proportions

We know that copper and oxygen together react to form two oxides-cuprous oxide and cupric oxide. The constant weights of two well cleaned and previously weighted porcelain boats are taken in the boat 1 some cupric oxide and in the boat 2 some cuprous oxide are taken they are placed in a hard glass tube, closed at both ends by corks and delivery tube. As shown in Fig. 4.2 this is the repitition of the experimental set up for the law of definite proportions.

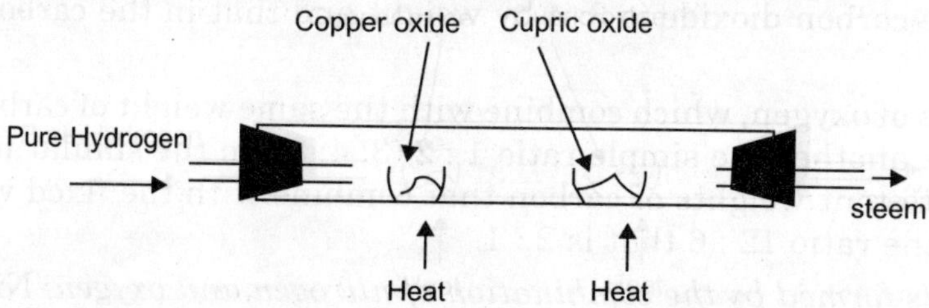

Fig. 4.2. The exp set up for the verification of Law of Multiple Proportion

Calculation

Let the weight of the porcelain boat 1 $= W_1$ g

Weight of the boat + cupric oxide $= W_2$ g

Weight of the boat + copper $= W_3$ g

The weight of the cupric oxide = $(W_2 - W_1$ g and weight of copper $(W_3 - W_1)$ g. Weight of oxygen in cupric oxide = $(W_2 - W_1) - (W_3 - W_1)$ or $(W_2 - W_3)$ g. That is to say, $(W_2 - W_3)$ g of oxygen combine with $(W_2 - W_3)$ g of copper to form cupric oxide.

$\therefore$ 1 g of oxygen will combine with $\dfrac{W_3 - W_1}{W_2 - W_3}$ of copper of x g of copper.

Similarly let the weight of boat 2 = a grams. Weight of the cupric oxide + boat 2 = b g. Weight of the boat + copper = c g of copper or x grams of copper.

$\therefore$ Weight of copper = $(c - a)$ g and the weight of the oxygen unite with $(b - c)$ g. So, $(b - c)$ g of oxygen unite with $(c - a)$ g of copper to form cuprous oxide.

$\therefore$ 1 g of oxygen will combine with $\frac{(c-a)}{(b-c)}$ g or y g of copper.

Experimental result show that the ratio of the different weights of copper (x and y g) in the two oxide combining with the fixed weight of oxygen will come out as 1 : 2. Thus, the law experimentally verified.

The Law of Reciprocal Proportions : The law states that the weight of two or more different elements, which separately combine with a fixed weight of another element are either the same as or are simple multiples or sub-multiples of the weights of these different elements when they combine with each other.

In other words the proportions in which any two elements combine with certain definite weight of a third element are the proportions or a simple multiple (or sub-multiple) of the proportions in which they combine amongst themselves.

Explanation : Let a and b g of two different elements X and Y react with c grams of another element Z. Now if X and Y combine together, the ratio by weight in which they do so will either be $a : b$ or a simple multiple of the ratio of the weights of X and Y which combine which combine with a constant weight of Z i.e. nb (where n and m are simple whole number).

Illustration : The element carbon and oxygen combine with hydrogen separately to form methane and water respectively. Chemical analysis that in methane, the ratio of the weights of hydrogen to carbon is 1: 3. Again, in water, the weight ratio of hydrogen to oxygen is 1: 8. Thus, in two compounds 3 parts by weight of carbon and 8 parts by weight of oxygen have separately combined with a constant weight of hydrogen (1 part). For easy remember see the arrow triangle below :

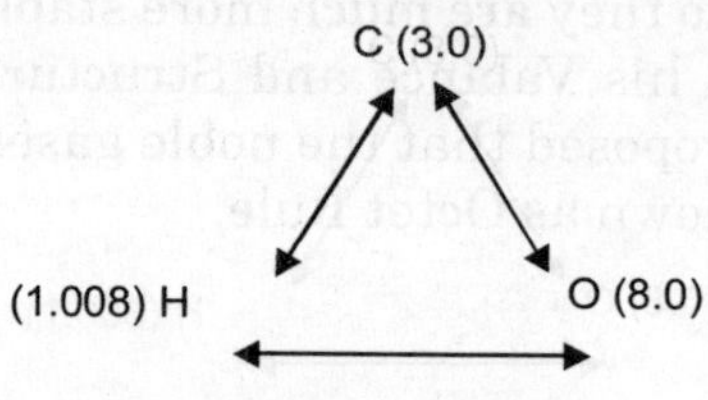

According to this law, if carbon and oxygen combine at all, they must do so either (1) in the ratio of 3 : 8 or (2) in its simple multiples or sub-multiples. Now carbon is known to combine with oxygen to give two different compounds-carbon dioxide and carbon mono oxide. In carbon dioxide the ratio of carbon to oxygen is 3 : 8 which is the same as predicted by the law. In carbon dioxide and carbon mono oxide the ratio is 6 : 8.

CHAPTER

5

Chemical Bonding

Introduction

As far as the word 'bonding' is concerned, it means that when a reaction occurs force or attraction occurs between the atoms of it as well as that of different elements or between the reactants, without a prefect chemical bond no chemical reaction can take place.

Actually *the bonding is the cement, which joins the bricks named atoms, To form houses named molecules or reactions.* In this chapter we shall be discussing about the chemical bonding, theory of chemical bonding, resonance, bond energy, oxidation number etc.

Chemical Bonding and Types of Chemical Bonding

Lewis octet theory

Atoms combine to form chemical bonds because of a decrease in potential energy with a corresponding increase in stability. Particles attract one another in such a position when they are in the state of lowest potential energy and also they are much more stable when bonded than separated. G.N. Lewis has established in his Valance and Structure of Atoms and Molecules, the modem Octet theory. He has proposed that the noble gases are extremely stable and they consist of 8 electrons, this is known as Octet Rule.

Chemical Bonding

According to some renowned chemists Chemical Bonding has been classified into two, the Ionic or Electrovalency, and Co-valency. Ionic or Electrovalent bonds are formed when the combining atoms *exchange* the valence electrons. Covalent Bonds are formed when the combining atoms *share* the valence electrons. Now we shall study the electrovalency and covalency.

Electrovalent and Ionic Bonding

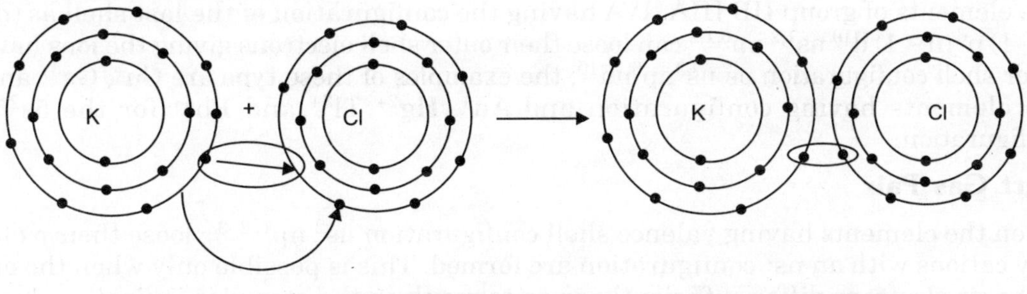

Fig. 5.1. One extra goes to chlorine makes it *Cl*⁻ and *K* remains

Theory of Electrovalent Bonding

We shall now study the details of electrovalent bonding or ionic bonding.

But Ions are formed?

Let us consider some elements like hydrogen, silver, cadmium and the inert gases.

> Hydrogen ion is the positive non-metal other non-metals are all electro-negative

Hydrogen ion is the only positive non-metal other non-metals are all electro-negative.

Formation of Ion in Hydrogen

As we all know that hydrogen atoms consists of only one ion of its type. The proton in nucleus of the hydrogen atom is positively charged ion H^+.

The Ions Found in Inert Gases

The inert gases having the configuration ns^2np^6 in their outer shell, except for n = 1, for which no p orbitals exist and the configuration is $1s^2$. There configurations are achieved by the representative element of group I, II and III. By the complete loss of their valance shell electrons and by the elements of group VII and VI and V acquiring 1, 2 and 3 electrons respectively in their outermost shells, some of the examples are shown in the table.

		H^+, He($1s^2$)	Li^+			
N^{3-}	O^{2-}	F^- Ne($2s^22p^6$)	Na^+	Mg^{2+}	Al^{3+}	Ti^{4+}
		Cl^- Ar($3s^23p^6$)	K^+	Ca^{2+}	Sc^{3+}	Zr^{4+}
		Br^- Kr($4s^24p^6$)	Rb^+	Sr^{2+}	Y^{3+}	Ce^{4+}
		I^- Xe($5s^25p^6$)	Fr^+	Ra^{2+}	Ac^{3+}	

Due to improper shielding of the nucleus by the $1s^2$ electrons, the $1s^2$ ions have properties differ from their rest of the numbers of the same group and tend to form more covalent bonds.

Formation of Ions of the Eighteen Elements

The elements of group IIB IIIA, IVA having the configuration of the last shell as (n – 1) s^2 (n – 1) p^6 (n – 1) d^{10} $ns^{1,2}$ $np^{0,1,2}$ can loose their outer shell electrons giving the ions having the outer shell configuration as $ns^2np^6nd^{10}$, the examples of these type are Cu^+, Ga^{3+} and Ge^{4+}, the elements having configuration and Au^+, Hg^{2+} Tl^{3+} and Pb^{4+} for the $5s^25p^65d^{10}$ configuration.

Inert Gas Pair

When the elements having valence shell configuration ns^2 $np^{1,\,2,\,3...}$ loose their p electrons only cations with an ns^2 configuration are formed. This is possible only when the energies of the ns electrons differ sufficiently so as to result in the stepwise ionization during the chemical bond formation. Therefore, only the post-transmission elements of group IIIA, group IVA and VA give such ions Ga^+ Ge^{3+} and As^{3+} for $4s^{2-}$ In^+ Sn^{2+} and Sb^{3+} for $5s^2$ and Tl^+, Pb^{2+} and Bi^{3+} for $6s^2$ ions.

Radius Ratio

The Structure of Ionic Compounds : As the cations and the anions in the ionic compounds are held by the static forces only, they try to arrange themselves in a geometrical arrangement such that the repulsion between the similar charges are minimized and the attraction between the opposite charges are maximized. This can be seen by considering the effect of the increase in size of the central ion. Qualitatively in a hypothetical octahedral arrangement of the ions; the spatial arrangement is seen almost conveniently through the sectional view of any four ion of one kind is the one where the anions touch each other and the cations also similarly. This is shown in Fig. 5.1.

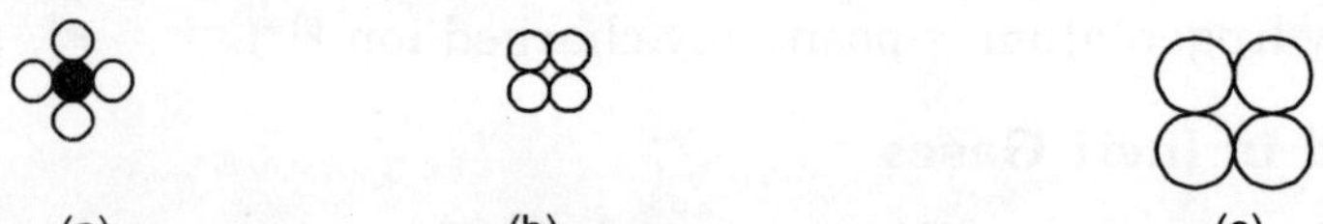

(a) (b) (c)

Fig. 5.2

If the size of the anion decreases the anion will no longer touch the other anions though they will be in contact with cation (a). If the relative size of the cation increases further more anions will able to fit around it giving greater stabilization with the change of co-ordination number and the structure. On the other hand, if the size of the anion is increased, it will no longer touch the cation and will increase the inter electronic repulsion which will push one of the anions away from the co-ordination sphere of the cation altering the structure.

The most common and stable ionic structures are tetrahedral, octahedral, and cubic forms corresponding to the co-ordination number 4, 6, 8 respectively which has been shown in Fig. 5.3.

The ionic bonding mainly occurs in the inorganic compounds, which are mainly in the crystalline form for examples are calcium chloride, zinc sulphide. As this configuration are not possible, the only exception is Au with $5s^2$ $5p^2$ $5d^{10}$ $6s^2$ configuration, found as *CsAu*.

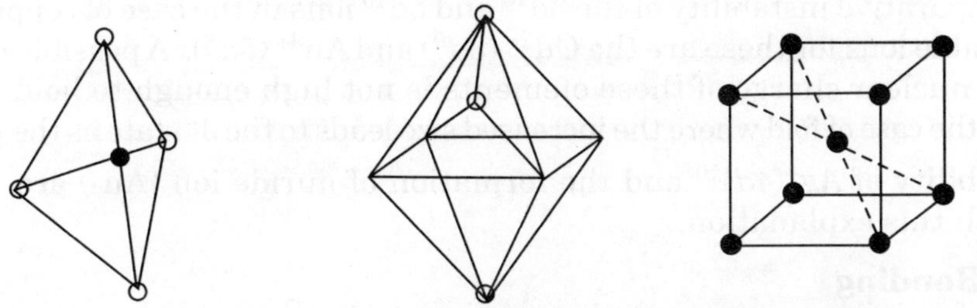

Fig. 5.3. The three common and stable ionic structures

The *d* and *f* Ions

The transition elements, which are having ions formed out of less outer valence electrons, they have the configuration as ns^2np^6dx (where x = 1 to 9) and this configuration has resulted without the ionization of d electrons, thus are classified as d ions. Because of similarities the examples can be given as Ti^{3+}, V^{2+}, Cr^{3+}, Fe^{3+} etc.

The *f* ions are derived from the inner transition elements and the following configuration $(n-1)\ S^2\ (n-1)\ p^2\ (n-1)\ d^{10}\ (n-1)\ f^{1\text{-}13}\ ns^2\ np^6$ where n= 6 for the actinons* and 5 for the Lanthenons.** These ions are very similar to each other, because they are similar to size, charge and electronic configuration in which the *f* electrons do not take part in chemical reaction.

Poly Atomic Ions

An element on combination with electronegative elements likes oxygen; fluorine or chlorine through bodings can loose their electro negativity and the formal charge. The formations of large number of oxoanions thioanions, oxocations, etc. are known as polyatomic ions. (These are very complicated to be discussed in this chapter).

Ions with Irregular Configuratons

These are certain ions that cannot be classified into any particular class, e.g. Hg^{2+}, Ge^{2+}, and Pb_9^{4-} etc. Ions from the above list can be noticed that the formation of them will become easy that is the following conditions are satisfied, the conditions are :

(*i*) The charge contained in these ion will be small.

(*ii*) Anions will be smaller in size than cations.

(*iii*) The ions when formed must have 8 electrons in the first orbit, also the second orbit must contain 18 electrons, the $6s^2$ pair, d or f ion (must be in order of decreasing stability.

---

*Actinons : The Actinium Ions.

**Lanthenons : The lanthanum Ions.

The comparative instability of the $3d^{10}$ and $5d^{10}$ ions in the case of copper and gold must be noted, stable ions for these are the Cu^{2+} ($3d^9$) and Au^{3+} ($5d^8$). A possible explanation may be that the nuclear charge of these elements is not high enough to hold the 18 electrons firmly, so in the case of fold where the increased size leads to the 3^+ state as the most stable state.

The stability of $Ag^+(4d)^{10}$ and the formation of auride ion (Au-) are, however, not in the line with this explanation.

Covalent Bonding

We shall now discuss about the covalent bonding in between chemical elements. As we all know that a covalent bond is formed when two atoms share their valence electrons. After sharing, the bonded electrons become indistinguishable and also if the sharing is equal then the bond becomes non-ionic.

In a covalent bonding the bonding can take place by sharing just single electron each by the valence shell electrons had originally belonged to one of the bonded atoms. These are known as co-valent bonds and are represented by an arrow, which indicates the acceptor atom and trails the donor atom.

The covalent bonding always belongs the unequal sharing thus resulting a positive charge in the donor atom and negative charge in the acceptor atom, e.g.

$$H_3N^+ \longrightarrow BCl_3$$

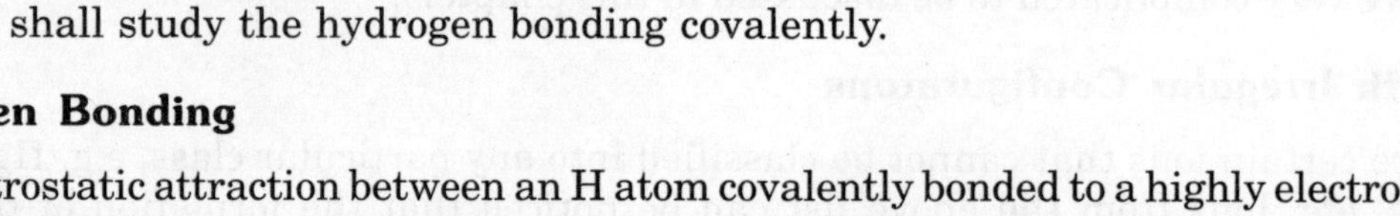

Fig 5.4. The donor and acceptor atoms in the covalent bonding in different elements.

Covalent bonding mainly occurs in the organic compounds as carbon and hydrogen. Here we shall study the hydrogen bonding covalently.

Hydrogen Bonding

The electrostatic attraction between an H atom covalently bonded to a highly electronegative atom and a lone pair of electrons on that electronegative atom of other molecule is called hydrogen bonding.

It is said that the hydrogen bonding is the weakest bonding among the other covalent bondings. The bond energy is less than 10Kcal/mol, while other covalent bondings have an energy of 120Kcal/mol. Hydrogen bonding appears in long chain like clusters of a large numb of associated molecules like as many as tiny magnets. Like covalent bonding hydrogen bond h a preferred bond direction also. This proves the fact that the hydrogen bonding occurs through the *p* orbitals which consists of the log pair of electrons of any electronegative atom. This also states that the hydrogen bonding takes place in a straight line. Few examples of hydrogen bonding are shown below in Fig. 5.5 :

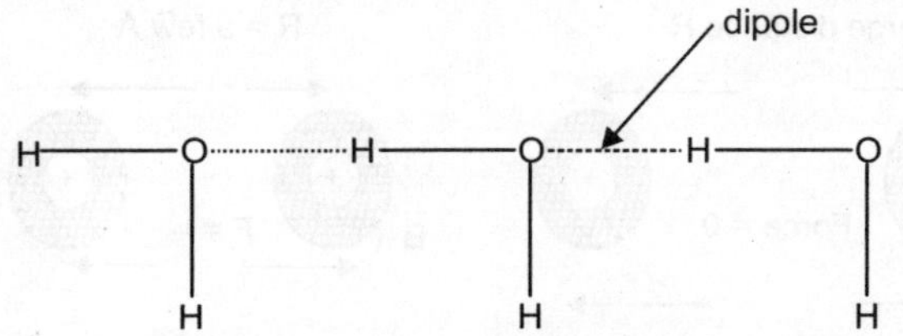

Fig. 5.5. Hydrogen bonding

Conditions for Formation: (1) Only *O*, *N* and *F*, which have very high electro negativity and small atomic size, are capable of forming hydrogen bonds.

(2) Hydrogen bond is longer and also weaker as said earlier.

(3) Hydrogen bonding must have a polar band, as shown below:

$$\overset{\delta^-}{\mathrm{O}}\text{———}\overset{\delta+}{\mathrm{H}} \text{ or, } \overset{\delta^-}{\mathrm{N}}\text{———}\overset{\delta+}{\mathrm{H}} \text{ or } \overset{\delta^-}{\mathrm{F}}\text{———}\overset{\delta^+}{\mathrm{H}}$$

(4) The electronegative atom (O, N or F) in the above polar bands must have unshared electron pair. Some of the hydrogen bond compounds are, water, ammonia, methane etc.

Van Der Waal's Forces

Dutch scientist named as Johannes Diderik Van der Waals (1873) introduced the concept of Van der Waals forces or bonds. He proposed a principle with a simple logical explanation. He said that there are two faulty assumptions in the kinetic molecular theory of gases:

1. The molecules of a gas are simply *point masses,* or in other words they occupied no volume this implied that the actual volume occupied by the gas molecules is negligible as compare, with the total volume of the gas.
2. The first moving molecules of a gas do not have attractive interactions or that the intermolecular attractions are completely absent in gases. Van der Waals also pointed out that in real gases, the molecules do have a volume and also exert intermolecular attractions. This is more, so at high pressure and low temperature the molecules are drawn fairly close together. Under these conditions, the individual molecular volume can no longer be ignored as compared to the total volume and the molecular attractions become much pronounced to be overlooked. Thus van der Waals advanced principle did help to correct the theories of the gasses.

Van der Waals principal also help us to understand to study the valency orbitals, the intermolecular forces. About the intermolecular forces which we are going to discuss now. The van der Waals forces are absent when the molecules are far apart from each other in a gaseous substance, but when the pressure is increased and the molecules are brought closer, the van der Waals forces are able to hold the molecules together in the liquid and solid state. Thus a gas is liquefied by allowing it to expand suddenly from high pressure. During expansion the gas overcomes the intermolecular forces of attraction. The energy required for attraction is obtained from the gas itself, thereby lowering the temperature and causing the liquification. Figure 5.6 belowing showing the van der Waals forces.

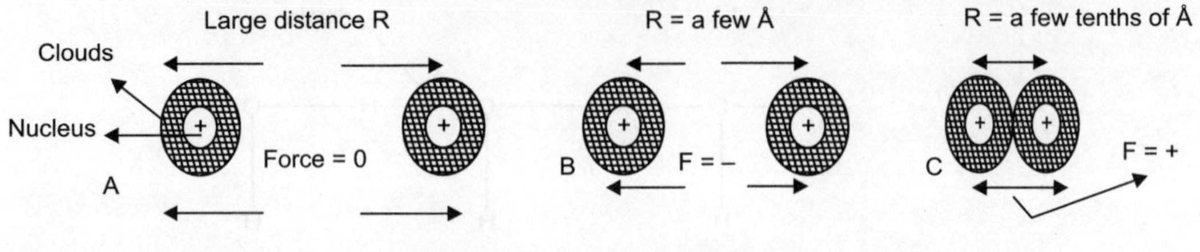

Fig. 5.6. Van der Waals forces

Octet Rule

The number and nature of the bonds formed by an element can be determined by octet rule. This rule states that the gain, loose or share electrons so as to eight (octa) electrons in the outermost shell, and in the innermost shell that number will be two.

Ionic bonds are formed by the elements having one, two or three electrons or more or less than the inert gas elements by the loss or gain of the requisite number of electrons. Elements having excess electrons tends to get rid of them by ionization forming cations having lower inert gas configuration, where elements have six or seven electrons in their valence shell tend to accept one or two electrons to give anions having the next inert gas shell configuration.

If the element has four or five electrons in the valence shell, it cannot acquire the stable octet (the eight-electron shell of the inert gas element having the configuration of $ns^2\ np^6$ by the loss or by the gain of the required number of electrons, the ionization energy or the electron affinity of the elements would prohibit the formation of ions. These elements attain the stable octet by mutual sharing of the valence electrons to the required extent.

Thus the elements of group *IA IIA* and to small extent *IIIA* from cations by the loss of one, two or three electrons respectively. The elements of group *VI* and *VIIA* and to some extent nitrogen and phosphorous tend to form monoatomic anions by gaining one, two or three electrons respectively. The elements of group *IIIA*, *IV A* and *V A* tend to form covalent bonds, by sharing the valence shell electrons and thus acquire the octet.

Elements with two or three electrons in the valence shell are sometimes not able to attain the octet rule either by the loss or gain of electrons or through mutual sharing (especially if the size of the atom is small e.g. beryllium, boron, or aluminium as due to large ionization energies cation formation is inhibited. If they combine by sharing electrons in the valence shell as in beryllium chloride, boron trifluoride or aluminium trichloride, they are fall short of octet by two or four electrons) such compounds are termed as electron deficient compounds. Accepting one or two pairs of electrons from donor atoms can complete the octet around the electron deficient atom and hence they form co-ordinate bonds with donor atoms.

Limitations of Octet Rule

The octet rule is no doubt useful but in spite its merits there are some demerits also, which we are going to discuss under this sub-head.

$$(BeCl)^{2-}\ (BF)^{4-} \text{ and } (AlCl_4)^-$$

1. Octet rule is only applicable for light elements like boron, scandium etc.
2. Transition metal ions obtained by loss of s and d electrons are quite stable and also they show multiple valences thus disobeying the octet rule.
3. When double or treble bond is formed by sharing four and six electron respectively the energy remains as equal to single bond.
4. The elements representing the groups III, IV, V and VI have more than 8 electrons and also the energy remains as equal to the single. The examples are $Al(H_2O)_6^{3+}$, SiF_6^{2-} PCl_5 PCl_6^- also SF_6 and IF, having 12, 12, 10, 12, 12 and 14 electrons respectively in their valence shell.
5. It is impossible to explain for the octet rule, the single electron bonds, e.g. Wand also 3-electrons bonds as NO, O_3 or NO_2. Octet rule cannot also explain metallic bonds.

Theory of Covalent Chemical Bonding

The theory of chemical bonding (here covalent bonding) can be explained with help of four different bond theories as:

1. Valence Bond Theory,
2. The Crystal field theory,
3. The Legand Field Theory,
4. The Molecular Orbital theory,
5. The Warner's Theory.

Valence Bond Theory

The famous American scientist Linus Pauling (1901-1997) established this theory. The theory mainly supports qualitative explanations and also stereochemistry, it also explains the magnetic properties of octahedral complexes, as shown below (Fig. 5.7):

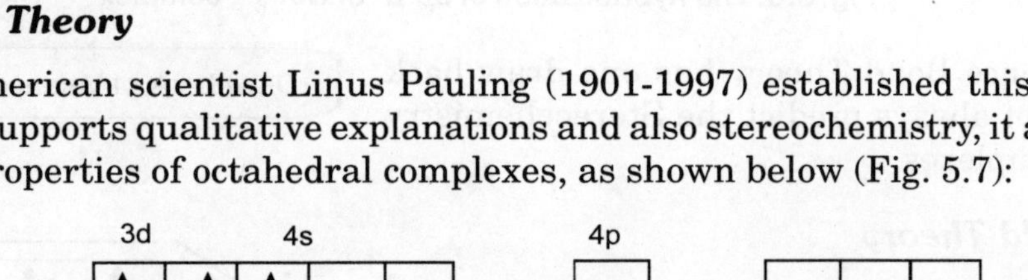

Fig. 5.7

Here it should be noted that each of the six legands in an electron pair donor. In order to accept the six electron pairs, the ion provides a set of equivalent 6 orbitals with their lobes directed to their apices of an octahedron. Such a set of orbitals is constructed by hybridization of the following group of orbitals.

S, Px Py Pz $d_x{}^2 - y^2$ and $d_z{}^2$.

It shows that a set of d^2sp^3 hybrid orbitals are formed and these are occupied by electron pairs (*xx*) donated by six NH_3 legands the complex will be paramagnetic corresponding to the presence of these unpaired electrons. This has been experimentally verified.

Next we may consider the CoIII (Cobalt III) complex Co $(NH_3)_6{}^{3+}$. The Co^{3+} ions has six 3d electrons with four unpaired electrons (see Fig. 5.7 above). In order to provide the required d^2sp^3 hybrid orbitals it is necessary to clear two of their electrons either by promoting them higher orbitals above 4*p* orbitals or by pairing within the 3d orbitals. The former is not energetically possible so the later is used.

The complex in Fig. 5.8 shows its diamagnetic with all the pairs of electrons. Similarly all octahedral complexes of Co (III) are diamagnetic. But there are few exceptions, among them one is $CoF_6{}^{3-}$ ion. This has four unpaired electrons as the free Co^{3+} ions. In such a case the metal legand bond is ionic, the outer 4d orbital, instead of 3d orbitals being involved in hybridization.

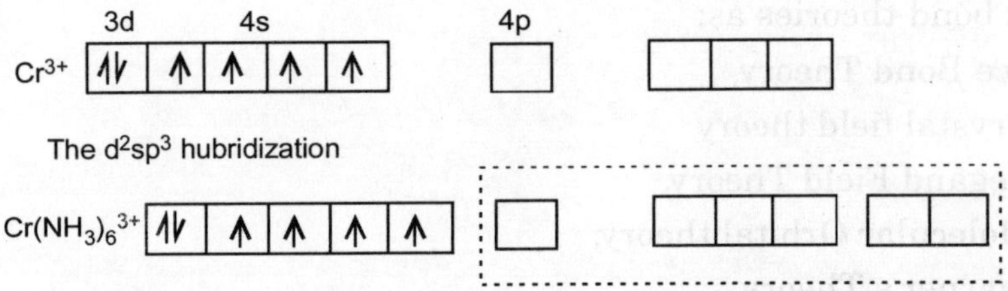

Fig. 5.8. The hybridization of ***sp³d²*** **of** $CoF_6{}^{3-}$ **complex**

Draw back of Valence Bond Theory

The Valence Bond Theory has one draw back that it cannot always predict the Stereochemistry of higher complexes.

Crystal Field Theory

This is purely electrostatic theory, which was first established by Bethe (1928) according to him there is a pure electrostatic attraction, which occurs between the central ion and the legands (here the legands are treated as point charges). The *d* orbitals on the metal are all of the same energy in the free state, that is degenerated and this degeneracy is destroyed when a complex is formed.

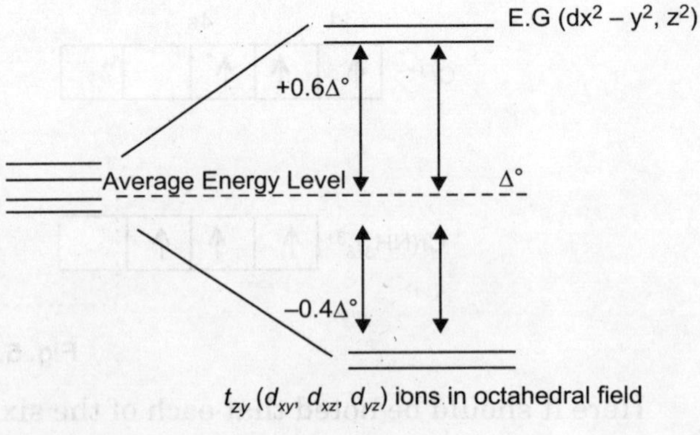

Fig. 5.9. The splitting of ***d*** **orbital in two different energy levels in the ocahedral field**

In an octahedral complex the ion is at the centre of the octahedron and the legands are at the six comers the e_g orbitals are directed along the axis ($d_x{}^2 - y^2$ and $d_z{}^3$ orbitals) and z. It is obvious that the approach of the six legands along x, y and z, $-x$ $-y$ and $-z$ direction will increase the energy of the $dx^2 - y^2$ and $d_z{}^2$ orbitals which are located along the axis much more than the d_{xy} and d_{xz} orbital which point in between the axis. Thus under the influence of an octahedral legand field the d orbitals split into two groups of different energy. (See Fig. 5.9).

The difference in the energy between the two d levels is given by Δ_0 or $10Dq$. It may be noted that e.g. orbitals are $+0.6\Delta_0$ above the energy and the t_{2g} orbitals are $-0.4\Delta_0$ which is below the average. The size of the energy gap Δ_0 between the t_{2g} and e_g levels can measured easily by recording the $U - V$ spectrum of the complex. Thus d electrons system e.g. $Ti\ (H_2O)_6{}^{3+}$ shows adsorption spectrum of the complex as shown in Fig. 5.10.

The spectrum is having a maximum and minimum peak at 20,300 cm^{-1} or 247kJ mol^{-1}. Thus the value of Δ_0 for $Ti\ (H_2O)_6{}^{3+}$ is 243kJmol^{-1}.

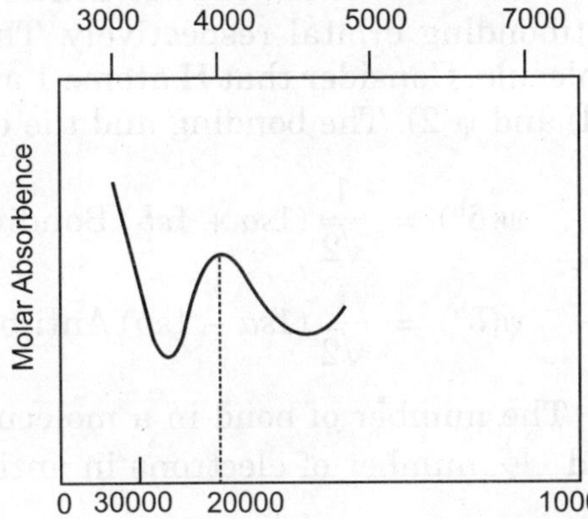

Fig. 5.10

For the same transition common legands can be arranged in order of strength (~) and the resulting series in is known as spectrochemical series.

$I <$ Br < Cl < NO_3 < OH < H_2O < NH_3 < NO^- < CN^- (strong legand)

The magnitude of the Δ_0 increases when the charge of the ion increases. In the tetrahedral legand field the splitting pattern of d-orbitals is different In this the direction of approach of the legands does not coincide exactly with either the e, or by orbitals. The t_{og}, in between the x, y and z axis) are closes to the direction of the legands than orbital so that their energy level is raised. This is just the opposite of the splitting pattern in an octahedral complex. The magnitude of the tetrahedral crystal field theory, based on simple electrostatic considerations, which offer a simple and clear model for setting examples for the magnetic and spectral properties of complexes fairly well. It fails to give the correct picture, as these are the deviations from the ideal situations represented by the model.

Legand Field Theory

Physicists developed this between 1930 and 1940 mainly *Van Vleck* and his associates and again rediscovered in 1950 by several theoretical chemists. It is the modified form of the Crystal Field Theory; it refers to the entire theoretical conditions, which are used to understand the bonding and its related electronic properties (magnetic, spectroscopic also) of the complexes of the transition elements.

In the real co-ordination complexes the interactions between positive ions and surrounding legands are not strictly electrostatic as stated by the crystal field theory. Experimentally, it has been proved that positive ion orbitals mix with the legand atom orbital. The degree of mixing values from one complex to another. Generally such an overlap is moderate and crystal field theory is modified to counter the overlap. Hence, it is named as Adjusted Crystal Field Theory.

Molecular Orbital Theory

Under the present sub-head we will discuss the Molecular Orbital Theory (M.O.T.). The basic idea of this theory is that each bonded atom pair is held together by an electron pair, which are located between the pairs of nuclei.

In the M.O.T. the valence electrons are considered to be associated with nuclei. In their simplest approximation molecular orbitals are obtained by linear combination of atomic orbitals (L.C.A.O). The method is also abbreviated as L.C.A.O.M.O. Molecular orbitals results from addition or subtraction of individual atomic orbitals leading to bonding and antibonding orbital respectively. The simplest example is that of H_2 that is hydrogen molecule. Consider that H atoms 1 and 2 have atomic orbitals described as wave function ψ(l) and ψ(2). The bonding and the orbitals are expressed by :

$$\psi(\delta^b) = \frac{1}{\sqrt{2}}(1sa + 1sb) \text{ Bonding}$$

$$\psi(\delta^b) = \frac{1}{\sqrt{2}}(1sa - 1sb) \text{ Anti-bonding}$$

The number of bond in a molecule is given by 1 (Number of electrons bonding M.O. and $\sqrt{2}$ number of electrons in anti-bonding M.O). Thus the number of bonds for H_2 is $\frac{1}{2}$ and for H_2^+ it is $\frac{1}{2}$.

In H_2 molecule both the electrons occupy the bonding orbital (δ) which results in saving of 2Δ equivalent to bond energy of the molecule. The molecule becomes stabilized by bond function. Overlap of orbitals (S – S, S – p etc.) along the axis floating the nuclei produces 0 molecular orbitals. This has been illustrated in Fig. 5.12.

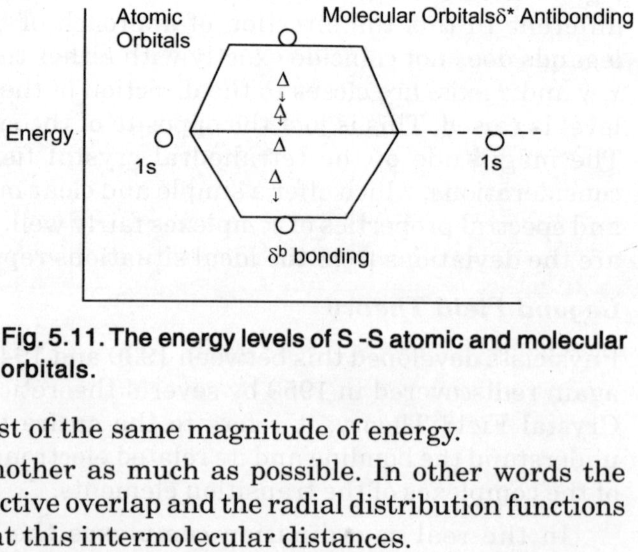

Fig. 5.11. The energy levels of S -S atomic and molecular orbitals.

The rules governing LCAO-MO are listed below :

(*a*) The atomic orbitals must be almost of the same magnitude of energy.

(*b*) The orbitals must overlap one another as much as possible. In other words the atoms must be close enough for effective overlap and the radial distribution functions of the two atoms must be similar at this intermolecular distances.

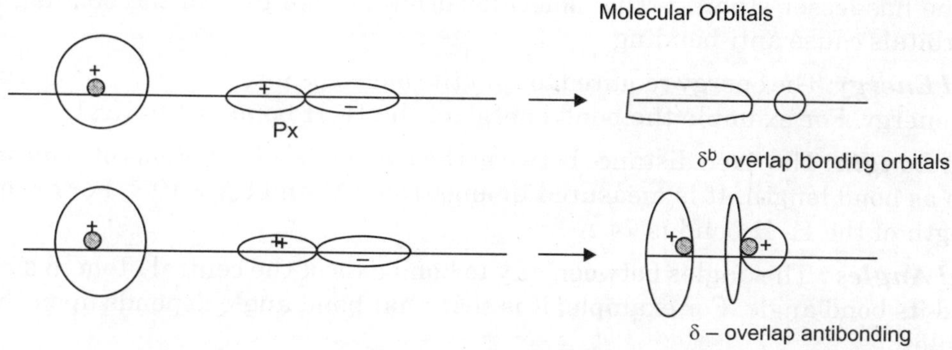

Fig. 5.12. The combination of orbitals in atoms giving 8 bonding (s – p)

(*c*) For bonding and anti-bonding M.O. either symmetry of the atomic orbitals must remain unchanged when rotated about the internuclear line or both atomic orbitals must change symmetry in an identical manner. The above-explained theories are applied in the different chemical reactions both inorganic and organic.

We shall now study the concept of resonance.

Concept of Resonance

The phenomenon of resonance is used to explain the chemical bonding in an empirical fashion. Resonance can be defined as stability of the bonds between the compounds due to dispersion of π electrons over a large region. The best example of the resonance is the cyclohexane and benzene (Fig. 5.13). Resonance is also termed as Mesomerism.

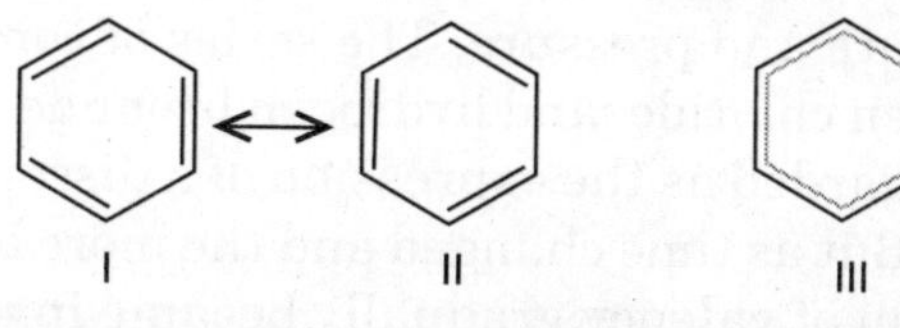

Fig. 5.13. Resonance in Cyclohexane and Benzene.

Resonance mainly occurs in those compounds having conjugated double bonds that is an alternate double and single bond (as in benzene). Benzene does not have both the double and single bond. The carbon-carbon bond lengths are same (1.39 Å) which is intermediate between a C–C bond length (1.54 Å) as also found in ethane. The double bonded arrow in Fig. 5.13 indicates double bonds.

We shall now study some of the short notes on some important terms of chemistry, regarding the chemical bonding:

Anti-bonding : When the covalent bonding takes place the electrons, which has taken part in bonding, they does not remain to their individual atoms, but the associate themselves with the molecule as a whole. Thus, the covalent bond is formed by the overlap of two atomic orbitals one from each atom. The molecular orbitals have greater stability and the atomic one has lesser stability, the molecular orbitals take part in the bonding and the atomic orbitals cause anti-bonding.

Bond Energy: The energy required to split the molecule into respective atoms is known as Bond energy. For example the bond energy of the H–H bond is 1 04Kcal.

Bond Length: The fixed distance between the two nuclei of the atoms of same molecule is known as bond length. It is measured in angstrom (Å) and 1 Å = 10^{-8}. For example the bond length of the H–H bond is 74 Å .

Bond Angles : The angles between any to bond axis of the central atom in a molecule are called its bond angle. For example, it is seen that bond angle depends upon the shape of molecules.

Dihedral Angle or the Angle of Torsion: In tetrahedral bonding the angle between the two bases is known as dihedral angle (*Di* = two hedral = base) or angle of torsion (as because it originates from rotation that is torsion about a single bond). This is denoted by phi = <ϕ>.

Werner's Theory

Valency, originally introduced as a numerical conception to express the combining ratio of atoms, was often confused with the combining power or affinity itself and was used as a substitute for the latter. In other words there was no difference between the valence strength and the strength of the bond. It seemed to be believed that the affinity or the binding force of the atom operates like valency in a discontinuous manner in integral units. Though it was not unknown that the stability of the compounds of one and the same element varied with the nature of the other element with combination and also with external physical conditions such as temperature and pressure. The series of compounds of hydrogen, such as hydrogen fluoride, hydrogen chloride, and hydrogen bromide, may be used for examples. However, the valency was regarded as the expression of a discontinuous chemical bonding force or, affinity of the atom. But as time changed and the more and more facts of chemistry were revealed, this conception of valency gradually became invalid. The conception of the partial valency came into light. In this light we can also see the theory of Thiele on partial valency, which was more improved in the year 1898 to support the behaviour of some complex organic compounds having conjugated double bonds.

A more significant development of this idea is found in the Co-ordination Theory of Complex Compounds, which was formulated by Alfred Werner (1893) to account for the combination of two or more simple compounds. According to his view, the affinity of the binding power of an atom in not completely saturated by union with another atom~ a certain amount of residual affinity remains unsaturated on each other after the combination. Werner therefore classified chemical compounds into two classes they are:

1. *Compounds of the First Order :* They are also regarded as simple compounds.
2. *Compounds of the Second or the Higher Order :* They are also known as the molecular or complex compounds.

Werner also distinguished two types of valences they are: *primary or the principal valency'* and *secondary, residual or auxiliary valences.*

The primary valence electron unites atoms to form simple compounds while second order valence electrons forms higher compounds. The valence force thus lost its original invariable and integral character, which attributed a definite integral value of the valence electrons or at least a definite maximum value. Werner introduced a new measure for the binding power of the atom, which was also expressed in integral units and was better known as co-ordination number. Each atom or ion was given a separate co-ordination number according to this view certain. According to this view certain atoms after or as a result of the satisfaction of their primary valency have the power to attract or hold around themselves a definite number of ions, groups or molecules. Giving rise to a closely associated complex in the form of a distinct name known as co-ordination complex. The groups, molecules or ions, thus held around the central atom, are said to be co-ordination number the co-ordinated units are arranged symmetrically around the central atom, e.g. the planar, tetrahedral, octahedral, and cubic. They cannot appear as ions in solutions. The binding force of the central atom was believed to be gravitational in nature and extended uniformly in all directions. The co-ordination number of the central atom giving the number of units associated with it, was therefore, supposed to be determined only by geometrical consideration, i.e. the amount of space available around it. As the elements of the transitional group are characterized by having very low atomic volumes. Werner also pointed out that they should naturally serve as centres of co-ordination complex.

Werner's Theory, is however, applicable only to a certain section of the compounds of the higher order i.e. only to the complex salts, where as the formation of a host of double or molecular compounds lies beyond its scope. However, Werner's Theory achieved a great triumph in successfully predicting the existence of isomers and asymmetric forms possessing the optical activity among the complex compounds.

CHAPTER

6

Air and Its Composition

Introduction

A vast jacket of a physical composition of different gases and vapours surrounds the Earth, which is known as the air or atmosphere. The total existing depth about 100 miles. This is the most important matter for all leaving organisms. In *Vedas* the atmosphere has been regarded as *the medium of life.*

Here in this chapter we shall study in detail the composition, physical and chemical properties and isolation of gases from air after liquification.

Classification of Atmosphere

Atmosphere has been classified rather has been divided into layers. The layers are from the surface of the earth is *Troposphere, Stratosphere,* then comes *Ionosphere,* and last one is *Exosphere.*

Here in this chapter we are going to discuss the physical composition of stratosphere and the troposphere.

Properties of Atmosphere

If we try to find out the properties of the atmosphere or air we must go back to the time of Aristotle (384-322 B.C) up to the eight century it was believed that air is one of the five elements of earth (*vayu*).

In 1673 Robert Boyle demonstrated that no burning could take place in the absence of air. He also showed that gunpowder got burnt in the absent of air. He thought that one of the ingredients of gun powder the potassium nitrate did the work of air.

M. V Lomonosov (1756) proved that a part of air combined with metals on heating those metals red hot.

In the year 1773 the most important classification of air was done by a famous chemist. Antoine Laurent Lavoisier (1743-1794). He conducted a very important experiment about the composition of air. By this experiment he was the first scientist to prove that air was not at all an element but a mechanical mixture. That whole of the air is not for respiration, it is that part which combine with other elements is the respirable portion of the air. It was also proved that no substance could burn or combine with other without this portion of the air, which respire. We are now going to discuss about the experiment of Lavoisier.

Lavoisier's Experiment

Lavoisier conducted this experiment mainly to investigate the composition of I = *p*~ Air. In this experiment he demonstrated that only one-fifths of the total volume of the air is respirable and the other four-fifths is named by him as 'Azote' which means non-respirable or inert.

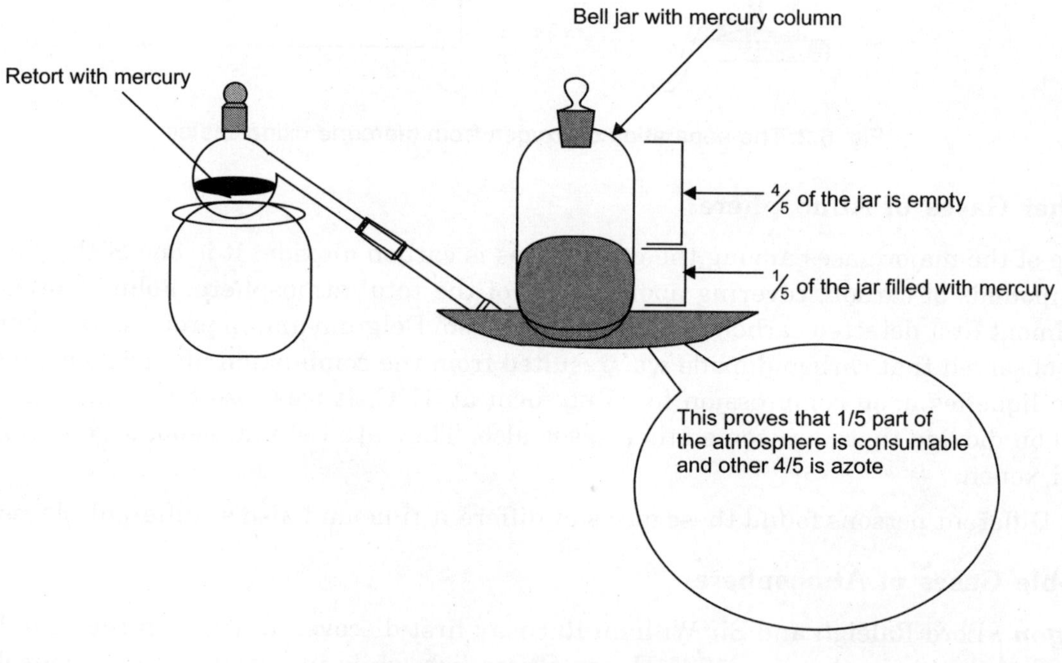

Fig. 6.1. The Laviosier's Experiment

The experiment of Lavoisier proved that atmosphere did contain a mixture of mainly two gases called Azote (later named as nitrogen) and oxygen. One-fifth of the atmosphere was full of oxygen and the other 4/5th was full of azote. When the azote and oxygen was mixed together the normal air was obtained. So from these two experiments Lavoisier concluded that:

1/4 was full of oxygen and other 4/5th was azote

(*a*) Air is a mechanical mixture of two gases namely Azote and oxygen:

(*b*) Oxygen supports combustion and life, but azote does not.
(*c*) Metal produces Calces (basic oxide) on absorption of oxygen.
(*d*) Combustion is the result of reaction oxygen.

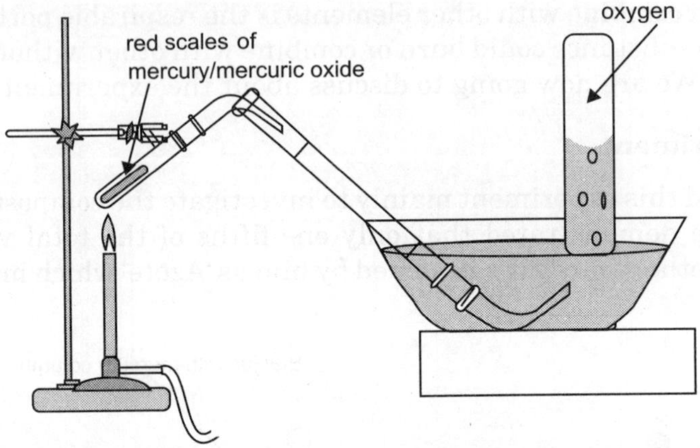

Fig. 6.2. The separation of oxygen from mercuric oxide scales

Other Gases of Atmosphere

One of the major gases among the other gases is carbon dioxide. It is one of the gaseous compounds of carbon, covering about .032% of the total atmosphere. John Baptist Van Helmont first detected carbon dioxide. He was from Belgium and in profession a chemist. He observed that carbon dioxide was resulted from the combustion and fermentation. It also liquefies upon compression to 75 kg/sqcm at 31°C. It was also found that not only carbon dioxide there are some other gases also. They are helium, neon, argon, krypton and xenon.

Different persons found these gases at different time and also at different places.

Noble Gases of Atmosphere

Argon : Lord Raleigh and Sir William Rarnsay first discovered Argon in the year 1894. But before this in the year 1795 Henry Cavendish while investigating on atmospheric nitrogen (phylogestical air) concluded that not more than 1/120th part of air might be some inert constituent. His work was forgotten until Lord Raleigh, a century later (1894) found that nitrogen prepared be removing oxygen from air is always about 0.5% more dense than as ammonia. The heavier gas remaining after both oxygen and nitrogen had been removed from air was the first of the noble gases to be discovered on the earth and was named as argon because of its chemical inertness (argoes = lazy).

Argon constitutes 1.3% by weight and .044% by volume and is found occluded in rocks. A major portion of the terrestrial argon has been produced since the Earth's formation, in potassium containing minerals by decay of the rare naturally occurring argon is a mixture of three stable isotopes argon-36, (.34%) argon-38 (0.06%) and argon-40 (96.60%). Argon is chemically inert because it has a complete valence shell, that is 8 electrons in the outermost orbit.

Argon can be converted into a colourless liquid at a temperature of –185°C and to a crystalline solid at a temperature of –189°C.

Helium : Helium was discovered (1868) in the gaseous atmosphere surrounding the sun by Joseph Norman Lockyer, who observed a yellow line in the solar spectrum which did not show any similarity with ant known substance in earth. Lockyer used the Greek name Helio's which means "for sun" in naming the element. Sir William Ramsay discovered this element on Earth in the year 1895 in a uranium bearing mineral (cleavite), a unique bright yellow line in the spectrum which matched the same yellow line observed by Lockyer.

Helium is also present in the Earth's atmosphere but only about .0005%. Helium can be obtained from natural gas by liquefying the other components at low temperature and high pressure. Helium can be liquefied at –286°C and 25 atmospheric pressure. Helium also gets solidified at 2.17°K (absolute temperature).

Neon : Sir William Ramsay and Marris W Travers discovered Neon in the year 1898 as a component most volatile of fractionation of crude Argon in liquid state, obtained from air. The gas is produced industrially by fractional distillation of air.

Neon liquefies at –246.048°C and freezes at –247°C when under low pressure it emits a bright orange light if an electric current is passed through the gas. In Earth's atmosphere the existence of neon is very little about 0018% by volume when the air is in dry state.

Krypton and Xenon : Krypton and Xenon are both available in the heavier fraction of the crude Argon. When the crude Argon is fractionally distilled the lowest fraction consists of krypton and the higher fraction consists of xenon. Krypton in atmosphere is present in 0.0001% by volume, xenon is present in the atmosphere about .0000009% by volume. The total contents of the atmosphere are shown in the bar graph (Fig. 6.3).

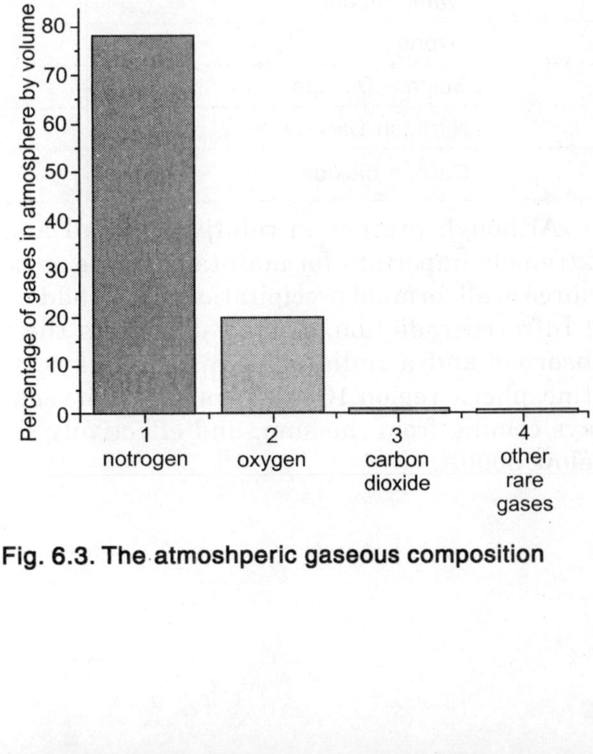

Fig. 6.3. The atmoshperic gaseous composition

The total amount of the atmospheric gases are given in the table below :

Name of the gases	Volume of the gas in percentage
Nitrogen	78.084
Oxygen	20.946
Hydrogen	0.00005
Methane	0.0002
Nitrous oxide	0.00005
Helium	0.000524
Argon	0.934
Neon	0.00018
Krypton	0.000114
Xenon	0.00000009

The uniformity of composition is maintained by mixing associated with atmospheric motions but above a height of about 90 km. The diffusion is much more and the lighter gasses like hydrogen and helium becomes available.

Of the gases present in variable concentrations water vapour, ozone, carbon dioxide, sulphur dioxides are principal importance. The typical concentration ranges of these gases (in percentage by volume) are stated in the table below:

Name of the other gases	Volume percentage
Water Vapour	0 - 7% average (0.032%)
Ozone	0 - 0.1%
Sulphur Dioxide	0 - 0.001%
Nitrogen Dioxide	0 - 000002%
Carbon dioxide	0 - 0.1%

Although present in relatively small amounts these variable constituents may be extremely important for maintaining life on the surface of the earth. Water vapour is the source of all forms of precipitation and in addition is an important absorber and also emitter of Infrared radiation. Carbon dioxide is the main agent for photosynthesis and a good absorber and a radiator of infrared rays. Ozone, which is present primarily in the atmospheric region 10 to 15 km above the sea level, is an effective absorber of ultraviolet rays coming from the sum, and effectively shields the earth from all kinds of radiation below 3000A.

CHAPTER

7

Hydrogen

Introduction

From Greek words Hydrous Genes, hydrogen has been derived, which means water producer. We will see in this chapter afterwards whether hydrogen is water producer, now we discuss its occurrence, properties and preparation (both lab and industrial).

Invention of Hydrogen

Henry Cavendish, famous English scientist, invented hydrogen in the year 1764; he called it *inflammable air* as it burnt in a pale blue flame when fire was introduced to hydrogen gas.

Occurrence

Hydrogen in earth is not very much available in free state, as it is the lightest element of the elements in the periodic table. It is also the lightest of all the gases in the atmosphere. Hydrogen mainly occurs in the form of compounds among which the water is the main (H_2O) and is also in some formations and compounds, which are going to be studied in this chapter in detail.

Preparation of Hydrogen

Hydrogen can be prepared very easily in the laboratory by a simple chemical process under normal temperature and pressure (room temperature and pressure).

Theory : Hydrogen is prepared from the reaction of granulated commercial zinc and dilute sulphuric acid the gas developed is collected in the gas jar with downward displacement of water.

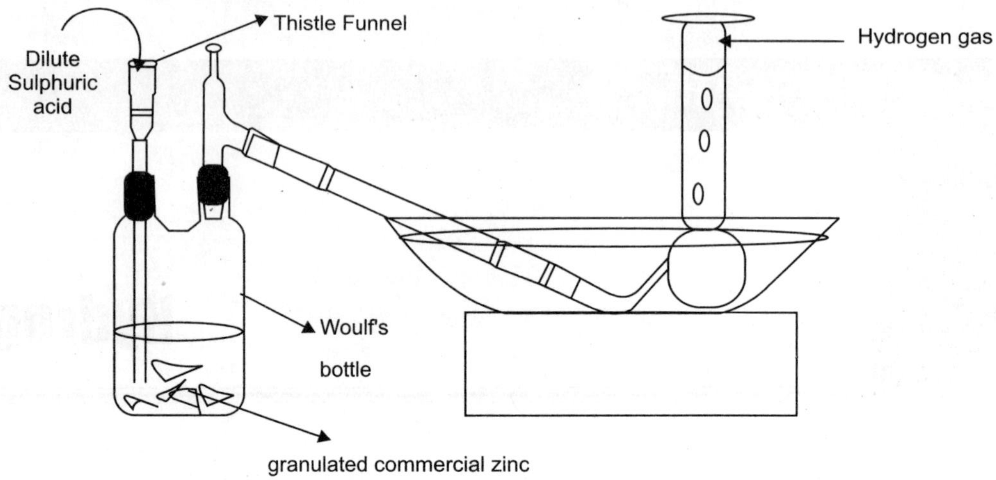

Fig. 7.1. The Laboratory preparation of Hydrogen Gas

Apparatus required	Chemicals required	Temperature	Pressure
Woulf's Bottle, Thirstle funnel, gas jar, delivery tube, corks etc.	Granulated Commercial zinc, Dilute Sulphuric Acid	Room temperature	760mm of Hg

$$\text{Zinc} + \text{Sulphuric Acid} \longrightarrow \text{Zinc Sulphate} + \text{Hydrogen}$$
$$Zn + H_2SO_4 \longrightarrow ZnSO_4 \Downarrow + H_2 \Uparrow$$

Flammable Gas

Precaution : During the experiment some of the precaution measure must be taken for the success of the experiment, they are as follows:

1. The apparatus should be air tight, because if there is any leakage the hydrogen gas will escape through the leak. To check this at first some water is poured in the bottle after all fitting, now instead of connecting the gas jar air is blown through the bottle and the open end of the delivery tube is closed with thumb the water level risen in the thistle funnel will, stay there, if not there is a leakage, to make it easily first most the rubber cork and then fix.
2. Stem of the thistle funnel must dip in the acid solution inside the Woulf's bottle.
3. Gas must be collected after 30 to 60 seconds after the evolution otherwise the gas will be filled with other unwanted gasses in the Woulf's bottle.
4. There should be no necked flame near the experimental set up because hydrogen can catch fire.

The above measures are very simple to carry out and if they are carried out properly there will be no trouble during the experiment.

Other Methods of Hydrogen Preparation

Hydrogen can be prepared by some other chemical processes also it is mainly prepared from water.

I. Decomposition of Water is done by some active elements, sodium potassium etc. When water is added (few drops only) to sodium or potassium metal a vigorous reaction takes place with the evolution of hydrogen gas

$$\text{Potassium} + \text{water} \longrightarrow \text{Potassium Hydrogen} + \text{Hydrogen}$$

$$2K + 2H_2O \longrightarrow 2KOH\Downarrow + 2H_2\Uparrow$$

$$\text{Sodium} + \text{Water} \longrightarrow \text{Sodium Hydroxide} + \text{Hydrogen}$$

$$2Na + 2H_2O \longrightarrow 2NaOH\Downarrow + H_2\Uparrow$$

Here water gets decomposed into hydrogen due to hydrolysis

II. At Boiling Point: Magnesium powder, zinc powder breaks up when it is boiled with water giving hydrogen and the respective metallic oxide.

$$\text{Magnesium} + \text{Water} \longrightarrow \text{Magnesium Oxide} + \text{Hydrogen}$$

$$Mg + H_2O \longrightarrow MgO\Downarrow + H_2\Uparrow$$

$$\text{Zinc} + \text{Water} \longrightarrow \text{Zinc Oxide} + \text{Hydrogen}$$

$$Zn + H_2O \longrightarrow ZnO\Downarrow + H_2\Uparrow$$

III. By Passing Steam over Heated Metals: Heated metals like magnesium, iron, zinc, cobalt nickel etc react with water in the form of steam. The reaction is reversible in the case of iron.

$$\text{Iron} + \text{Steam} \longrightarrow \text{Ferric Oxide} + \text{Hydrogen}$$

$$3Fe + 4H_2O \longrightarrow Fe_3O_4\Downarrow + 2H_2\Uparrow$$

IV. By the Action of Hydrides: The metallic Hydrides can also decompose water at ordinary temperature, e.g. calcium Hydride (hydrolyth) reacts with water accordingly:

$$\text{Hydrolyth} + \text{Water} \longrightarrow \text{Calcium Hydroxide} + \text{Hydrogen}$$

$$CaH_2 + 2H_2O \longrightarrow Ca(OH)_2\Downarrow + 2H_2\Uparrow$$

V. From Hydroxides of Sodium, Potassium: When zinc aluminium, tin, etc. metals react with sodium or potassium hydroxide, hydrogen is produced

$$\text{Zinc} + \text{Sodium Hydroxide} \longrightarrow \text{Sodium Zincate} + \text{Hydrogen}$$

$$Zn + 2NaOH \longrightarrow Na_2ZnO_2\Downarrow + H_2\Uparrow$$

$$\text{Tin} + \text{Potassium Hydroxide} \longrightarrow \text{Potassium stannate} + \text{Hydrogen}$$

$$Sn + 2KOH \longrightarrow K_2SnO_3\Downarrow + H_2\Uparrow$$

Purification of Impure Hydrogen

Type of Impurities	Removal Process
Oxygen (from air)	Can be removed over hot iron.
Sulphur Dioxide, Carbon dioxide Oxides of Nitrogen	Is generally removed by passing hydrogen through potassium hydroxide solution.
Compounds of Phosphorous, Sulphur and Arsenic,	By passing at first through potassium permanganate solution and then through silver nitrate solution to remove arsenic.
Atmospheric Moisture	By drying over calcium chloride or phosphorous pentoxide.

Generally pure hydrogen is prepared by the electrolysis of barium hydroxide solution by using nickel electrodes, the gas is at first passed through hot platinum gauge (not through platinum asbestos because this reacts with hydrogen producing silicon hydride) to burn any traces of oxygen in it. Then the gas is passed through potassium hydroxide solution and sulphuric acid, because the acid will get reduced forming a little sulphur dioxide, then the gas is passed through phosphorous pentoxide to remove the atmospheric moisture. Ultimately the gas is then passed through palladium because only palladium absorbs hydrogen and all other inert gases are left behind, this is generally done at an elevated temperature. Then the hydrogen is again recovered by strongly heating the palladium.

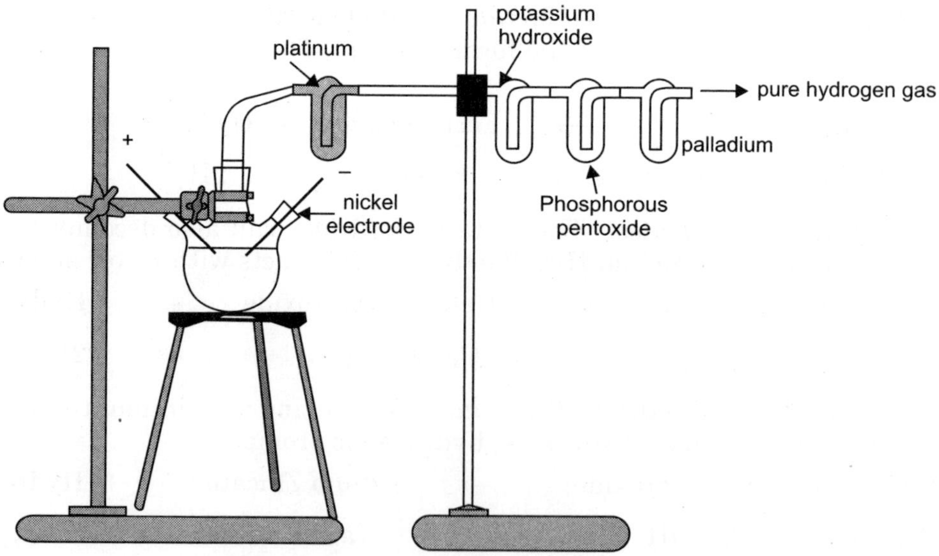

Fig. 7.2. The Apparatus set up for the preparation of Pure Hydrogen from the Electrolysis of barium hydroxide with nickel electrodes

Properties of Hydrogen

Physical Properties : Hydrogen is physically a colourless, odourless and tasteless gas, it is very much less soluble in water. Among all the gases hydrogen is the lightest gas known, one litre of hydrogen weights only 0.9g at Normal Temperature and Pressure (N.T.P).

Hydrogen becomes liquid at a temperature below –253°C and gets solidified at –250°C. Hydrogen has an interesting property of adsorption or occlusion by the metals like platinum and palladium. When the surface of a palladium or platinum metal is exposed to hydrogen gas, the gas is soaked to the surface. The detail physical properties of hydrogen gas is given in the table below:

Property	Value
Melting Point	–259.34°C
Boiling Point	–252.87°C
Density of Solid Hydrogen at –259°C	0.0858g/cm^3
Density of Liquid at –252.87°C	0.0708g/cm^3
Critical Temperature	–240.17°C
Critical Pressure	12.9792 Pascal
Critical Density	0.01312g/cm^3
Specific Heat at Constant Pressure	
Gas at 25°C	34.2cal/(g)(°C)
Liquid at –256°C	1.93cal/(g)(°C)
Solid at –259.8°C	0.63cal/(g)(°C)
Heat of Fusion at –259.8°C	13.9 col/g
Heat of Vaporization at –252.87°C	107cal/g
Thermal conductivity	0.000444cal/(cm)(s)(°C)
Viscosity at 25°C	0.00892 centipoise.

Chemical Properties: Hydrogen has various types of chemical properties to discuss them a separate book will be necessary so a very few of them will be discussed in this chapter.

Valence capacity of Hydrogen Atom (1) Electrovalency in forming the hydride ion W which is hydrated in water as the hydronium or hydroxonium ion, H_3O^+. The relative activity of aqueous solution depends upon the concentration of the hydrogen ion.

(2) Electrovalency in forming the hydride ion, as in compounds NaH, aaH_2 etc.

(3) Hydrogen is also efficient in forming Covalent bonds as in hydrochloric acid and water.

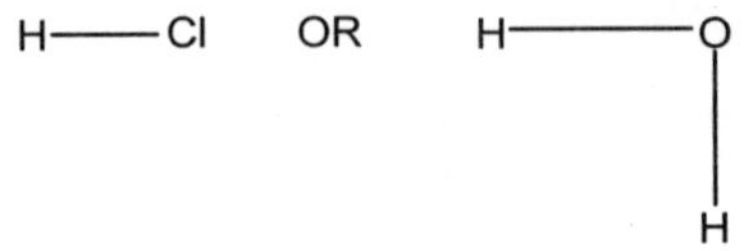

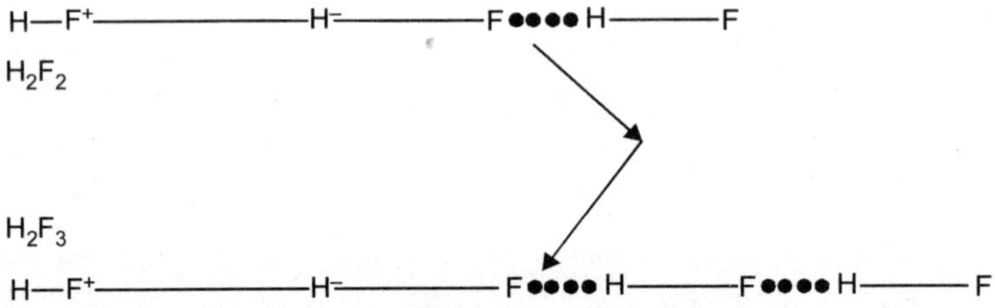

Hydrogen also shows bonding with acetic cetic acid and carbon

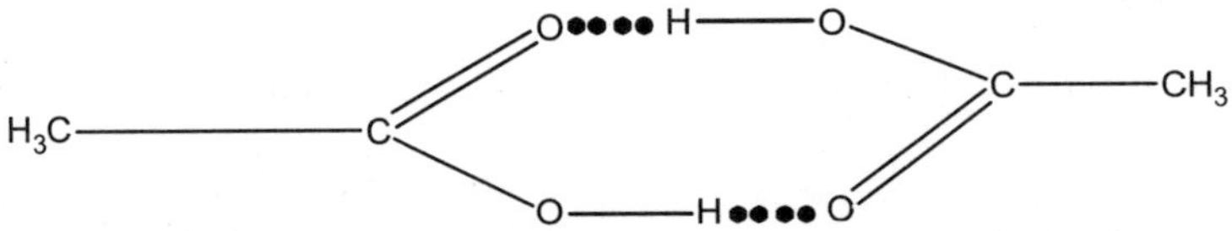

We shall study some more chemical properties, which are mainly experimental.

1. *Action on Litmus:* We shall see that there is no change in either red or blue litmus paper when exposed to the gas jar containing hydrogen gas, in moist condition, this means that hydrogen is neutral to litmus.
2. *Combustion of Hydrogen Gas* is not a supporter of combustion but is self-combustible. When a glowing splinter is used in the gas jar a loud Pop sound is heard and the gas burns with a pale blue flame.
3. *Exp: For the demonstration of Combination Property of Hydrogen Gas:*

HARMFUL

Students are requested to perform these experiments before a teacher

As we all know that hydrogen is known as water producer. Henry Cavendish named it, he to supported the statement and performed the experiments below:

Exp. 1 Hydrogen gas is produced from a source (hydrogen Generator) and passed through calcium chloride collected in the bulb to remove moisture, then the gas is made to escape through the jet tube and the gas is lighted when it comes in the contact of the bulb. Droplets of clear liquid collects on the bottom outer wall of the flask, which turns blue, the anhydrous white copper sulphate (Fig. 7.3).

Exp. 2 (Fig. 7.4.) To prove that Hydrogen is a combustible gas but not a supporter of Combustion

When a glowing splinter is introduced in a gas jar containing hydrogen the splinter is extinguished and the gas burns with a pale blue flame by making a loud "pop" sound at the mouth of the gas jar.

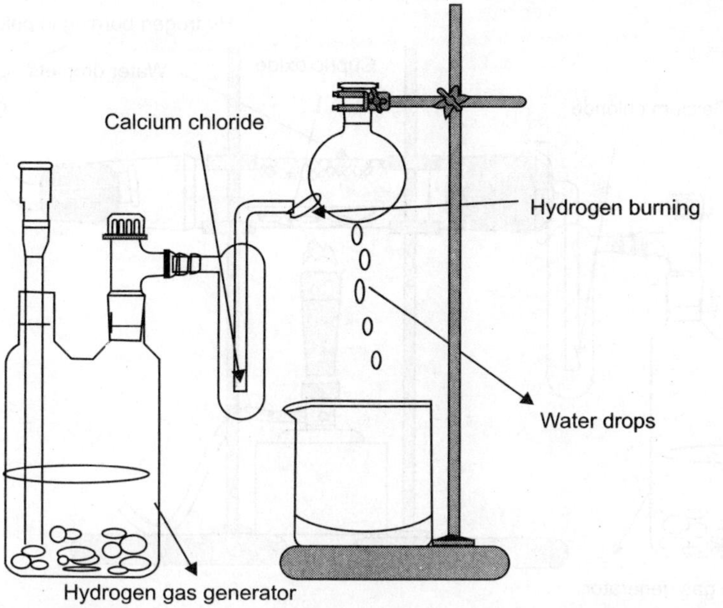

Fig. 7.3. Experimental sepup that Hydgrogen Gas is a water producer

Reproduction Property

Hydrogen in the chemical laboratory is used as a reducing agent because of its excellent oxygen removing properties. When hydrogen gas is passed over a red-hot metallic oxide, the oxygen gets combined with hydrogen and the metal is left behind its original form.

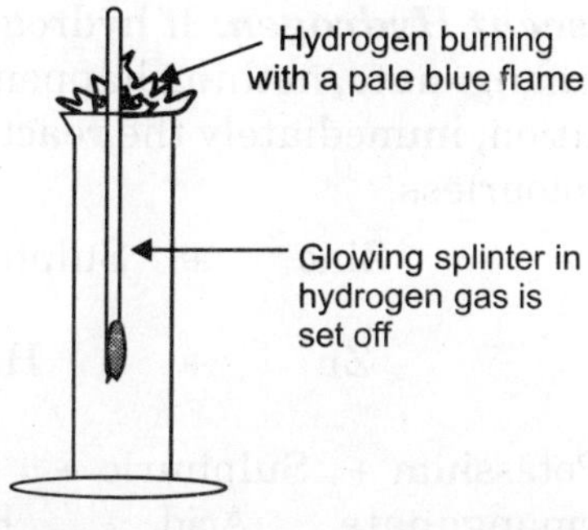

Fig. 7.4. To prove that Hydrogen is a combustible gas but not a supporter of Combustion

From the source of hydrogen gas a long hard glass tube of 3 to 4 inch diameters is connected by a delivery tube, on the other end of the glass tube a small jet is connected. Inside the glass tube some copper oxide is kept. The whole apparatus is checked for air tightness. At first hydrogen gas is passed through the tube to remove the air inside the tube then the glass tube is heated strongly at the place where the copper oxide is kept. After some times it will be seen that the colour of the oxide is changing from black to brown, also some fine droplets of pure liquid is seen collected on the inner walls of the tube. The clear liquid is water which can be identified by introducing some anhydrous copper sulphate, which will turn blue immediately when comes in contact with the liquid. Here when the copper oxide is heated its oxygen gets reacted with the hydrogen producing water and the metal remains as it was.

Cupric Oxide + Hydrogen ⟶ Water + Copper

$$CuO + H_2 \longrightarrow H_2O + \downarrow Cu$$

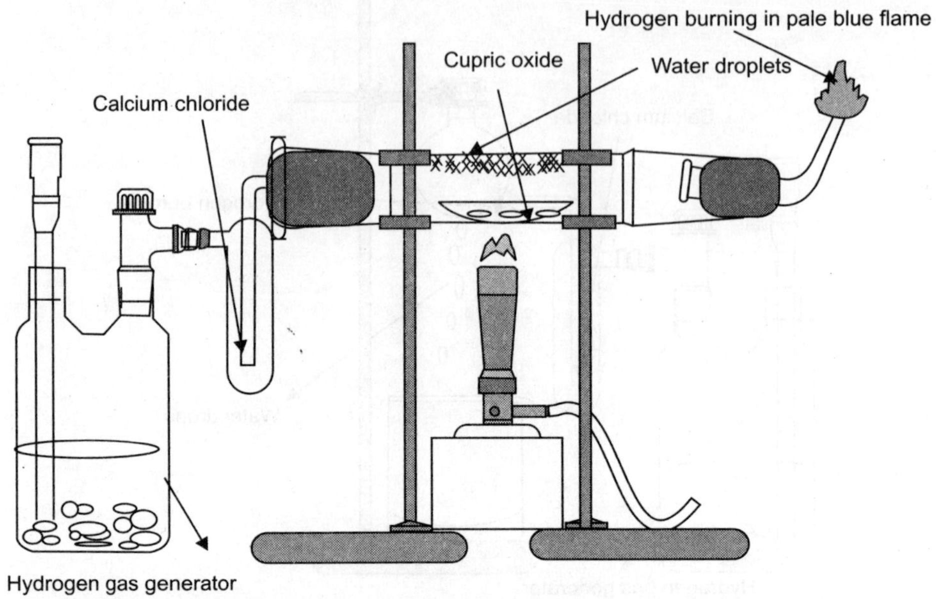

Fig. 7.5. Experimental setup that Hydrogen Gas is a water producer

Types of Hydrogen

Nascent Hydrogen: If hydrogen gas is bubbled through an acidic solution of potassium permanganate, nothing happens, if some granulated zinc are placed in the permanganate solution, immediately the reaction starts and the colour of the solution changes from pink to colourless.

Zinc + Sulphuric Acid ⟶ Zinc Sulphate + Nascent Hydrogen

$$Zn + H_2SO_4 \longrightarrow ZnSO_4\Downarrow + 2H\Uparrow$$

Potassium Permanganate + Sulphuric Acid + Nascent Hydrogen ⟶ Potassium Sulphate + Manganese Sulphate + Water

$$2KMnO_4 + 3H_2SO_4 + 10H \longrightarrow K_2SO_4\Downarrow + 2MnSO_4\Downarrow + 8H_2O$$

Yellow Ferric chloride also turns green when introduced to nascent hydrogen by above process.

Ferric Chloride+Nascent Hydrogen ⟶ Ferrous Chloride+ Hydrogen Chloride

$$FeCl_3 + H \longrightarrow FeCl_2 + HCl$$

Atomic Hydrogen: When hydrogen is electrically discharged at room temperature and under very low pressure active hydrogen is produced. Active hydrogen readily combines with lead and sulphur to form their respective hydrides and it can be also reduce oxides from metals at room temperature.

Preparation of Hydrogen in Kipp's Apparatus

This apparatus is commonly known as hydrogen generator, it is widely used in the modern chemical laboratories, for quick availability of the hydrogen gas.

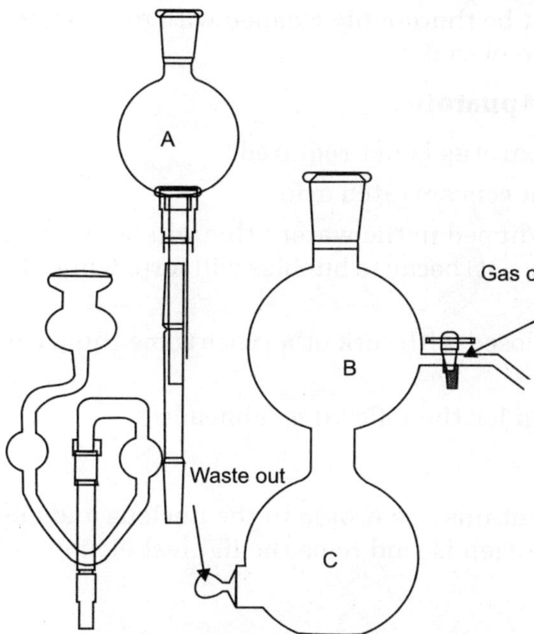

Fig 7.6A. Dismantled Kipp's Apparatus

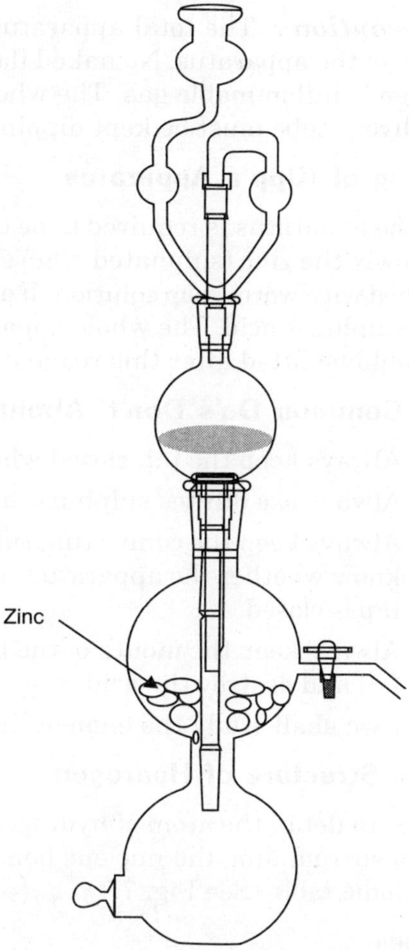

Fig 7.6B. Kipp's Apparatus producing hydrogen

Kipp's Apparatus is made up of glass consists of three domes A, B and C the top one (A) consists of a long stem going down up to the dome up to the dome (C) the middle dome (B) consists a tap through which the gas escapes. The top dome can be removed from the total assembly of the other two domes. The middle and the bottom cannot be separated. In the middle dome (B) zinc dust is kept and the top dome is placed in the assembly. From the top dome (A) through a meandering thistle funnel dilute sulphuric acid is poured in the apparatus in such a way that the acid reaches in the bottom dome(C) and it comes in the middle dome (B). This start the reaction, which produces hydrogen gas, this gas, escapes through the tap. When the tap is closed the gas inside creates pressure on the acid which is pushed back from the second dome to the dome (C) and finally to dome (A), thus stopping the reaction and the production of gas.

Precaution : The total apparatus must be air-tight because any leak will stop the working of the apparatus. No naked flame should be brought near the apparatus, because hydrogen is inflammable gas. The whole gas must be kept in a cool and ventilated place. The delivery tube must be kept dipping in the water.

Cleaning of Kipp's Apparatus

When the apparatus is required to be cleaned, at first the acid should be pippeted out and then slowly the zinc (sulphated zinc) should be removed and the whole apparatus should be washed with worm soap solution. If any patch is present it should be removed by applying a little sulphuric acid. The whole apparatus must be thoroughly cleaned with distil water and should be fitted after this reagents should be placed.

Some Common Do's Don't' About Kipp's Apparatus

- ✓ Always keep the tap closed when the apparatus is not required.
- ✓ Always use dilutes sulphuric acid and not concentrated acid.
- ✓ Always keep the connecting rubber tube dipped in the water (this will help you to know whether the apparatus is leaking or not) because bubbles will form while the tap is closed.
- ✓ Always keep the mouth of the top dome closed with cork otherwise some thing can fall and destroy the acid.

Now we shall study the element hydrogen under the following subheads.

Atomic Structure of Hydrogen

If we see in detail the atom of hydrogen gas, it contains one proton in the nucleus and one electron surrounding the nucleus hence the hydrogen is said to be the lightest element of the periodic table (See Fig. 7.7).

Isotopes

The word 'isotope' is a Greek word meaning *Iso* = Same and *Tope* = place, that means the substance having same atomic number but different atomic masses and they aslos stay at the same place in the periodic table. Hydrogen has two well-known isotopes: 1. Deutreum and 2. Tritium.

According to the classical chemistry isotopes of an element are expected to have same chemical properties. Therefore, it is believed that the behaviour of the isotopes of an element are essentially the same. However, a change in the mass of an atom in a molecule produces changes in the vibrational, rotational and also in the transitional motions of the molecule. In diffusion of gasses it has long being known that the velocity of the molecules changes with the mass of the molecule.

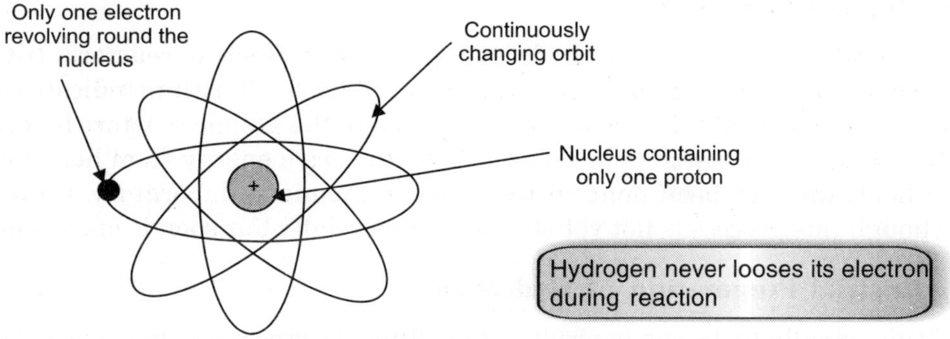

Fig. 7.7. The Diagrammatic representation of the atomic structure of hydrogen.

The isotopes of hydrogen has different properties, they combine with oxygen mainly, when deuterium combines with oxygen deuterium oxide (D_2O) is formed, it is also known as heavy *water.* Tritium also reacts with oxygen like deuterium; it is called tritium oxide (T_2O). The properties of these two oxides of the isotopes are tabulated below:

Properties	Hydrogen	Deuterium	Tritium
Relative Atomic Mass	1.0078267	2.01439	3.017
Nuclear Spin Quantum number	1/2	1	1/2
Nuclear Magnetic Moments (Nuclear Magnetrons*)	2.79268	0.857386	2.98
Nuclear Quadrapole Moment ($e \times 10^{-28}$ m^2)	0	2.77×10^{-3}	0
Radio Active Stability	Stable	Stable	

We shall also see the physical properties of the isotopic molecular hydrogen in the table below:

Properties	Hydrogen**	Deuterium	Tritium
Melting Point	–259.04°C	–254.2°C	–252.4°C
Boiling Point	–252.61°C	–249.3°C	–247°C
Heat of Fusion (kJ mol^{-1})	0.117	0.197	0.250
Heat of Vapcurization (kJ mol^{-1})	0.904	1.22	1.393
Critical Temperature (°K)	33.19	38.35	40.0
Critical Pressure	1315.20Kpa	1664.77Kpa	1834Kpa
Heat of Dissociation at 25°C	435kJ/mol	18.5 kJ/mol	15.1kJ/mol
Inter molecular Distance	74.17pm	74.17pm	74.17pm***

-Magnetrons - All of this data refer to the mixture of ortho and para-hydrogen that are in equilibrium.
*These data refer to Hydrogen at normal isotopic composition.
***That is containing 156 ppm of Deuterium predominantly at HD.

Isotopes of Hydrogen

Generally these isotopes of hydrogen are used in nuclear reactors, tritium is generally used in detection of atmospheric conditions and excess of tritium indicates heavy continental air. Deuterium oxide heavy water is required in the atomic reactors for cooling the reactor. Now-a-days a research is in progress that to derive energy from heavy water in the form of heat; this has been done in the nuclear chemistry laboratory, California University. Though this process is not yet started commercially but soon it may come to use.

Industrial Preparation of Hydrogen

Commercially hydrogen is produced by different processes they are as follows:

$$\text{Ethane} + \text{Steam} \xrightarrow{\text{Nickel}} \text{Carbon Monoxide} + \text{Hydrogen}$$

$$C_2H_6 + H_2O \xrightarrow{\text{Nickel}} 2CO\Uparrow + 5H_2\Uparrow$$

Catalytic Reaction: Hydrogen is produced in a large scale by using hydrocarbons natural gas (Methane), petrol, diesel, etc. Water vapour (steam) reacts with volatile hydrocarbon in the presence of nickel at 750°C-1100°C to produce carbon monoxide.

$$\text{Carbon Monoxide} + \text{Steam} \xrightarrow{\text{Iron Oxide}} \text{Carbon Dioxide} + \text{Hydrogen}$$

$$2CO + H_2O \xrightarrow{\text{FeO}} CO_2\Uparrow + H_2\Uparrow$$

Here in the above reaction ethane is used as the main hydrocarbon source.

The above process is conducted under much lower temperature than the nickel catalytic process which is (280°C-360°C). This process is also known as steam hydrocarbon process, which has replaced the once most popular industrial process which required much higher temperature and extremely expensive reagents like mono-ethylamine.

Other Propertiess

Hydrogen is also prepared by the reaction of steam with iron at (780°C-820°C) the reaction is shown below in Fig. 7.8.

$$\text{Ferrosoferric Oxide} + \text{Carbon Monoxide} + \text{Hydrogen} \rightleftharpoons \text{Iron} + \text{Iron Oxide} + \text{Carbon} + \text{Steam}$$

$$Fe_3O_4 + CO + H_2 \rightleftharpoons Fe + 2FeO + CO_2 + H_2O$$

$$\text{Iron} + \text{Iron Oxide} + \text{Steam} \rightleftharpoons \text{Ferrosoferric Oxide} + \text{Hydrogen}$$

$$Fe + 2FeO + 2H_2O \rightleftharpoons Fe_3O_4 + 2H_2$$

The manufacture of hydrogen on an industrial scale involves the reaction between steam and iron. The process is known as *Lane Process.*

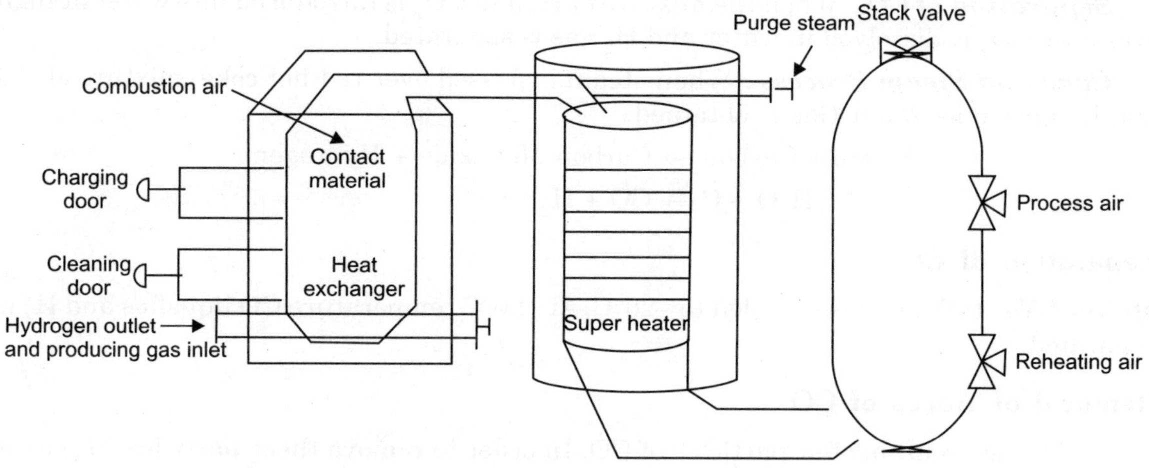

Fig. 7.8. The Process of Industrial Preparation of Hydrogen by Lane's Process

Spongy iron from the reduction of spathic iron ore (ferrous carbonate) is heated to redness and steam passed over it. The hot ferrosoferric oxide, (Fe_3O_4), is then reduced with water gas. Water gas is made by passing steam over red hot carbon and it consists of a mixture of carbon monoxide and hydrogen, with a smaller amount of carbon dioxide:

By the Cracking of Methane

When natural gas is heated in the absence of air up to 70°C, it decomposes to produce hydrogen.

$$\text{methan} \longrightarrow \text{Carbon} + \text{Hydrogen}$$

$$CH_4 \longrightarrow C(\text{solid}) + 2H_2$$

During cracking of methane "Carbon Black" is obtained as by-product.

By Thermal Decomposition of Ammonia

When liquid ammonia is vapourized and heated up to 100°C in the presence of a catalyst, it decomposes to produce nitrogen and hydrogen.

$$\text{Ammonia} \rightarrow \text{Nitrogen} + \text{Hydrogen}$$

$$2NH_3 \rightarrow N_2 + 3H_2$$

Separation of Nitrogen : Mixture of N_2 and H_2 is cooled to –196°C. At –196°C, N_2 liquefies and H_2 is separated in gaseous state.

Steam/ alcohol method : When methyl alcohol is passed over steam at 250°C, a mixture of hydrogen and CO_2 is obtained.

$$\text{methyl Alcohol} + \text{Steam} \longrightarrow \text{Carbon Dioxide} + \text{Hydrogen}$$

$$CH_3OH + H_2O \longrightarrow CO_2 + 3H_2$$

Separation of CO_2: When the mixture of H_2 and CO_2 is introduced into water at high pressure, CO_2 is dissolved in water and H_2 gas is separated.

Coke and Steam Process : When steam is passed over red-hot coke, mixture of CO and H_2 known as Water Gas is obtained.

Steam + Carbon → Carbon Monoxide + Hydrogen

$$H_2O + C \rightarrow CO + H_2$$

Separation of CO

Method No. 1: Water gas is cooled to –20°C. At –20°C temperature CO liquefies and H_2 is separated.

Removal of Traces of CO

H_2 so obtained contains few particles of CO. In order to remove these particles. H_2 gas is treated with NaOH.

Carbon Monoxide+Sodium Hydroxide ⟶ Sodium formate

$$CO + NaOH \longrightarrow HCOONa$$

Method No. 2 : Water gas is heated with steam in the presence of iron oxide or cobalt oxide. Due to this reaction CO is converted in to CO_2.

Hydrogen + Carbon Monoxide ⟶ Carbon Dioxide + Hydrogen

$$H_2 + COH_2O \longrightarrow CO_2 + 2H_2$$

Separation of CO_2 : When a mixture of H_2 and CO_2 is dissolved in water at high pressure, CO_2 is dissolved in water and H_2 gas is separated.

Electrolysis

This is the process where hydrogen is separated electrically from water. In this process steady D.C. electricity is applied by two different electrodes at the positive electrode oxygen is liberated and at the negative electrode hydrogen is produced. The diagram of the whole experimental set up is shown in Fig. 7.9.

We shall now see a table consisting the properties of the three oxides of hydrogen.

Properties	Hydrogen Oxide	Deuterium Oxide	Tritium Oxide
Molecular Weight	18.015	20.028	220.032
Freezing Point	0°C	3.81°C	---------
Boiling Point	100°C	101.42°C	101.51°C
Triple Point	0.01°C	3.82°C	4.49°C
Critical Temperature	374.15°C	370.09°C	---------
Temperature of the Maximum density of Liquid	3.984°C	11.185°C	13.403°C

Properties	Hydrogen Oxide	Deuterium Oxide	Tritium Oxide
Delectric constant of the liquid at 20°C	81.5	80.7	—
Viscosity of the liquid at 20°C (metepɔise)	1.002	1.248	—
Ion Product of Liquid at 25°C (mol dm^{-3})	1.0×10^{-14}	0.03×10^{-14}	—
Lattice Constant of the Solid (pm)	735.2	735.2	—
Refractive Index of the liquid at 20°C	1.33300	1.32830	—

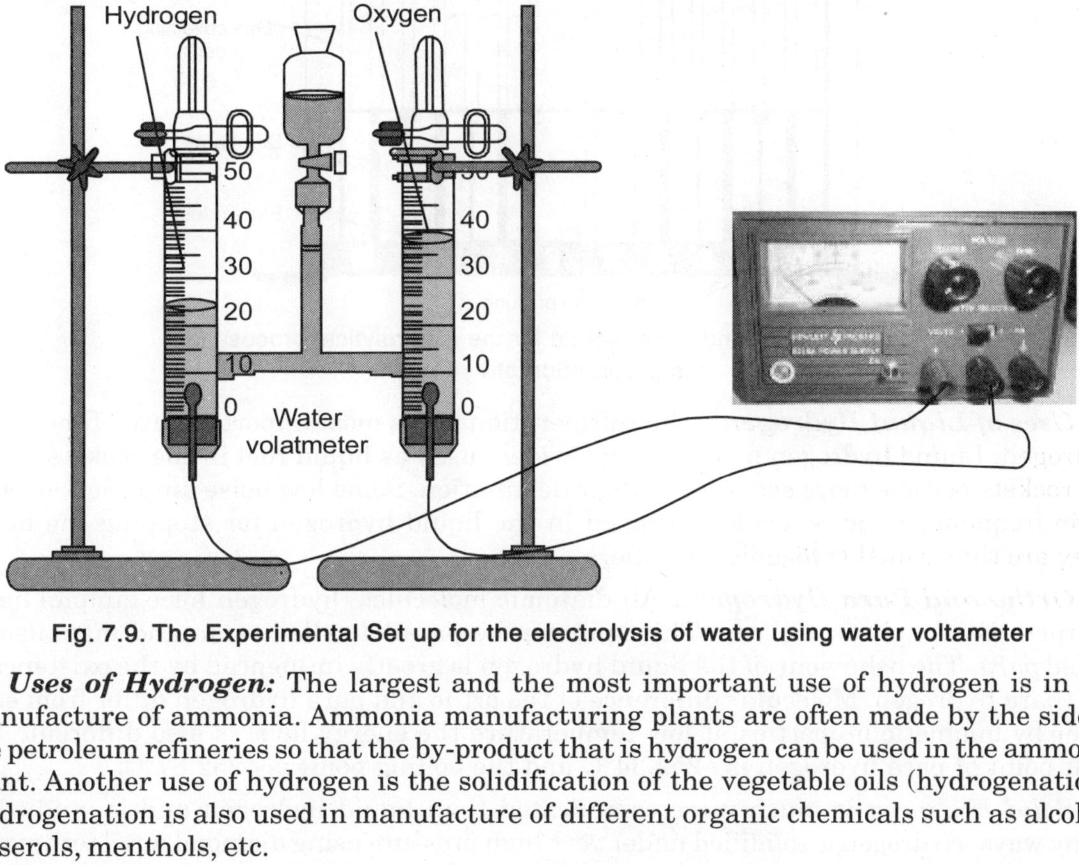

Fig. 7.9. The Experimental Set up for the electrolysis of water using water voltameter

Uses of Hydrogen: The largest and the most important use of hydrogen is in the manufacture of ammonia. Ammonia manufacturing plants are often made by the side of the petroleum refineries so that the by-product that is hydrogen can be used in the ammonia plant. Another use of hydrogen is the solidification of the vegetable oils (hydrogenation). Hydrogenation is also used in manufacture of different organic chemicals such as alcohol, gleserols, menthols, etc.

Hydrogen is used to manufacture hydrochloric acid. Hydrogen is also used in the space research organization (ISRO) for the production of fuel for rockets.

Liquid Hydrogen : James Dewar first produced liquid hydrogen in the year 1898, he was able to retain its liquid state because of famous Dewar's flask. In earlier days hydrogen was liquefied by applying a high pressure, which was a risky process because there was a chance of explosion. Modern liquifiers, however, use close refrigeration by inert gases by inert gases (like helium) which condense hydrogen at low pressure and thus the hazards of explosion can be avoided.

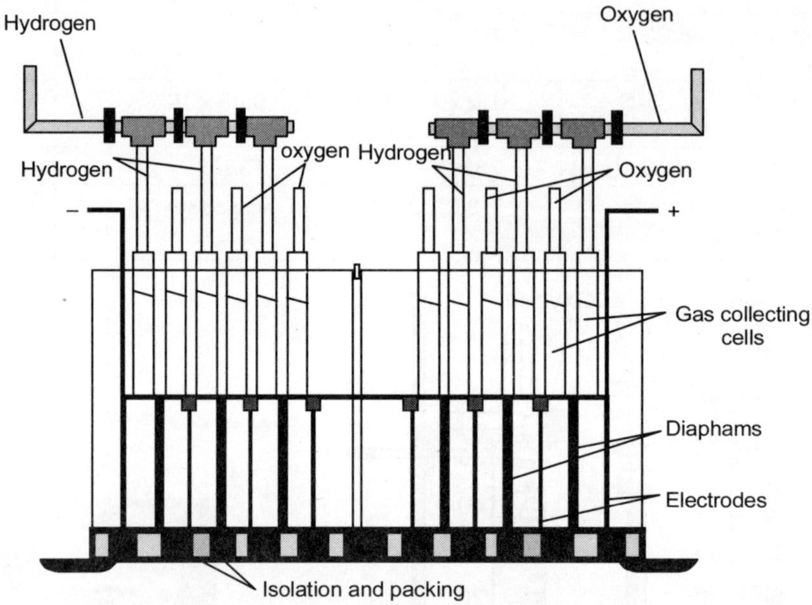

Fig. 7.9A. The Industrial set up for the electrolytical process for the preparation of hydrogen

Uses of Liquid Hydrogen : The refrigeration is the most important use of the liquid hydrogen. Liquid hydrogen is now-a-days widely used as liquid fuel in the rockets where the rockets become more accurate to their destination. Some low noise amplifiers used in radio frequency circuits are kept dipped in the liquid hydrogen for stopping the noise. They are thus called cryogenic amplifiers.

Ortho and Para Hydrogen : All diatomic molecules (hydrogen for example) has a characteristics spin isomerism. This spin can be parallel called *ortho* and may also be called *para*. The behaviour of the liquid hydrogen is greatly influenced by the existence of the (para hydrogen. Molecular difference of the ortho and para hydrogen differ from each other by magnetic properties at low temperature the energy level is also different. The melt point of para hydrogen is –259.34°C and the boiling point is –252.87°C.

Solid Hydrogen is much more complicated than liquid hydrogen and is studied in many ways. Hydrogen is solidified under very high pressure using diamond anvil techniques and the use of high explosive shock waves, when hydrogen gas molecules contract to form solid the press becomes 5.7 GigaPascals. The experiments have concluded with microscopic samples with diamond faces.

The major technological interest in solid hydrogen (or more likely of the solid isotopes deuterium and tritium) as the raw material of the nuclear reactors for production of power generators.

Metallic Hydrogen : During a very high-pressure experiments it is very difficult to maintain systems that whether the hydrogen will remain as same or become atomic or rather metallic. This actually happens when the pressure is sufficiently high and also the distance between molecules is more or less same as the distance between the electron and the nucleus of the hydrogen atom. The pressure applied for converting metallic hydrogen from ordinary hydrogen is 200 to 700Gpa, though it is not possible in all labs to carry out such experiments by the metallic hydrogen is one of the major constituents of the planets Jupiter and Saturn. The physical properties of this type of hydrogen have been the subject of interest and research but unfortunately there was no real calculations for instance whether or not it will be a liquid at temperatures and pressures or will become monoatomic solid. It is a solid, which has been proved by some theoretical chemists and also in a considerable high transition temperature can become super conductors. However, the quantities of the metallic hydrogen, which would be produced in any considerable process, will be so small that they would have no practical applications.

CHAPTER

8

Oxygen

Introduction

After Lavoisier, in the year 1773 P.B. Scheele recognized oxygen as the part of the atmosphere, which supported the combustion and respiration. According to Lavoisier oxygen consisted only 1/5th of the total atmosphere. Scheele was first to isolate oxygen but he reported his work in 1777. On the other hand in the year 1774 Joseph Priestely prepared oxygen gas from mercuric oxide.

Lavoisier named the gas as oxygen because he believed that oxygen produced all the acids (Oxous = acid, Genus = producer), but this was proved wrong in later days.

Occurrence

Oxygen in Earth's atmosphere is available in free elementary state (21% by volume and 23% by weight approximately) water consists of 89% of oxygen by weight and earth's crust consists of 46% of oxygen by weight, oxygen is abundantly available from the trees due to their photosynthesis, where carbon dioxide is consumed and the oxygen is left as the by product, the equation is shown below:

$$\text{Carbon Dioxide} + \text{Water} \xrightarrow[\text{Chlorophyll}]{\text{Sunlight}} \text{Glucose} + \text{Oxygen} + \text{Energy}$$

It is a fact that an average a tree when fully grown produces a huge amount of oxygen as by product, which is not so easy and also economical for us to produce it in the factory. We shall now study about oxygen.

Catalysis is the action of catalysts in a chemical reaction

Catalysis

Catalysis is a process in which a chemical substance enhances a chemical reaction but itself, remains inactive. The chemical reagents performing catalysis are called catalysts.

The catalysts are of two type: (1) *Positive catalysts*, which brings down the reaction temperature and enhances the speed of the reaction thereby making the progress of the reaction fast. The second type (2) *Negative* Catalysts, these catalysts reduce the temperature but the time required for the reaction does not change, these catalysts are better known as *inhibitors.* Here are some of the names of the catalysts:

Catalyst	Process	Reaction	Type
Water	Reaction between ammonium nitrate ammonium chloride and zinc	$3NH_4Cl + 2NH_4NO_3 + Zn = Zn(NH_3)4Cl_3 + NO_2 + 4H_2O$	Positive
Acides of Alkalies	Conversion of cane sugar into Glucose	$C_{12}H_{22}O_{11} + H_2O = 2C_6H_{12}O_6$	Negative
Metals and Metallic Salts			
(a) Finely divided platinum or vanadium pentoxide	1. In contact process for the manufacture of sulphuric acid.	$SO_2 + O_2 \leftrightarrow 2SO_3$	Negative
	2. In oxidation of methyl alcohol	$2CH_3OH + O_2 \leftrightarrow 2H_2CO + 2H_2O$	Negative
(b) Platinum wire	In Ostwald process of nitric acid preparation form ammonia	$4NH_3 + 5O_2 = NO + 6H_2O$	Positive
(c) Finely divided Nickel	In hydrogenation of vegetable oils	Unsaturated esters (oils) + Hydrogen = Esters (Fats)	Negative
(d) Iron and Cobalt	Manufacture of hydrogen from natural gas.	$CH_4 + 2H_2O = 4H_2 + CO$	Positive
(e) Ferric and Aluminium oxide	Boseh Process for the manufacture of hydrogen from water gas.	$CO + H_2O = H_2 + CO$	Positive
(f) Cupric Chloride	Decans process of manufacturing of chlorine from hydrogen chloride	$4HCl + O_2(air) \leftrightarrow 2H_2O = 2Cl_2$	Positive

Characteristics of Catalysis

1. Catalyst agent remains unchanged in mass and composition at the end of the reaction.
2. A very small quantity of the catalyst can accelerate the change appreciably.
3. Action of catalyst is specific that is different chemical reactions requires different catalysts, for e.g.

$$\text{Formic Acid} \longrightarrow \text{Water} + \text{Carbon Monoxide}$$
$$H_2CO_2 \longrightarrow H_2O + CO$$

Formic Acid $\longrightarrow$ Hydrogen + Carbon Dioxide

$$H_2CO_2 \longrightarrow H_2 + CO_2$$

4. In *reversible reaction a catalyst accelerate equally the forward as well as reverse that IS the percentage of reactants at the presence of catalyst equilibrium is quickly established.
5. Actually capability of the catalyst is sometimes increased by the process of some other substance (which alone may not function as catalyst) such substances are called promoters or actuators. For example in Barber Process of manufacture of ammonia iron is the catalyst and the molybdenum is the promoter.

Theories of Catalysis

1. Intermediate Unstable Compound Theory : It explains many catalytic reactions but not all. In the preparation of sulphuric acid by chambers process the reaction is

Sulphur Dioxide + Oxygen $\rightleftharpoons$ Sulphur Trioxide

$$2SO_2 + O_2 \rightleftharpoons 2SO_3$$

[is catalyzed by nitrogen monoxide]

Here nitrogen monoxide (NO) is catalyst + Oxygen (O_2) (one reactant)- $2NO_2$ (in first step)

Nitrogen Dioxide + Sulphur Dioxide [Other reactant] $\longrightarrow$ Sulphur Trioxide + Nitrogen Monoxide

$$NO_2 + 2SO_2 \longrightarrow 2SO_3 + NO_2$$

This is A + catt $\longrightarrow$ A catt

A Catt + B $\longrightarrow$ AB + Catt (where A and B are two compounds and Catt is the catalyst).

2. Adsorption Theory: -The catalyst, especially when not finely divided can absorb on its surface both the reactants and thus brings them nearer and they react faster (law of Mass action). Modem view is that adsorption is not a mere physical phenomenon, it is believed that catalyst forms intermediate compound with adsorbed reactants. These compounds interact and resultant product evaporates living behind the catalyst.

$$\left.\begin{array}{l} \text{A + Catalyst} \rightarrow \text{A Catalyst} \\ \text{B + Catalyst} \rightarrow \text{B Catalyst} \end{array}\right\} \text{A catalyst + B Catalyst = AB Catalyst}$$

Experiment Laboratory Preparation of Oxygen

Theory: In the laboratory oxygen is mainly prepared by the decomposition of potassium chlorate in the presence of manganese dioxide, which acts as a catalyst. The gas is collected by the downward displacement of water (see Fig. 8.1).

---

*Promoters are catalysts like substances, which influence the catalysts at their presence.

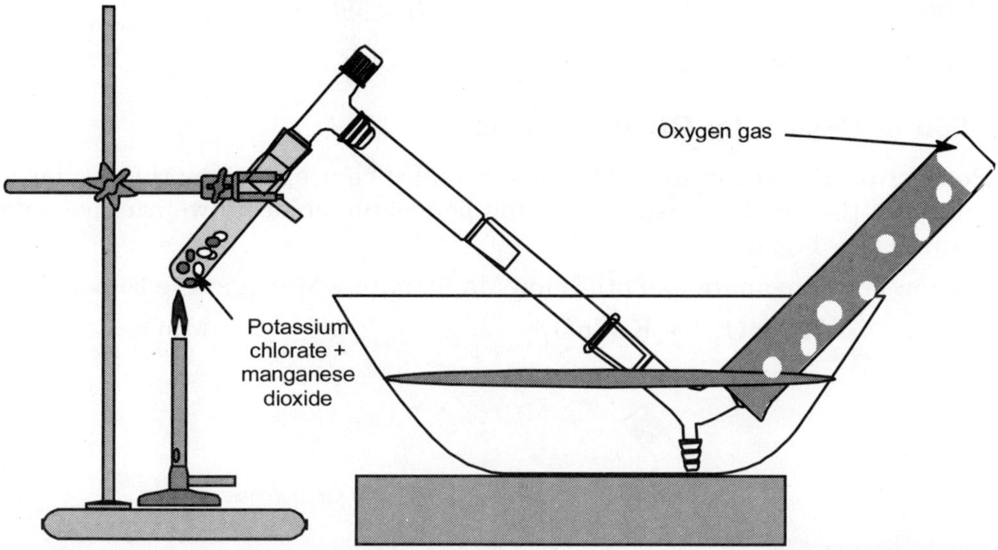

Apparatus	Reagents	Temperature	Pressure
Test tube, delivery tube, clamp stand, burner, gas jar, bee hive self through etc.	Potassium Chlorate, Manganese dioxide	(610°C) without catalyst (257°C) with catalyst	normal

Fig. 8.1. Laboratory Preparation of Oxygen

Reaction Equations

Potassium Chlorate $\xrightarrow{\text{at 380°C}}$ Potassium Perchlorate + Potassium Chloride

$$4KClO_3 \longrightarrow 3KClO_4 \Downarrow + KCl \Downarrow$$

Potassium Perchlorate $\xrightarrow{\text{above 610°C}}$ Potassium Chloride + Oxygen

$$KClO_4 \longrightarrow KCl \Downarrow + 2O_2 \Uparrow$$

Potassium chlorate $\xrightarrow{\text{above 610°C}}$ Potassium Chloride + Oxygen

$$4KCl_3 \longrightarrow 4KCl \Downarrow + 6O_2 \Uparrow$$

These reactions have been carried out without using catalyst

Potassium Chlorate	+	[Manganese Dioxide]	$\longrightarrow$	Potassium Chloride	+	Oxygen	+	[Manganese Dioxide]
$2KClO_3$	+	MnO_2	$\longrightarrow$	$2KCl \Downarrow$	+	$3O_2 \Uparrow$		

Preparation of Oxygen by Other Methods

From Potassium Permanganate: Oxygen can also be prepared by the similar method as said above in the lab, but instead of using potassium chlorate we can use potassium permanganate (see Fig. 8.2).

Potassium Permanganate → Potassium Manganate + Managanese Dioxide + Oxygen

$$2KMnO_4 \rightarrow K_2MnO_4 + MnO_2 + O_2$$

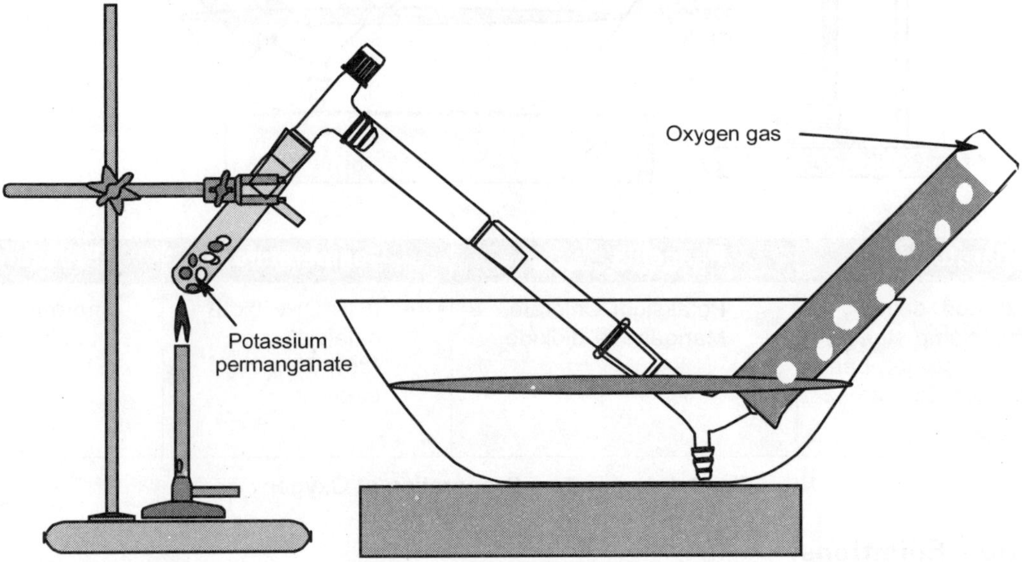

Fig. 8.2. The Preparation of Oxygen from Potassium Permangant

From Sodium Peroxide (Oxone) : Oxone, sodium peroxide mixed with little ferric and manganese salt can produce oxygen when water is added drop by drop from a dropping funnel. This process is unique because it does not require high temperature, or heat (See Fig. 8.3).

Collection of Oxygen

Usually oxygen is collected in a common gas jar and here little mishandling wastes much of the oxygen during experiment, there is also a scarce of leakage. Another most convenient process is to collect oxygen in a gasholder, it is easier to handle and much safer to handle than ordinary gas jars. Fig. 8.4. shows the diagram of a gasholder.

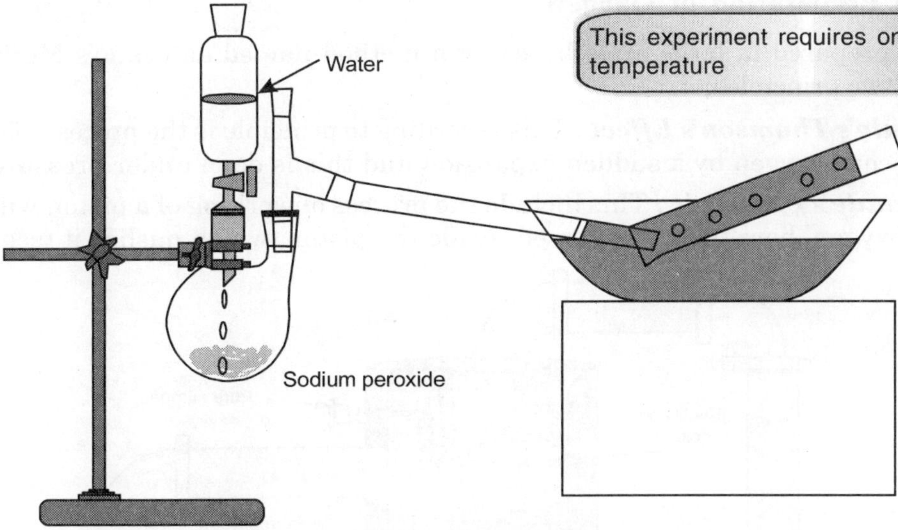

This experiment requires only room temperature

Fig. 8.3. Preparation of Oxygen from Oxone (sodium peroxide)

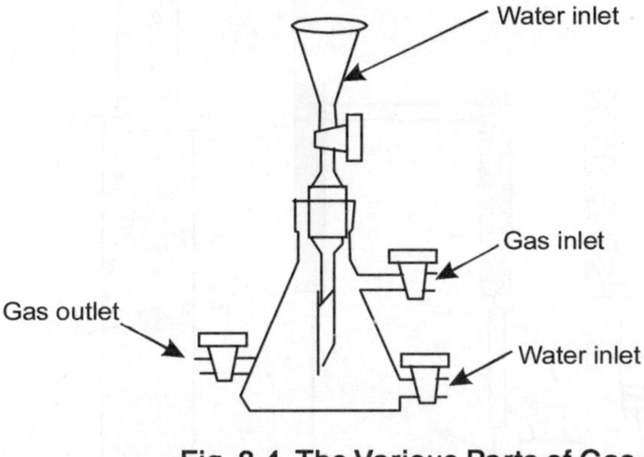

It is necessary because when the gas inside the holder escapes the partial pressure vacuum which is filled up by water, if it is not done, the gas holder will burst

Fig. 8.4. The Various Parts of Gas Holder used in Lab

Operation of Gas-holder and Safty Rules

The operation of the gas-holder is very simple as shown is Fig. 8.4 through the tap B from the reservoir water_enters to the holder. The tap connected to the gas outlet helps the gas to escape; there is also a water escape tube through which the water continuously runs out of the holder during collection of gas. During the collection of the gas the gas outlet tap and the tap B is closed and all the other taps are opened and during the use of the gas the gas inlet tap is closed and both the taps B and the water exit tap are simultaneously opened.

Industrial Preparation of Oxygen

Oxygen is prepared in large scale by using a method named as Claud's Method, which consists of two principles:

1. ***Joule's Thomson's Effect :*** This according to principle is the process of cooling the gaseous oxygen by a sudden expansion and this is done under pressure.
2. ***Claude's Principle :*** This include the process of working of a piston with the help of oxygen, here the gas is forced inside the piston, which pushes it (see Fig. 8.5).

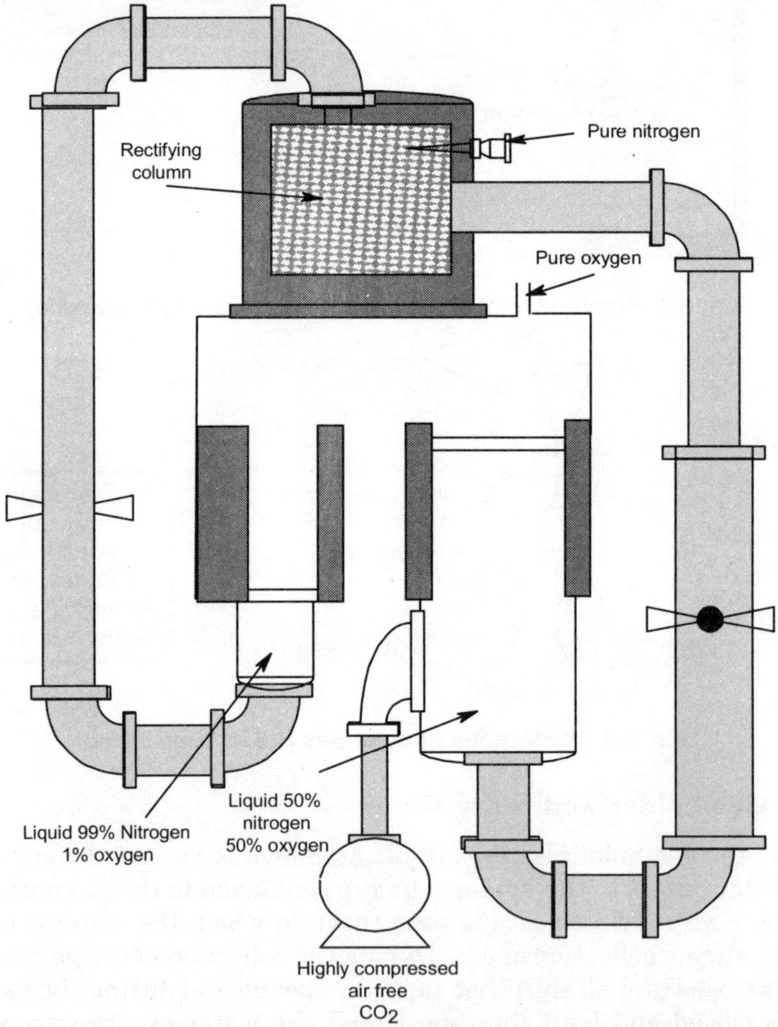

Fig. 8.5. The Industrial Preparation of Oxygen Gas by Claude's Process

Dry air free from carbon dioxide, compressed to 30 to 40 atmospheres and cooled to 15°C is send to heat exchanger where it is cooled below 0°C by out going oxygen and nitrogen. Next it does work in an expansion engine and cools to a very low temperature. Air enters at the base of the tower and passes up the central pipes surrounded by liquid oxygen when almost whole of oxygen and enough nitrogen gets liquefied and drain down as liquid (50% oxygen and 50% nitrogen). The unliquefied portions come down through the outer pipes and they are collected as liquid (99% nitrogen and 1% Oxygen), pressure of the vapour forces the liquid (50% nitrogen + 50% oxygen) to the middle of the column. From descending liquid, nitrogen boils off and escapes as pure nitrogen and the liquid collected finally is almost pure oxygen which are led to heat exchangers where they cool incoming air. Liquid air or oxygen is transferred into Dewar's Flask (see Fig. 8.6).

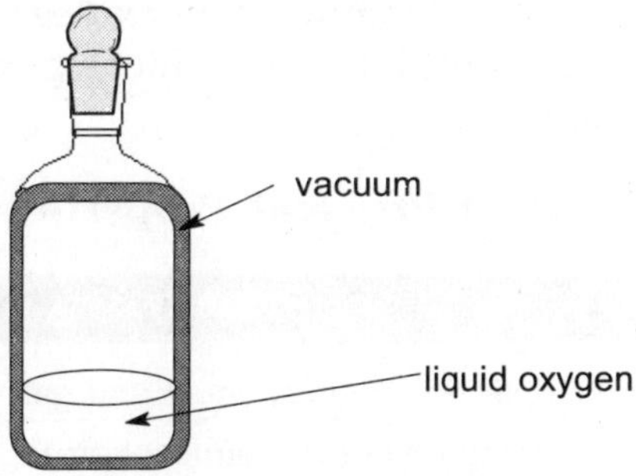

Fig. 8.6. Liquid Oxygen

Properties of Oxygen

Physical Properties : As we all know oxygen is the most important element for the living organism. In chemistry (periodic table) it has got a very sound position in the group VIA. Oxygen physically is a colourless odourless and taste less gas. Oxygen is a supporter of combustion but it self is not combustible.

Experiment : In a gas gar full of oxygen a glowing splinter is introduced (red hot match stick) the splinter bursts into flame, which shows that oxygen is the supporter of combustion, but oxygen does not split up which shows that it is not a combustible gas (See Fig. 8.7).

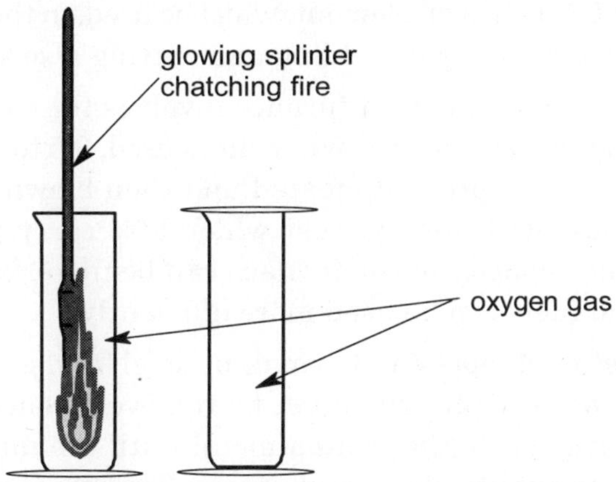

Fig. 8.7. The Experimental Set up to Prove that Oxygen is Supporter of Combustion and Also Combustible

Oxygen has a good place in our environment about 21% in the air, 85.7% in earth's crust and 46.6% in seawater. About three parts of oxygen by volume dissolve in 10 parts of fresh water at 20°C. Practically all chemical compounds except inert elements (the inert gases) forms compounds with oxygen and these compounds are called oxides.

Most elements form oxides when heated in an atmosphere of Oxygen. Many elements form more than one oxide, e.g. sulphur, iron, etc. The properties of oxygen are tabulated below:

Chart of Physical Properties

Property	Value
Triple Point (solid, liquid and gas at equilibrium)	–218.80°C
Boiling Point (at 1 atmosphere)	–182.97°C
Gas density at 0°C at 10^5 Pa pressure	1.4290 g/litre
Liquid density at the normal boilig point	1.142 g/ml
Solubility in water at 20°C per 1000g of water	30

Isotopes of Oxygen

Oxygen has seven isotopes, three of which are stable; the most abundant isotope is ^{16}O (99.759%). The next one is ^{18}O (0.204%) and the last is ^{17}O (0.37%). The four unstable are 140(17 sec)^{15}O (2.07 minutes)^{19}O (29 seconds) and ^{20}O(14 seconds).

Uses of Oxygen

Oxygen is widely used in a variety of applications beside respiration of the life form. They are as follows :

Metallurgical Uses : Oxygen is an element, which is used, in the different metallurgical process such as smelting. Terming, gas welding and cutting also surface conditioning.

Smelting: Smelting of ore in the blast furnace involves the combustion of about 1 ton of oxygen for each ton of material produced. When air is used, 3.5 tons nitrogen accompanies each ton of oxygen, which is compressed, heated and then blown in the blast furnace. A large number of heat is lost with the exhaust which also carry powdered ore and coke away as dust and limit the capacity of the furnace can be increased, less expensive fuels can replace some coke and fuels can be used more efficiently.

Metal Refining: In refining copper and in making steel from pig iron various impurities such as carbon sulphur and phosphorous must be removed from the metal by oxidation process. Injecting pure oxygen into the molten metal until the impurities are completely removes can produce better quality steels and copper. This process can also be utilized in the electric arc furnace also. This technique is called *basic oxygen process* the most common form of the steel manufacture is L-D process there this process is used.

Reheat Furnaces: With specially designed burners oxygen can be utilized without raising flame temperature with substantial fuel saving and less pollution furnace temperature is more easily achieved and controlled.

Welding, Cutting, Surface Conditioning: The high temperature flame of oxy-hydrogen or oxy-acetylene torch can generate as high as 3200°C temperature for welding (see Fig. 8.8).

Powerful jet oxygen is used with the acetylene mixture; this mixture is lighted which becomes the flame, for welding or cutting metals. The ratio of oxygen and the acetylene.

Fig. 8.8. The Oxy-acetylene welding torch

Hydrometallurgy: The mechanism by which metal values are leached from ores may involve oxidation. Oxygen gas is dissolved in the leach fluid to extract uranium from deeply buried ore deposits. Similarity copper is obtained from previously discarded mine waste.

Chemical Synthesis: Several syntheses in the chemical industry involve oxygen. They are partial oxidation of hydrocarbons.

Fuel synthesis: In fuel synthesis syn gas is prepared which is a mixture of hydrogen and carbon monoxide, this is formed during the reaction of fuel hydrocarbon and oxygen. This syn gas is required in the production of many organic materials called menthol, octane etc.

Oxygen is also used in the manufacture of pigments, making fuel for rockets the liquid oxygen is used. Liquid oxygen is also used for refrigeration. The solid fuel rockets were generally used which, required hydrocarbon based polymers, that produced ozone by self-combustion, which distract the trajectory of the rockets. Nowadays the liquid fuel rockets used liquid hydrogen and oxygen, which ignite easily and propel correctly and the trajectory does not change. The launch vehicles require about 3000 tons of fuel and in which 2000 tons are liquid oxygen.

> Syn-gas is the mixture of hydrogen and carbon dioxide

Biological application: Divers use nitrogen and oxygen or oxygen with any inert gas such as helium combination for respiration under water. Aviators of high altitude flying jet aircraft, require oxygen as their respiration gas. There are numerous such uses, which is impossible to discuss in such a short book like this.

Ozone

When sparks are passed through oxygen in air, a characteristic odour appears caused by the formation of new substances called Ozone (Greek *Ozein* means to smell). Ozone can be prepared from absolute dry oxygen. Hence, it is clear that oxygen is the only constituent of ozone. Ozone is named because of its pungent odour. The molecular mass of ozone is 48, as we all know that the atomic mass of oxygen is 16 (15.9994), therefore the combination of three atoms of oxygen forms ozone having a molecular mass of 16 + 16 + 16 = 48 (47.99820 with a formula of O^3.

Preparation of Ozone

As ozone is not freely available in the atmosphere it is prepared in the laboratory by different methods which are now going to be discussed. First very simply ozone can be prepared by an electrical discharge which is created inside a glass flask full of dry oxygen as shown in Fig. 8.9.

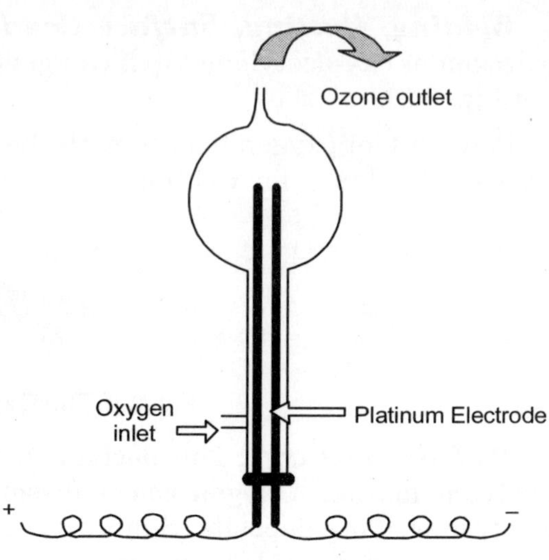

Fig. 8.9. Laboratory Ozonizer

Siemen's Ozonizer consists of two coaxial glass inside of inner and outside of the outer is coated with tin plates which are connected to a high voltage induction coil. A slow stream of cool dry oxygen is passed through these plated and as mixture ozone 20% and oxygen 80% escapes at the other end. Mixture is cooled to give dark bluer liquid ozone (at –112.5°C) (See Fig. 8.10).

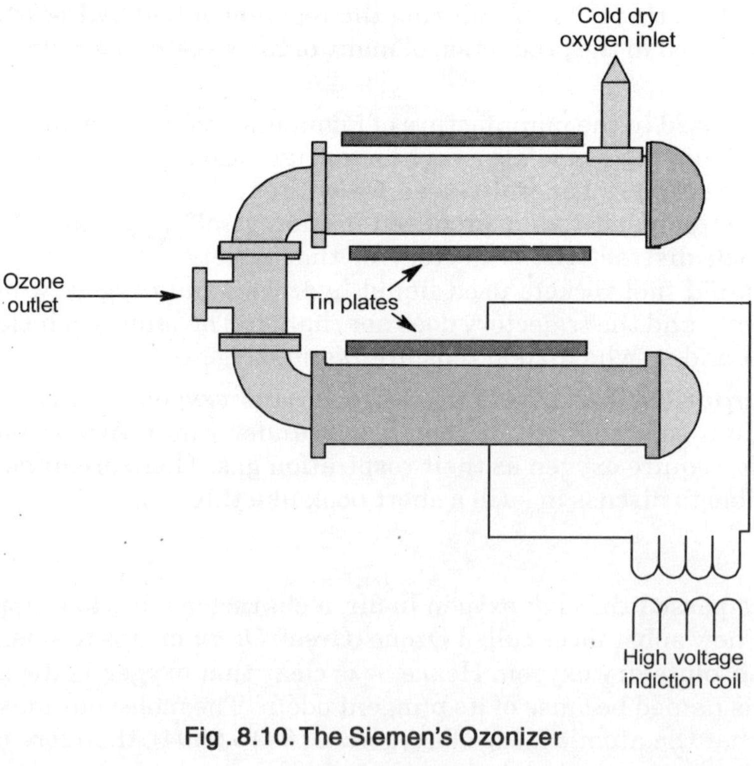

Fig . 8.10. The Siemen's Ozonizer

Brodie's Ozonizer

It is more or less similar to the Siemen's Ozonizer but instead tin plated electrodes it uses platinum electrodes are used and also dilute sulphuric acid is also used in which the electrodes are dipped in (Fig. 8.9). Oxygen is allowed to pass through which undergoes an electrical discharge and is partially converted to ozone.

Halske's Ozonizer

Here ozone is prepared from atmospheric oxygen. In chamber porcelain tubes are placed, aluminium rods are placed which acts as electrodes, which are connected to the high voltage source (10kV A.C). This is the main industrial process of ozone (see Fig. 8.11).

Properties of Ozone

Physical Properties: Among the physical properties of ozone, its solubility in water is greater than that of oxygen, the solubility is in 100 volumes of water ozone is soluble up to 49 volumes. The formation of ozone from oxygen can be expressed by the equation

$$\text{Oxygen} \rightleftharpoons \text{Ozine} \longrightarrow 285 \text{ kJ}$$

$$3H_2 \rightleftharpoons 2O_3$$

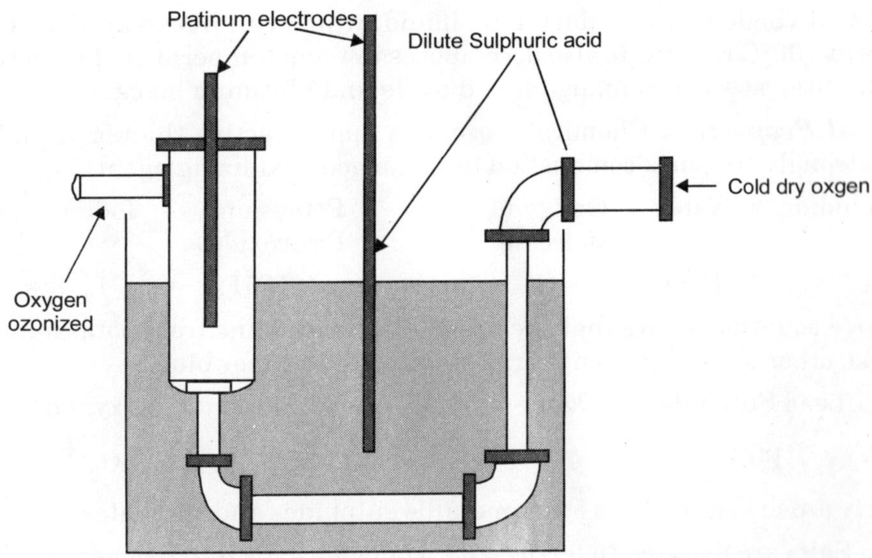

Fig. 8.11. The Brodie's Ozonizer

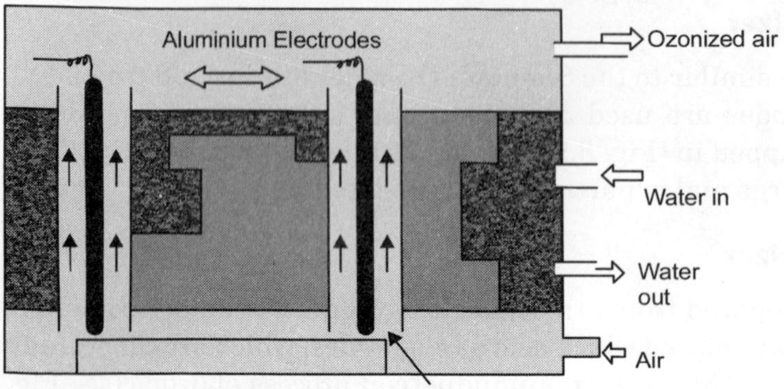

Fig. 8.12. The Halske's Ozonizer

'rom which it follows that the standard method of formation of ozone is positive and is
4 kJ/mol. In addition, as the cause of this reaction three molecule of the gas gets ›rmed into two that is the method of the system diminishes. As a result the standard in this reaction is also positive (163kJ/mol). Thus the reaction transformation of into ozone cannot processed spontaneously - it the expenditure of energy. The reaction - the decomposition of ozone proceeds spontaneously because it is attended ›crease is the energy level of the system. In other words, ozone is an unstable ıce.

·112°C it condenses to a dark blue liquid, which can freeze at –251.4°C. The gas ›oses at 100°C rapidly. It also decomposes in room temperature but in the presence table catalyst such as manganese dioxide and platinum black.

emical Properties: Chemically ozone is more reactive than oxygen, like oxygen ılso strongly supports combustion also is a good oxidizing agent.

ium Iodide + Water + Oxygen Molecule ⟶ Potassium Hydroxide + Iodine + Oxygen

$$2KI + H_2O + O \longrightarrow 2KOH + I_2 + O_2$$

› above equation shows that the ozone liberates iodine from potassium iodide (KI) ı, like other oxidizing agent turns starch iodide paper blue.

Lead Sulphide + Ozone ⟶ Lead Sulphate + Oxygen

$$PbS + 4O_3 \longrightarrow PbSO_4 \Downarrow + 4O_2 \Uparrow$$

ıilarly ozone also oxidizes other metallic sulphides and sulphates.

·rous Salts are oxidized to ferric salts. Halogen acids into halogens and waters.

'erous Sulphate + Sulphuric Acid + Ozone ⟶ Ferric Sulphate + Water + Oxygen

$$FeSO_4 + 2H_2SO_4 + 2O_3 \longrightarrow Fe(SO_4)_3 \Uparrow + 2H_2O + 2O_2 \Downarrow$$

$$\text{Hydrogen Chloride} + \text{Ozone} \longrightarrow \text{Water} + \text{Chlorine} + \text{Oxygen}$$
$$2HCl + O_3 \longrightarrow H_2O + Cl_2 + O_2$$

It converts potassium manganate into potassium permanganate

$$\text{Potassium Manganate} + \text{Water} + \text{Ozone} \longrightarrow \text{Potassium Permanganate} + \text{Potassium Hydroxide} + \text{Oxygen}$$
$$2K_2MnO_4 + H_2O + O \longrightarrow 2KMnO_4 + 2KOH + O_2$$

Elements like sulphur, phosphorous, iodine gets oxidized to their oxyacids such as sulphuric acid phosphorous acid, iodic acid etc. Metals like silver is converted into silver oxide.

On exposure to ozone mercury looses its mobility and sticks to the wall of the vessel it is kept, this phenomenon is called Tailing of mercury

$$\text{Silver} + \text{Ozone} \longrightarrow \text{Silver oxide} + \text{Oxygen}$$
$$2Ag + O_3 \longrightarrow Ag_2O \Downarrow + O_2 \Uparrow$$
$$\text{Iodine} + \text{Water} + \text{Ozone} \longrightarrow \text{Iodic Acid} + \text{Oxygen}$$
$$I_2 + H_2O + 5O \longrightarrow 2HIO_3 + 5O_2$$
$$\text{Mercury} + \text{Ozone} \longrightarrow \text{Mercuric Oxide} + \text{Oxygen}$$
$$2Hg + O_3 \longrightarrow Hg_2O + O_2$$

In a few reactions all the three atoms of oxygen are used up and so no oxygen is liberated, as shown in the example below:

$$\text{Stannous Chloride} + \text{Hydrochloric Acid} + \text{Ozone} \longrightarrow \text{Stannous Chloride} + \text{Water} + \text{Chlorine}$$
$$3SnCl_2 + 6HCl + O_3 \longrightarrow 3SnCl_2 + 3H_2O + 3Cl_2$$
$$\text{Sulphur Dioxide} + \text{Ozone} \longrightarrow \text{Sulphur Trioxide}$$
$$3SO_2 + O_3 \longrightarrow 3SO_3$$

Like other compounds that can be yield atomic, ozone acts as a bleaching agent, disinfectant and germicide.

Ozone can bleach the colour of indigo and other vegetable colours it *destroys rubber crocks etc.*

Ozonides

With unsaturated compounds ozone forms addition products called *ozonides.*

$$\text{Ethylene} + \text{Ozone} \longrightarrow \text{Ethylene Oxonide}$$
$$C_2H_4 + O_3 \longrightarrow C_2H_4O_3 \Downarrow$$

$$\text{Benzene} + \text{Ozone} \longrightarrow \text{Benzene Triozonide}$$
$$C_6H_6 + 3O_3 \longrightarrow C_6H_6(O_3)_3 \Downarrow$$

Reaction with Peroxides

Ozone also reacts with different peroxides, e.g. hydrogen peroxide, barium peroxide and sodium peroxide.

Hydrogen Peroxide	+	Ozone	⟶	Water	+	Oxygen
H_2O_2	+	O_3	⟶	H_2O	+	$2O_2 \Uparrow$
Barium Peroxide	+	Ozone	⟶	Bariuim Oxide	+	Oxygen
BaO_2	+	O_3	⟶	BaO	+	$2O_2 \Uparrow$

1. Particular smell, which is found during electrical sparking
2. Turning blue of starch iodide paper.
3. Sticking of Mercury to the glass walls.
4. Turning tetra-methyl base paper violet.
5. Benzedyne paper turns brown when come in contact with ozone.

Calculation of Formula of Ozone Emperiment

In this experiment a sample of ozone and oxygen mixture in two flasks which were kept over mercury. Ozone in one of them was absorbed by turpentine oil and the mercury level rose up which indicated indirectly the level of ozone. A flame to decompose ozone into oxygen heated the second flask, here the increasing volume pushed down the mercury level. When both the volumes in the flask were compared, it was seen that in the apparatus in the apparatus A. The volume was double from the original level and in the flask B. The level of the mercury was pushed down by same as the original level as shown in Fig. 8.13.

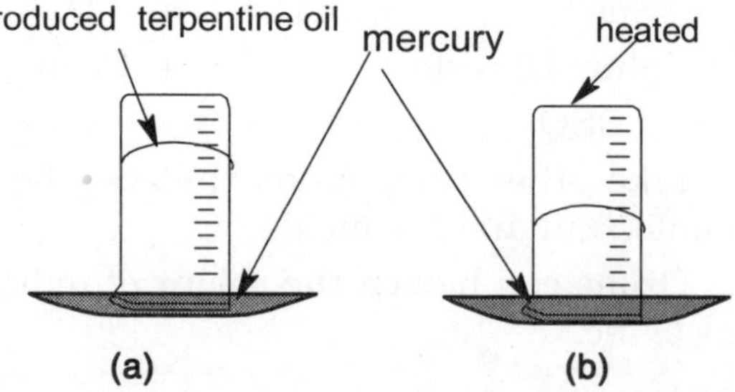

Fig. 8.13. The Industrial Preparation of Oxygen Gas by Claude's Process

It is proved that when V cc (let the volume of ozone absorbed by the turpentine in the flask A) changes into oxygen V + V/2 cc (let the volume of oxygen formed by the decomposition of ozone is V/2cc).

Hence, 2 volumes of ozone decompose into 3 volumes of oxygen (by Avogadro's law) 2 molecules of ozone yields 3 molecules of oxygen since the oxygen molecule is diatomic 2 molecules of ozone contains 6 atoms of oxygen.

∴ Molecular Weight is 48. Hence, the formula is O_3.

Structure of Ozone Molecule

The ozone at the vertex of the triangle is 120°, which reveals that the central oxygen is in the state of Sp^2 hybridization here. Accordingly from the view-points of valence bond theory, the formation of the ozone molecule is shown below :

The hybrid Sp^2 orbital of the central atom containing one electron (see Fig. 8.14) overlaps with the p. orbital of one of the extreme oxygen atoms the result being the formation of a sigma (0) bond the Pz orbital of the central atom. It is not participating in the hybridization is oriented at right angles to the plane of the molecule and also contains an unpaired electron. It overlaps with the similarity Pz orbital of the same extreme atom of oxygen, which leads to the electron pair occupying one of the hybrid sp^2 orbital orbitals the central oxygen atom in the donor acceptor way. Electron dot formula is shown below:

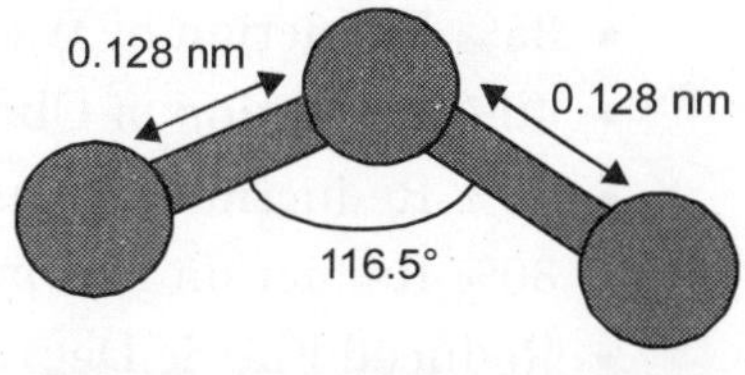

Fig. 8.14. The Ozone molecule

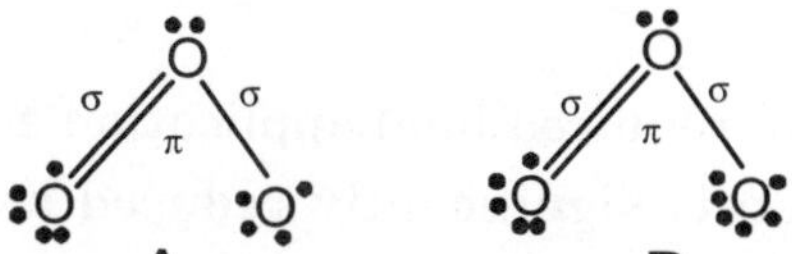

A B

According to formula A, the bonds of the central oxygen atom with the two extreme atoms are not equivalent oxygen atom with the two extreme atoms are not equivalent - one is double and one is single bonded. The identical lengths of these bonds however point to their equivalence. Hence, in addition to the formula A, formula B can correctly describe the structure of the ozone molecule. This signifies that the formula of the ozone molecules is intermediate between A and B so it is shown in the figure below :

Here the dashed line indicates that the delocalization of the 7[-electron pair that is its belonging to all three oxygen atoms. Consequently, the 7 bond in the ozone molecule is the three-centre one.

A consideration of the M.O. (Molecular Orbital Method) also leads to the conclusion thai a three-centred bonding molecular orbital is formed here.

Uses of Ozone

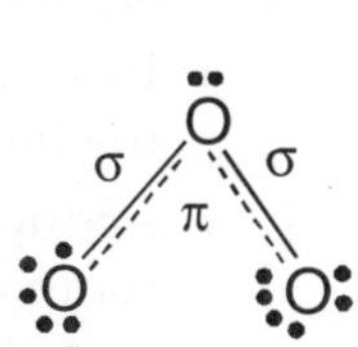

Scientists has developed capabilities for the treatment of water and wastewater containing a wide variety of contaminates. They continue to put significant effort into research and development to find new solutions and to refine existing methodologies in the areas of water treatment. Typically, the process uses a combination of chemical, bacteriological and mechanical technologies when creating a water treatment system as many waste streams have unique characteristics associated with their contamination.

BENEFITS OF OZONE

Laundry

- 100% Reduction of Hot Water Usage.
- 35% Reduction of Water/Sewer Usage.
- 40% Reduction of Chemical Usage.
- 35% Reduction of Wash Cycle Time.
- 30% Reduction in Drying Time.
- Reduced Fabric Degradation.
- Reduced Labour Costs.
- Improved Work Environment for Laundry Facility Employees.

Wastewater Treatment

Agricultural

- Significant reduction of required land application acreage.
- Ammonia volatilization is significantly reduced as the ammonia is oxidized to nitrates.
- Noxious odours are significantly reduced in the barns, at the treatment system site and in the spray fields.
- Health conditions inside the animal houses are greatly improved due to flushing the animal houses with ozone and aerobically treated wastewater.
- Water following treatment is clean enough to be used as drinking water for animals.
- In some applications, high grade fertilizer is a by-product of the treatment process.

Industrial

- Oxidizes residual organics following pre-treatment, lowering the BOD/COD prior to discharge.
- Can be used for bioremediation pre-treatment since ozone oxidizes residual long chain organics, which are toxic to bacteria and reduces them to shorter chain organics, which are then more easily consumed by bacteria.
- Increases the dissolved oxygen level in the waste stream as the ozone reverts back to elemental oxygen.
- Reduces residual color present in the waste stream resulting from the presence of compounds such as tannins, lignins, etc.
- Reduces residual odour/taste present in the waste stream resulting from the presence of compounds such as volatile fatty acids, hydrogen sulfides, ammonia, etc.
- Precipitates suspended solids since ozone is a micro-flocculent that allows for the physical removal of the resulting suspended solids via filtration.

- Increases the oxidation-reduction potential in the waste stream due to both the ozone and oxygen by-products that is beneficial for maintaining oxic conditions.
- Rapidly kills microorganisms including pathogens with minimal contact time at the proper concentration to create a sterile waste stream.

Aquaculture

- Proven performance in both fresh and saltwater applications.
- Reduced bacteria counts.
- Destroys organic contaminants, parasites and reduces disease.
- The byproduct of the ozone oxidation is oxygen, which is a necessary component for the healthy respiration of the fish.
- Results in added cost savings due to a reduced need for other treatment chemicals to control disease and fish health.

Vegetable and Fruit Processing and Meat Packaging

- Longer shelf life.
- Improved produce/products quality.
- Cleaner produce/products.
- Safe system operation.
- No chemical residual on the produce/products.
- Easier and more economical operation.
- Automatic operation.

Cooling Tower and Chillers

- Reduced pumping power needed to extract water from wells or reservoir and pump to water treatment facility.
- Reduced chemical, filtration, and maintenance costs associated with treating and purifying at the water treatment facility.
- Reduced pumping power for distributing the water from the water treatment facility to the end-user.
- Reduced pumping power and associated costs to transport wastewater (blowdown) to the sewage treatment plant.
- Reduced chemical and maintenance costs, and reduced pumping power associated with sewage treatment at the plant.
- Reduced costs associated with permits allowing the discharge of treated sewer water to a river or stream.

Ozone in Nature

In nature, there is a cycle of oxygen just like there is a cycle of water. Oxygen is released from plants on land and plankton in the sea during photosynthesis. The oxygen is lighter than air and floats upward in the atmosphere. At the 20-30 km region, strong ultraviolet radiation in the 185-200 nanometer wavelength bombards the oxygen and turns some of it into ozone. The ozone created exists as a thin layer in the atmosphere and it blocks out the small portion of the UV spectrum that it absorbs. The great majority of the UV reaches the earth allowing suntanning, which Dr. Michael Carpendale of the San Francisco Veteran's Administration Hospital has noted is useful in a very efficacious therapy developed in the early years of this century. We hear a great deal about the thinning of the ozone layer in the media, but the facts are otherwise. Ozone production in the upper atmosphere is dependent on the amount of energy coming from the sun. During peaks of solar activity, ozone is created at a greater rate. During lulls in the sunspot cycle, the ozone layer is thinner. The lowest level ever measured was in 1962. At night, on the dark side of the planet, the ozone layer disappears, in a few hours. The layer is reformed as the sun rises in the morning. There is no ozone over the poles in the winter because there is no sunlight. Ozone is produced constantly in the upper atmosphere as long as the sun is shining, and since ozone is heavier than air, it begins to fall earthward. As it falls, it combines with any pollutant it contacts, cleaning the air nature's wonderful self-cleaning system. If ozone contacts water vapor as it falls, it forms hydrogen peroxide, a component of rainwater, and the reason why rainwater causes plants to grow better than irrigation. Ozone is also created by lightning, and the amount produced in an average storm is often triple the allowable limit of .015 PPM as set by the US EPA. This ozone is what gives the air the wonderful fresh smell after a rain, and is of the highest benefit to anyone fortunate enough to be breathing it. Ozone is also created by waterfalls and crashing surf, which accounts for the energetic feeling and calm experienced near these sites. Another way ozone is produced is by photons from the sun breaking apart nitrous oxide, a pollutant formed by the combustion of hydrocarbons in the internal combustion engine. This ozone can accumulate in smog due to temperature inversions and is a lung and eye irritant.

These are the forms of ozone created by natural processes in the atmosphere.

Medical Ozone

The first ozone generators were developed by Werner von Siemens in Germany in 1857, and 1870 saw the first report on ozone being used therapeutically to purify blood, by C. Lender in Germany.

There is evidence of the use of ozone as a disinfectant from 1881, mentioned by Dr. Kellogg in his book on diphtheria. In October of 1893, the world's first water treatment plant using ozone was installed in Ousbaden, Holland, and today there are over 3000 municipalities around the world that use ozone to clean their water and sewage. In 1885, the Florida Medical Association published "Ozone" by Dr. Charles J. Kenworth, MD, detailing the use of ozone for therapeutic purposes. In September 1896, the electrical genius Nikola Tesla patented his first ozone generator, and in 1900, he formed the Tesla Ozone Company. Tesla sold ozone machines to doctors for medical use, the same thing we are doing 100 years later, with a design based on one of his from the 1920s. We have seen one of these 75-year old generators, and it still works perfectly. Tesla produced ozonated olive oil and sold it to naturopaths, and we do, too. In 1898, the Institute for Oxygen Therapy was started in Berlin by Thauerkauf and Luth. They injected ozone into animals and bonded ozone to magnesium, producing Homozon. Beginning in 1898, Dr. Benedict Lust, a German doctor practicing in New York, who was the originator and founder Naturopathy, wrote many articles and books on ozone. In 1902, J.H.Clarke's *A Dictionary of Practical Materia Medica,* London describes the successful use of ozonated water in treating anemia, cancer, diabetes, influenza, morphine poisoning, canker sores, strychnine poisoning and whooping cough. In 1911, *A Working Manual of High Frequency Currents* was published by Dr. Noble Eberhart, MD. Dr. Eberhart was head of the Department of Physiologic Therapeutics at Loyola University. He used ozone to treat tuberculosis, anemia, chlorosis, tinnitus, whooping cough, asthma, bronchitis, hay fever, insomnia, pneumonia, diabetes, gout and syphilis.

In 1913, the Eastern Association for Oxygen Therapy was formed by Dr. Blass and some German associates.

During World War I, ozone was used to treat wounds, trench foot, gangrene and the effects of poison gas. Dr. Albert Wolff of Berlin also used ozone for colon cancer, cervical cancer and decubitis ulcers in 1915.

In 1920, Dr. Charles Neiswanger, MD, the President of the Chicago Hospital College of Medicine published *Electro Therapeutical Practice*. Chapter 32 was entitled "Ozone as a Therapeutic Agent."

In 1926, Dr. Otto Warburg of the Kaiser Institute in Berlin announced that the cause of cancer is lack of oxygen at the cellular level. He received the Nobel Prize for Medicine in 1931 and again in 1944, the only person to ever receive two Nobel Prizes for Medicine. He was also nominated for a third.

In 1929, a book called *Ozone and Its Therapeutic Action* was published in the US listing 114 diseases and how to treat them with ozone. Its authors were the heads of all the leading American hospitals.

In 1933, the American Medical Association, headed up by Dr. Simmons set out to destroy all medical treatments that were competitive to drug therapy. The suppression of ozone therapy began then, and it continues in the US to this day.

The Swiss dentist E.A. Fisch was using ozone in dentistry before 1932, and introduced it to the German surgeon Erwin Payr who used it from that time forward.

Aubourg and Lacoste were French physicians using ozone insufflation from 1934-1938.

In 1948, Dr. William Turska of Oregon began using ozone, employing a machine of his own design, and in 1951, Dr. Turska wrote the article "Oxidation" which is still relevant today, and is included in our booklet. Dr. Turska pioneered injection of ozone into the portal vein, thereby reaching the liver. From 1953 onward, German doctor Hans Wolff used ozone in his practice, writing the book *Medical Ozone,* and training many doctors in ozone therapy. In 1957, Dr. J. Hansler patented an ozone generator which has formed the basis of the German expansion of ozone therapy over the last 35 years. Today over 7000 German doctors use ozone therapy daily. In 1961, Hans Wolff introduced the techniques of major and minor autohemotherapy. In 1977, Dr. Renate Viebahn provided a technical overview of ozone action in the body. In 1979, Dr. George Freibott began treating his first AIDS patient with ozone, and in 1980, Dr. Horst Kief also reported success treating AIDS with ozone. In 1987, Dr. Rilling and Dr. Viebahn published *The Use of Ozone in Medicine,* the standard text on the subject. In 1990, the Cubans reported on their success in treating glaucoma, conjunctivitis and retinitis pigmentosa with ozone. In 1992, the Russians revealed their techniques of using ozone bubbled into brine to treat burn victims with astounding results.

Today, after 125 years of usage, ozone therapy is a recognized modality in many nations: Germany, France, Italy, Russia, Romania, Czech Republic, Poland, Hungary, Bulgaria, Israel, Cuba, Japan, Mexico, and in five US states.

Ozone Concentration

Medical ozone is produced in varying concentrations. The quantity of ozone in comparison with the quantity of oxygen in the gas stream is called percent concentration. It is measured in micrograms (μg) of ozone per millilitre (or cc) of the mixture. A litre of oxygen weighs 1.4 grams.

Therefore: 0.5% × 1.4 gm = 7 μg/cc 1.0% × 1.4 gm = 14μg/cc 1.5% × 1.4 gm = 21ug/cc 2.0% × 1.4 gm = 28ug/cc 2.5% × 1.4 gm = 35ug/cc 3.0% × 1.4 gm = 42ug/cc 3.5% × 1.4 gm = 49μg/cc 4.0% × 1.4 gm = 56ug/cc 4.5% × 1.4 gm = 63ug/cc 5.0% × 1.4 gm = 70ug/cc 5% or 70 mg/cc is considered to be the upper limit of concentration for internal use of medical ozone.

Dr. Greenberg of the Kief Clinic has shown, in vitro, that at concentrations of 90 μg/cc there was crimping of red blood cells, which was definitely harmful. Experiments by F. Sweet et al have shown inhibition of growth in healthy cells at concentrations above 72 μg/cc. If we stay below that level, we will have no problems. Ozone Therapy is Safest Known Therapy Ozone has been found to be an extremely safe medical therapy, free from side effects. In a 1980 study done by the German Medical Society for Ozone Therapy, 644 therapists were polled regarding their 384,775 patients, comprising a total of 5,579,238 ozone treatments administered. There were only 40 cases of side effects noted out of this number which represents the incredibly low rate of .000007%, and only four fatalities. Ozone has thus proven to be the safest medical therapy ever devised.

Dosage and Frequency

When it comes to dosage and frequency of administration, there is some difference of opinion. Dr. Carpendale believes that a high concentration is necessary to kick-start the immune system initially, followed by much lower concentrations. He believes that continued high concentrations may be immunosuppressive, based on T-4 cell counts. Other doctors, such as Dr. Turska, recommend initial medium concentration doses, three times per week, followed by twice weekly at lower concentration, followed by weekly injection as long as necessary. Dr. Stan Beyrle recommends injection every four days at medium concentration. Dr. Wang has been giving daily injections at medium concentration, and direct injection into breast tumors. Dr. Freibott recommends very high concentrations at low dosages, with the emphasis on observing the patient's blood saturation. Dr. Rillings classic, *The Use of Ozone in Medicine,* recently reprinted, gives many recommendations on dosage and concentration. There is no evidence that long term treatment on a daily basis has any detrimental effect. Doctors who have used it for decades have only positive results to report. Ozone is blatantly non-toxic. There is no evidence of free radical damage; in fact, ozone is the best free radical scavenger there is. Ozone also stimulates production of superoxide dismutase, catalase, and glutathione peroxidase, which are the enzymes in the cell wall which protect the cell from free radical damage, so ozone actually helps prevent free radical damage. Dr. Horst Kief of Germany recommends taking Vitamin A and Vitamin E supplements when receiving ozone treatments. It is know that Vitamin C is antagonistic to ozone, and persons taking megadoses of Vitamin C should maintain a 12 hour spread between ingestion and the ozone treatment, although ozone does not break down Vitamin C in the body. This effect of Vitamin C can be used to advantage in intravenous administration. Sometimes a patient will have a lot of coughing caused by ozone outgassing in the lungs from having had a bit too much too fast from the IV. If the coughing continues longer that 30 minutes, it can be stopped by administering 5000 mg of Vitamin C orally. The ozone reaction will end quickly and the patient will be more comfortable and have a better attitude toward the therapy. The rate of injection should be very slow, about 10 cc per minute. Since intravenous injections are 95-98 % oxygen and 2-5 % ozone gas, some doctors have expressed concern about embolism. However, there is no danger of embolism from injections of oxygen and ozone. Only nitrogen forms a dangerous gas bubble, as when divers get the bends from surfacing too fast. The human body runs perfectly well on 100% oxygen; consider the fighter pilots who breathe 100% oxygen daily for years — they have the highest reflexes, visual acuity and level of general health of any group of humans.

Ozone and Magnets

Doctors have reported that they can enhance ozone therapy by using magnet therapy simultaneously. Permanent magnets can be used with the north pole facing upward, toward the patient, on the underside of the treatment table. Magnets cause a polarization of red blood cells, which have iron in them. The polarization causes red blood cells to unclump and become more flexible, so that they can bend and get through the finest capillaries, improving microcirculation and preventing literally hundreds of diseases. Therefore, there is a synergistic effect between ozone and magnets.

Ozone for Prevention

Ozone is a powerful therapeutic tool for curing disease, but it is equally important for *prevention* of disease. The hundreds of different diseases named by allopathy are but symptoms of one underlying cause. That cause, as proven by two-time Nobel Prize winner Dr. Otto Warburg, is hypoxia, or oxygen starvation at the cellular level. This is the cause of degenerative disease (arthritis, atherosclerosis, multiple sclerosis, rheumatism, cancer, etc.). Ozone both treats and prevents most communicable disease as well (mumps, measles, influenza, cholera, tropical fevers, etc.) Regular use of ozone in the home can provide high levels of immunity from most common diseases, and relegate immunization to the dustbin of history. Our present allopathic health system is disintegrating under financial stress, and it can easily be replaced by prevention through use of ozone, supplemented by ozone injection for serious cases, and emergency room hospitals for accident victims. This system will be far less expensive than our present system, where 95% of our health dollar is spent in the last year of life, trying to undo a lifetime of toxic buildup.

Ozone and Water in the Body

The human body is 2/3 water. Of that, 90% is lymph and 10% is blood. The cell functions by burning sugar in oxygen to provide energy. The waste products are carbon dioxide and water. If there is insufficient oxygen at the cellular level, the burn will be incomplete, and carbon monoxide and lactic acid will be formed. The body cannot easily rid itself of monoxide; it prevents hemoglobin from picking up fresh oxygen, and the body temperature is lowered. The lactic acid will build up in the system, clogging the nerve pathways, eventually calcifying and causing degeneration. More oxygen is required to come in and oxidize these toxins, but if it is not available, they build up. The blood will carry a heavy load of sludge, and toxins will be deposited in the fat. The water that composes the body gets dirtier and dirtier. Disease is the result. This is where ozone shines — in eliminating toxicity from the body. Ozone taken on a daily basis will, over time, clean all the fluid of the body, safely. Ozone has been used to clean water for large cities for over 100 years. The water engineers have a value that they use to measure the effectiveness of ozone in cleaning water. This is the CT value. It is a product of concentration *x* time. (C*x*T).

Ozonated Water

For prevention, a major benefit can be derived from regularly drinking ozonated water. Water is a fascinating substance, and we all take it for granted. Chemically it is considered to be on oxygen atom bound with two hydrogen atoms. The bond angle between the two hydrogen atoms is known to be variable, depending on the amount of energy in the molecule. Research has shown that water whose bond angle is 101 degrees is 'dead' water, bereft of life-giving energy. When water is distilled the bond angle expands to 120 degrees upon evaporation, but collapses to 101 degrees upon condensation, and is therefore dead. A bond angle of 103 degrees corresponds to average water. A bond angle of 106 degrees produces activated, energized water, and is attainable by placing a magnet, north pole inward, against the water container. The highest energy obtainable in liquid water is a bond angle of 109.5 degrees, and this is attainable only ozonating water at 4 degrees C. Ozone will not stay in water for very long, even at 4 degrees C. To hold the ozone in the water over long periods, it is necessary to add a few drops of Concentrace, which is a solution of trace minerals from the Great Salt Lake with the sodium, cadmium, copper and lead removed. The ozone hangs on to the minerals without oxidizing them and remains available over many months. In general, unmineralized water should not be consumed.

Drink water that has gone through reverse osmosis, carbon filtering, and is then ozonated. There are some contaminants that will pass through R/O and carbon, such as fluorine. Ozonating the water removes all such contaminants and energizes the water until the bond angle reaches 109.5 degrees. Ozonate water for 15 minutes per litre, about five minutes per glass.

Ozonating the Lymph

Women have an advantage, in that vaginal insufflation requires no preparation, and can be administered for very long periods of time, hours in fact. The gas will usually find its way into the uterus, out the Fallopian tubes, and then into the abdominal cavity. Liver problems and pelvic inflammatory disease (PID) can be addressed in this way. This is also a good way of getting ozone into the lymph system. For men, cleaning the lymph system is not as easy, and requires use of a body suit or a steam cabinet. The body suit is a less that popular aesthetic experience. The Saunette steam cabinet, however, is a pleasurable experience. Because of the moist heat, the pores are open, and the capillaries are dilated. The ozone enters and oxidizes toxins in the fat, the lymph and the blood. The skin is the largest organ of elimination. The person sweats the oxidized toxins back out, avoiding the dump of toxins to the liver and colon which can bring on the symptoms of toxic shock overload. Instead, the person emerges from the steam cabinet feeling extremely relaxed and mellow, and ready for bed. This is an ideal way of counteracting the stress of the day.

Breathing Ozone

Ozone is safe to breathe when it is bubbled through extra virgin olive oil. This is an excellent therapy for asthma and bronchitis and pneumonia, especially when combined with magnetic therapy. Breathing of ozone has been practiced in North America for over 90 years. When ozone is bubbled through olive oil continuously for weeks, the oil starts to change. First it loses its color, then it begins to foam, and eventually it becomes a gel, although the oil is not oxidized. If it is kept refrigerated at 40 degrees F, this gel will hold on to its ozone for more than ten years. This gel applied to the skin has many uses: on cuts, scrapes and burns; insect bites, diaper rash, eczema, impetigo, herpes, etc. When massaged vigorously into the body by a trained therapist over time it has enabled multiple sclerosis patients to regain the use of their limbs. Ozonated gel is 95% as active as ozone gas. The ozonated gel liquefies as soon as it reaches skin temperature. It is an excellent lubricant for intercourse and provides more protection the highly-touted condom for the prevention of disease, due to the bactericidal, virucidal and fungicidal action of ozone. It is an excellent product for your pet as well.

What Does Ozone Do?

Ozone: inactivates viruses, bacteria, yeast, fungus and protozoa. stimulates the immune system cleans arteries and veins, improves circulation' purifies the blood and lymph normalizes hormone and enzyme production reduces inflammation reduces pain, calms the nerves stops bleeding prevents shock prevents stroke damage reduces cardiac arrhythmia improves brain function and memory oxidizes toxins, allowing their excretion chelates heavy metals; it works well in conjunction with EDTA prevents and reverses degenerative diseases prevents and treats communicable diseases prevents and eliminates auto-immune diseases

How does Ozone Work?

1. *Inactivation of bacteria, viruses, fungi, yeast and protozoa :* Ozone disrupts the integrity of the bacterial cell envelope through oxidation of the phospholipids and lipoproteins. In fungi, ozone inhibits cell growth at certain stages. With viruses, the ozone damages the viral capsid and disrupts the reproductive cycle by disrupting the virus-to-cell contact with peroxidation. The weak enzyme coatings on cells which make them vulnerable to invasion by viruses make them susceptible to oxidation and elimination from the body, which then replaces them with healthy cells.
2. *Enhancement of circulation :* In circulatory disease, a clumping of red blood cells hinders blood flow through the small capillaries and decreases oxygen absorption due to reduced surface area. Ozone reduces or eliminates clumping and red cell flexibility is restored, along with oxygen carrying ability. Oxygenation of the tissues increases as the arterial partial pressure increases and viscosity decreases. Ozone also oxidizes the plaque in arteries, allowing the removal of the breakdown products, unclogging the blood vessels.
3. *Stimulation of oxygen metabolism :* Ozone causes an increase in the red blood cell glycolysis rate. This leads to the stimulation of 2,3-diphosphoglycerate (2,3-DPG) which leads to an increase in the amount of oxygen released to the tissues. There is a stimulation of the production of the enzymes which act as free radical scavengers and cell wall protectors: glutathione peroxidase, catalase, and superoxide dismutase. Ozone activates the Krebs cycle by enhancing oxidative carboxylation of pyruvate, stimulating production of ATP. Ozone also causes a significant reduction in NADH and helps to oxidize cytochrome C. Prostacyclin, a vasodilator, is also induced by ozone.

Ozone and Atmosphere

Why are we interested in radicals?

Radicals, having an unpaired electron, are highly reactive. This makes them the driving molecules in air chemistry. They cleanse the atmosphere, are responsible for the formation of ozone (O_3) and can also initiate aerosol formation. One of the prime radicals for air chemistry is the hydroxyl radical, OH. Simple in structure, it is one of the most reactive molecules in the atmosphere.

It is mainly formed by water molecules, ozone and sunlight, which are available almost everywhere. This ubiquitous production source makes it the most important oxidizing molecule in the atmosphere. A close chemical relative, the hydroperoxyl radical, HO_2, is a major source of tropospheric ozone. It reacts with the NO radical to form NO_2, which is then photolyzed, producing ozone. In addition, OH and HO_2, together called HO_2, play a role in the production of aerosols through the formation of low volatility compounds like organic acids or sulfuric and nitric acid.

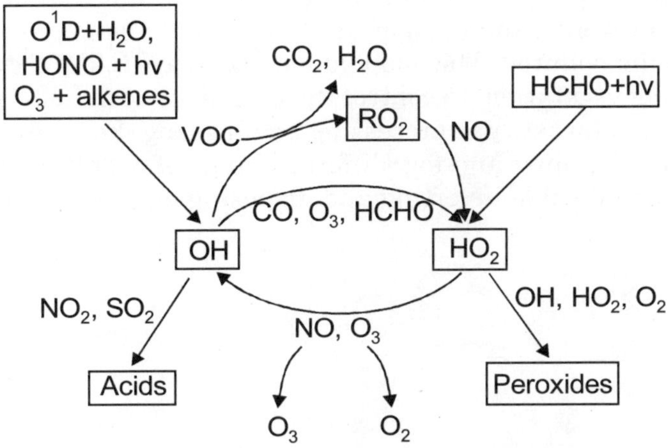

Fig. 8.15.

The destruction of ozonosphere and the increase of carbon dioxide in the air caused many environment disasters. The direct cause of Greenhouse Effect is discharging too much carbon dioxide in to the air while burning coal, petroleum and natural gas. Since the Industry Revolution in 18th century, people made use of coal, petroleum in a large amount, and the annual discharge amount of carbon dioxide to the atmosphere increased from 90 million t. in 1860s to 5 billion t. in 1950s. In the recent hundred years, carbon dioxide in the atmosphere has increased by 25 percent, and two thirds of the forest that can regulate and absorb carbon dioxide was felled. And the agricultural activities of overusing petrochemical materials, pesticides and chemical fertilizer caused a large amount discharge of sulfur dioxide and nitrides, which can float in the air several thousand kilometers away, producing chemical reaction with vapor and form acid rain. At present, acid rain has swept over the whole world. At least 3600 thousand square km forest was destroyed by acid rain, over 4000 lakes have become dead lakes. In China, a fairly large scale of acid rain belt has formed in southern area of the Yangtse River. About 61.8 per cent of acid rain occurred in the cities of southern area, and the area of acid rain accounts for 30 per cent of the territory. Acid rain lasted for some time in the summer of 1982, and russet spots emerged on the leaves of crops in a large area after raining and most seedlings of cereal crops died, which caused severe reduction of crop output. Ozonosphere can protect the people and other animals from being radiated excessively by UV rays, but now there is a hole in ozonosphere, which is as large as the territory of America, as deep as Everest. If ozonosphere reduces by 10 per cent, the young fish in 10 metres deep of the sea will die in 15 days. If ozonosphere remains only 2 per cent, it will be the deadline of the lives on the earth.

There are three shapes of atmospheric polluted materials: solid, liquid and gas. Solid polluted materials refer to the dust particles of different size and character; liquid polluted materials refer to all kinds of acid fog and organic solvent drops; gas polluted materials refer to sulfide, nitrides, carbon dioxide and hydrocarbon and some organic gas, such as nitrogen dioxide. The specific content of atmospheric environment protection and construction are to prevent wind erosion and implement the engineering of preventing atmosphere from being polluted. The adopted measures are to build cropland shelterbelt, to protect the existing vegetation, to control the discharge of industrial waste gas, and to construct the facilities of waste gas processing engineering. For example, different kinds of equipment for cleaning dust and fog, different kinds of desulfuration and denitrating equipment, and carbon dioxide inverting and disposing engineering equipment.

CHAPTER 9

Oxides and Peroxides of Hydrogen

Introduction

Water and hydrogen peroxide are the two oxides (H_2O) and hydrogen peroxide (H_2O_2).

WATER

Occurrence

Water is the most important constituent of the earth's surface ($\frac{1}{4}$ land and $\frac{3}{4}$ water). Water it self (Liquid) and water vapour (Vapour State). Water is considered as the most useful thing of the earth's surface, the 'life of earth's surface. Water is also present in other planets and also in human body.

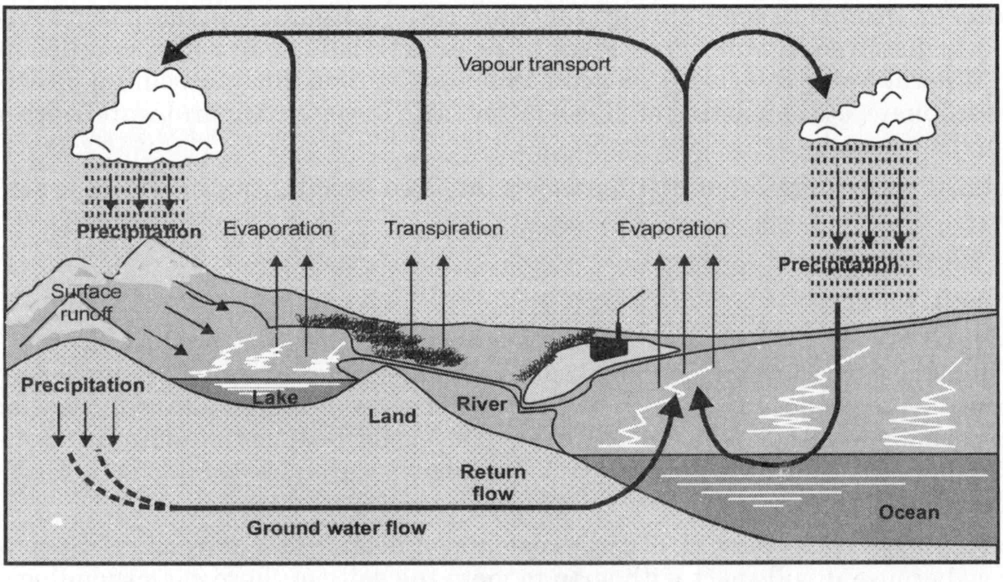

Fig. 9.1. The Water Cycle

The three forms of water are interchangeable, that is when the temperature of the liquid state reaches at 0°C the water (Liquid) is transferred to ice (Solid), also if the temperature of the water is increased to 100°C the vapour is produced in the form of steam. This has been clearly shown in Fig. 9.1.

Natural Sources of Water

There are many sources of water provided by our nature, they are rain, spring, river and mineral water and we shall now discuss in a brief about these sources.

Rain Water : Rain water is considered as the purest form of water, but due to pollution it has gone far from purity, many dissolve gases such as carbon dioxide, nitrogen dioxide, sulphur dioxide etc. have made the water impure. It also consists of dust particles and other substances, which are regularly discharged from the factories and automobiles.

Spring Water : When the rainwater falls on the surface it penetrates through the surface of the earth and appears as springs. These spring dissolves in it different matters of the earth's surface. It also gets attracted with carbonic acid. A small portion of heavy metals like iron, etc. also get dissolved.

But rainwater when penetrates through the different layers of rocks in the earth's surface, where it is collected in a large volume and due to pressure rises up to the surface in form of spring. These springs appear through different layers of the earth.

River Water : Number of springs gives rises to rivers, also the snow peak mountains, during summer when snow melts the molten snow also gives rise to many rivers. The river whether originated from snow or rain, has to pass through many regions and has to collect much undesired matter, including organic wastes. River passes through many cities, towns and villages hence it carries a considerable amount of water along with it.

Mineral Water : In some springs there are certain organic and inorganic substances which gets dissolved in water gives rises to a characteristic taste which is called mineral water. Mineral water can contain a good amount of carbon dioxide, sodium bicarbonate, sodium sulphate, etc. imparting or give a bitter taste in water. Mineral waters are used in medicinal purpose.

Hard and Soft water

Natural waters are often distinguished into two broad groups. They are :

1. Hard Water, and
2. Soft Water.

1. Hard Water : Water is said to be hard when it does not produce enough lather when added to soap solution. Soap molecules actually reduce the surface tension of water and thus able to produce bubbles, when the water soap mixture is stirred these bubbles stay in permanent lather, which does not subside easily. Soap is the sodium or potassium salt of higher fatty acids, which is very much soluble in water. Its formula is $C_{17}H_{33}COONa$ (sodium Oleate).

Maturality of water consists of excess calcium or magnesium or iron ions. Soap will not lather as because it will react with soap to yield the salts of those corresponding metals elements.

For examples : When Ferric ion reacts with soap the takes place as shown below:

Ferric Ion	Soap		Ferric Oleate	Sodium Ion
Fe^{++}	$2C_{17}H_{33}COONa$	$\longrightarrow$	$(C_{17}H_{33}COO)_2Fe$	$2Na^+$

The degree of hardness is determined by the reaction water with a standard soap solution the hardness of water is expressed in parts of calcium carbonate per 100,000 parts of water (parts per million = PPM). When the water does not contain or contain in very little amount of calcium, magnesium or ferric ions is termed as soft water.

Hard water is unsuitable for boiler and for laundry. In laundry there is a lather formation problem and in boiler heavy calcium and magnesium ions gets deposited on the inner surface of the boiler as *boiler scales* being poor conductor of heat reduces the efficiency of the boiler and also the surface of the boiler gets damaged.

The surface of the boiler and the layer of the boiler scales have a difference in concentration and relaxation texture, which creates flaws in the boiler material. Hard water is, therefore, softened before using in the boiler and in the other important purposes. Hardness of water is generally distinguished as *permanent hardness* and *temporary hardness.*

Temporary Hardness : This is because of the presence of dissolved bicarbonates of calcium and magnesium. This can be removed by boiling the insoluble precipitates of carbonates. This is the reason why such hardness is termed as *temporary hardness.*

Permanent Hardness : This hardness occurs when the water contains dissolved chlorides and sulphates of calcium and magnesium. These compounds cannot be removed only by boiling that is why the water is termed as *permanent hard* water.

Softening of Water

As we know that hard water is not suitable for boiler and for laundary purpose, so there are some process of removing temporary hardness can be done by boiling which can be expressed by the following equation:

Calcium Bicarbonate (insoluble) $\longrightarrow$ Calcium Carbonate (soluble) + Water

Temporary hardness also can be removed by the addition of a calculated amount of slaked lime (calcium hydroxide)

Calcium Hydroxide + Calcium Bicarbonate $\longrightarrow$ Calcium Carbonate + Water

$$Ca(OH)_2 + Ca(HCO_3)_2 \longrightarrow 2CaCO_3 + 2H_2O$$

Magnsium Bicarbonate + Calcium Hydroxide $\longrightarrow$ Calcium carbonate + Water + Magnesium Hydroxide

$$Mg\,(HCO_3)_2 + Ca(OH)_2 \longrightarrow 2CaCO_3 + 2H_2O + Mg(OH)_2$$

Permanent and temporary hardness both can be removed by passing water through zeolite or hydrated sodium aluminium silicate called permutit. The permutit has a property of ion exchange; it exchanges its negative ions with the positive magnesium and calcium ions. The actual reaction is shown below :

Sodium Zeolite	+	Calcium Ion	$\longrightarrow$	Calcium Zeolite	+	Sodium Ion
Na_2Z	+	Ca^{++}	$\longrightarrow$	CaZ	+	$2Na^{++}$

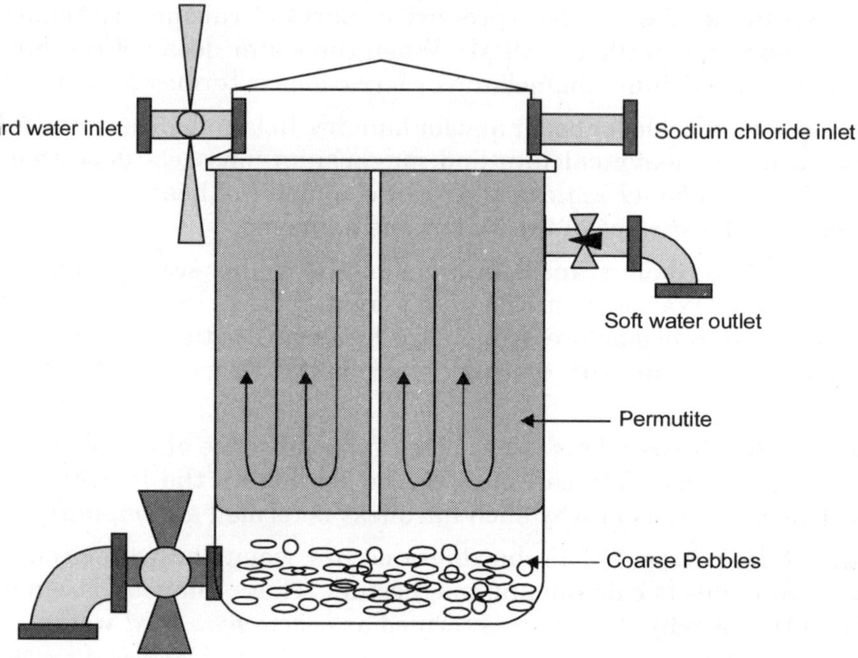

Fig. 9.2. The Process of Water Softening by Permutite Process

The reaction is reversible as seen from the above. The permutite bed when gets exhausted can be reused again by moistening with a solution of sodium chloride (common salt). But now-a-days another much reliable process is being used. The name of the process is Ion Exchange. Where the deionization of water is done by using ion exchange resins. These are granular substances and are often more efficient than zeolite, these ion exchange resins are organic compounds they contain a sulphonic acid group (SO_3H). The resin is first converted into sodium sulphonate with the treatment of sodium chloride and then sodium chloride and the softening resin takes place as shown in the equation below.

Sodium Resin	+	Calcium Ion	$\rightleftharpoons$	Calcium Resin	+	Sodium Ion
2Na.R	+	Ca^{++}	$\rightleftharpoons$	R_2Ca	+	Na^+

here R represents resins

In small scale or domestic water is also softened by the use of sodium hexametaphosphate, with calcium ion it forms a complex anion, which does not precipitate soap.

Sodium Hexametaphosphate	+	Calcium Sulphate	$\longrightarrow$	Sodium Calcium Hexametaphosphate	+	Sodium Sulphate

$$Na_2Na_4(PO_3)_6 \quad + \quad CaSO_4 \longrightarrow Na_2CaNa_2(PO_3)_6 \Downarrow + \quad Na_2SO_4$$

Also polymanino acid like ethylenediaminotetraacetic acid, commonly known as EDTA also used in scale water softening process.

Pure and Distil Water

The process of distillation does the ultimate purification of water. Here the process is very simple, the water is heated on a furnace up to a temperature nearby 101°C and the steam generated is cooled by a series of condensers and the water is obtained. If necessary this water is cooled and further distilled, sometimes the water is mixed with a little of potassium permanganate an then boiled.

All the soluble mineral matters of water are now removed without using the process of distillation. But using the process of ion exchange resins, the water is first passed through a bed of negative ion exchanger, which is a bed of granular organic acids. The acid character sometimes appears which in terms reduces the calcium and magnesium ions by exchange.

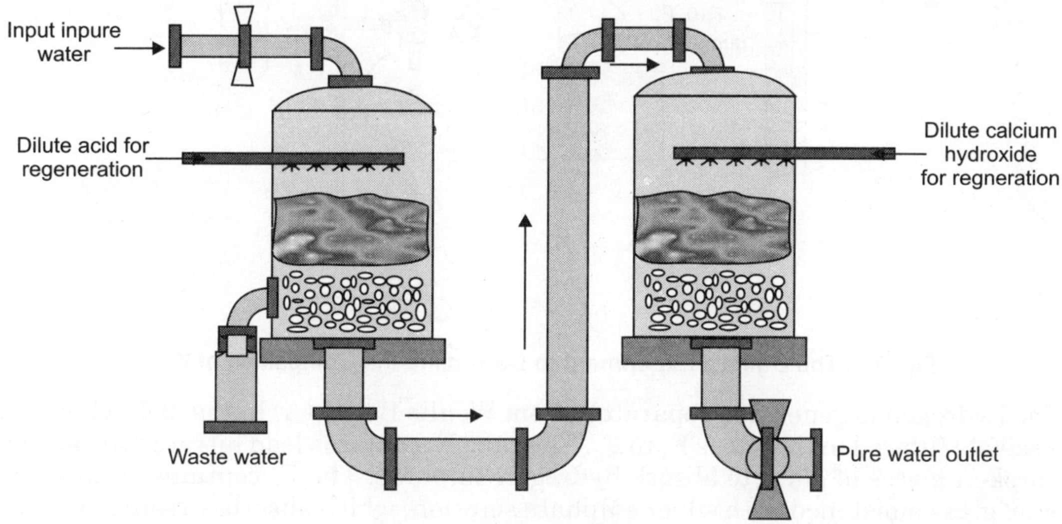

Fig. 9.3. The Water Purification by Ion-Exchange Method

Composition of Water

Composition of water can be of two types - the physical and the chemical composition. At first we will discuss the physical composition of water.

Physical Composition : Physical composition mainly means those constituents which is dissolved in water physically e.g. dust particles, dissolved gasses, e.g. oxygen, carbon dioxide, etc. The water also contains some chemical elements e.g. calcium, magnesium, iron etc. These physical components get easily separated by different methods of filtration.

Chemical Composition : To discuss the chemical composition of water we shall have to discuss a long story. In the year (1781-1784) Henry Cavendish, the famous English scientist first seperated water into two different elements - the hydrogen and oxygen (2 volumes of hydrogen and one volume of oxygen). Lavoisier also supported this theory. Cavendish performed the experiment of electrolysis and got these two gases. The following processes can determine the chemical composition ofwater, which is discussed below (Fig. 9.4).

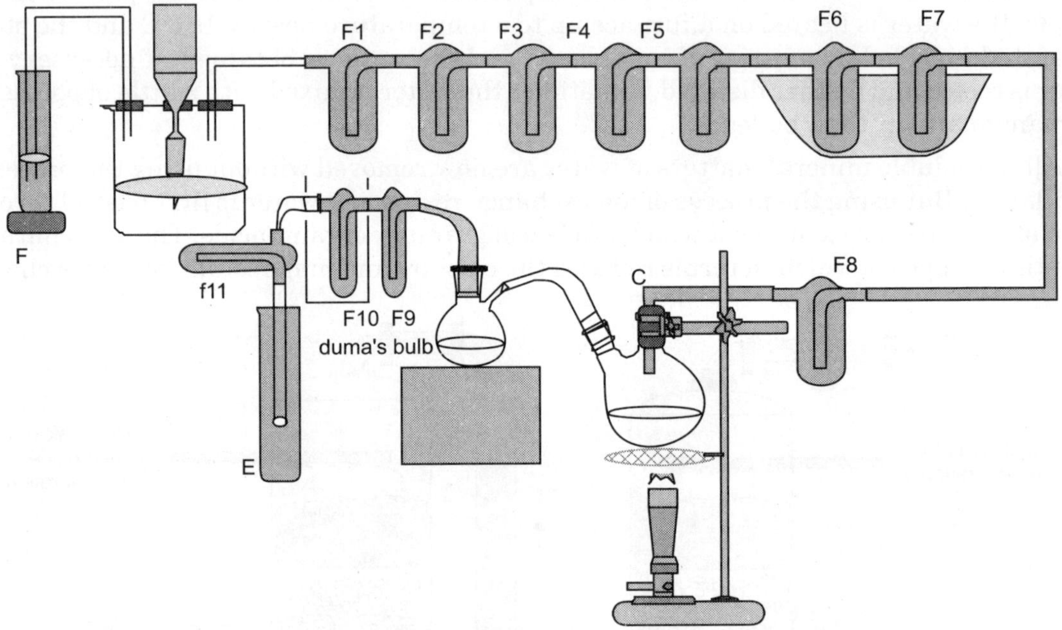

Fig. 9.4. The Duma's Experiment to Determine the Compsition of Water

The hydrogen is generated separately from Woulfs Bottle (A) in Fig. 9.4, which is then successively filtered in the tubes F_1 to F_7. The tube F_1 contains lead nitrate solution spread over broken pieces of glass to absorb hydrogen sulphide. The F_2 contains similar broken pieces of glass moistened with silver sulphate solution, which absorbs arsenic hydride, the next tube F_3 to F_5 contains potassium hydroxide, which removes acid vapour F_3 contains pumice stone soaked with potassium hydroxide solution, F_4 contains solid pieces of potassium hydroxide. F_6 and F_7 contain phosphorous pentoxide or pumice stone soaked with sulphuric acid to remove moisture. These are placed in freezing mixture and are followed by a smaller tube F_5 filled with phosphorous pentoxide, which serves as a guard tube. Dry copper oxide is placed in a hard glass tube C, connected by means of a delivery tube to receiver D, a bulb provided with a long neck containing calcium chloride. The bulb D collects the water formed, it is followed by a series of drying tube four in number F_9– F_{12}, F_9 consists of solid potassium hydroxide F_{10} consists of phosphorous pentoxide or sulphuric acid in pumice stone which is kept inside with freezing mixture. F_{11} are guard tubes containing phosphorous pentoxide or sulphuric acid, which serves as the moisture absorber from the outside of the apparatus system. The residual hydrogen escapes by bubbling through the sulphuric acid in cylinder E. The cylinder F in the extreme left of the apparatus set up consists of mercury, serves as safety valve. The bulb c with copper oxide was heated inside the bulb C by a spirit lamp for 10 to 12 hours, as the hydrogen was steaming through.

After the experiment was over the apparatus set up was allowed to cool in a current of hydrogen, which was then displaced by air. The bulb C and the receiver D and the absorption system was then detached exhausted and weighted.

From the mean of the several experiments it was fond that 7.98 part by weight of oxygen combined with 1 part of hydrogen by weight to produce 8.98 part of water by weight. This experiment was done by J.B dumas.

Morley's Experiment

In the year 1826 Morley did a remarkable experiment on the gravimetric composi-tion of water. The experi-ment is as follows, in a sealed tube that has two openings as shown in the Fig. 8.5. He passed an accurately measured amount of hydrogen and oxygen in pure and dry condition. This tube has two platinum terminals marked negative and positive for electrification a reservoir tube was attached with this apparatus, which was kept in a freezing mixture. When electricity was passed the hydrogen and oxygen reacted to form water. When sufficient quantity of water was collected, the apparatus was further cooled, after disconnection, the water in the reservoir got solidified. This water in the form of ice was carefully taken out and ratiocinated*. The apparatus was carefully weighted before and after the experiment.

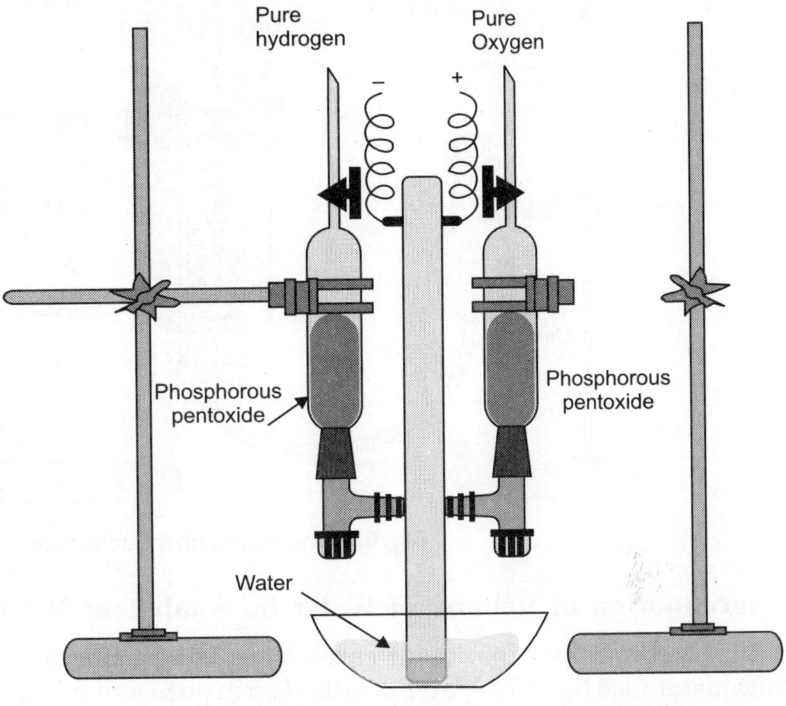

Fig. 9.5. The Morley's Apparatus and Emperiment

From the data, Morley found the ratio of oxygen and hydrogen is 7.9396 to l or 8: 1.

Volumetric Composition of Water : The volumetric method is used to determine the

*Ratiocination : Analysis.

volumetric composition of water. The apparatus used for this purpose is Hofmann's Eudiometer (see Fig. 9.6). In this experiment carefully measured hydrogen and oxygen of known volume is mixed and this mixture is allowed to explode by applying electrical current through the gas mixture. The volume of unused hydrogen was then measured after explosion. The result observed was those two volumes of hydrogen combines with one volume of oxygen to form water. All gases were measured under same condition of temperature and pressure over mercury. The volumetric composition of steam was similarly determined by exploding two volumes of hydrogen and one volume of oxygen within the closed limb of an eudiometer, surrounded by a jacket of amyl alcohol vapour, so that the water forms remains as steam. The gases and the steam formed are measured.

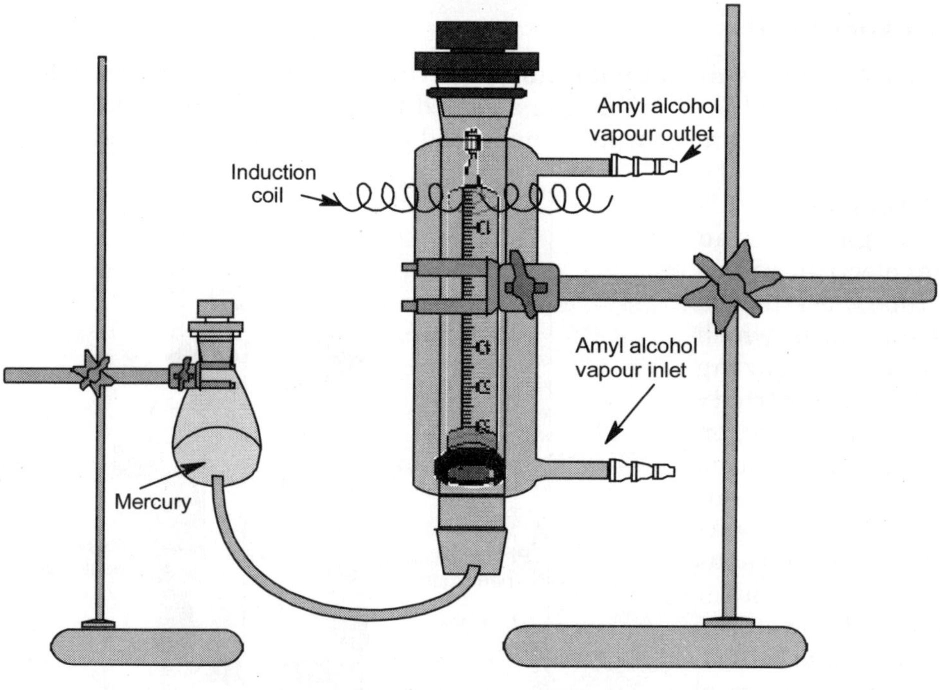

Fig. 9.6. The Hofmann's Eudiometer.

Determination of Volume of Water by Analytical Method

In this method the experiment carried out is by using an apparatus called Hofmann's Voltameter (see fig. 9.7). Water acidified with pure and dilute sulphuric acid is used as the testing solution, through which an electrical current is allowed to pass. Hofmann's Voltameter consists of two graduated each provided with a tap at top and platinum electrode at the bottom. Also there is a reservoir situated at the centre. Water is poured in the reservoir till it fills both the tubes. Taps are closed then, the current is switched on. The observation was like this, both the tubes did collect some gas, the tube with the negative end of the battery collects the gas which is double in volume with the gas collected in the other tube. After analysis it was found that the gas collected in the negative tube was hydrogen and the gas collected in the positive tube is oxygen.

From the above experiment Hoffmann found that exactly 2 volumes of hydrogen reacts with one volume of oxygen, to form two volumes of steam.

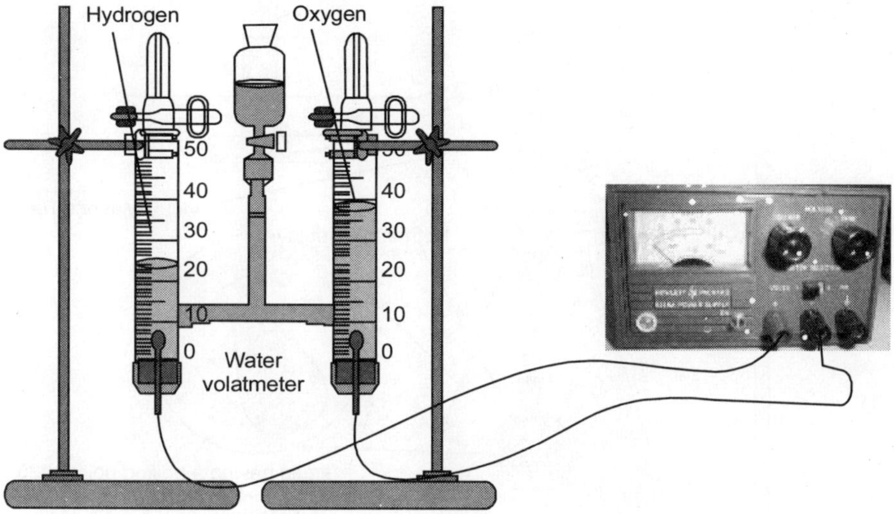

Fig. 9.7. The Hofmann's Voltameter

Complicated Chemical Concepts

Vapour Density: This is the ratio of equal volumes of vapour and the oxygen (the later being taken as 16). There are several methods of determining the vapour density. They are going to be discussed. Vapour density is required for the determination of molecular weights of different chemical substances.

Calculation of Vapour Density by Victor Mayor's Method

Principle: A known weight of a volatile substance solid or liquid is vapourized in a long tube (see Fig. 9.9) to displace a volume of air, equal to volume of vapour formed. Noting the volume of vapour displaced and its pressure and temperature, we can calculate the vapour density of substance:

Let X gm be the weight of substance vapourized and V cc be the volume of air (displaced) at N.T.P. (= volume of vapour formed) that is, X g of vapour at N.T.P. occupy V cc.

$$\therefore \quad \text{Vapour Density} = \frac{\text{Weight of Vapour}}{\text{Weight of Equalvolume of Hydrogen at N.T.P.}}$$

$$\text{or,} \quad \text{Vapour Density} = \frac{X}{V \times 0.00009}$$

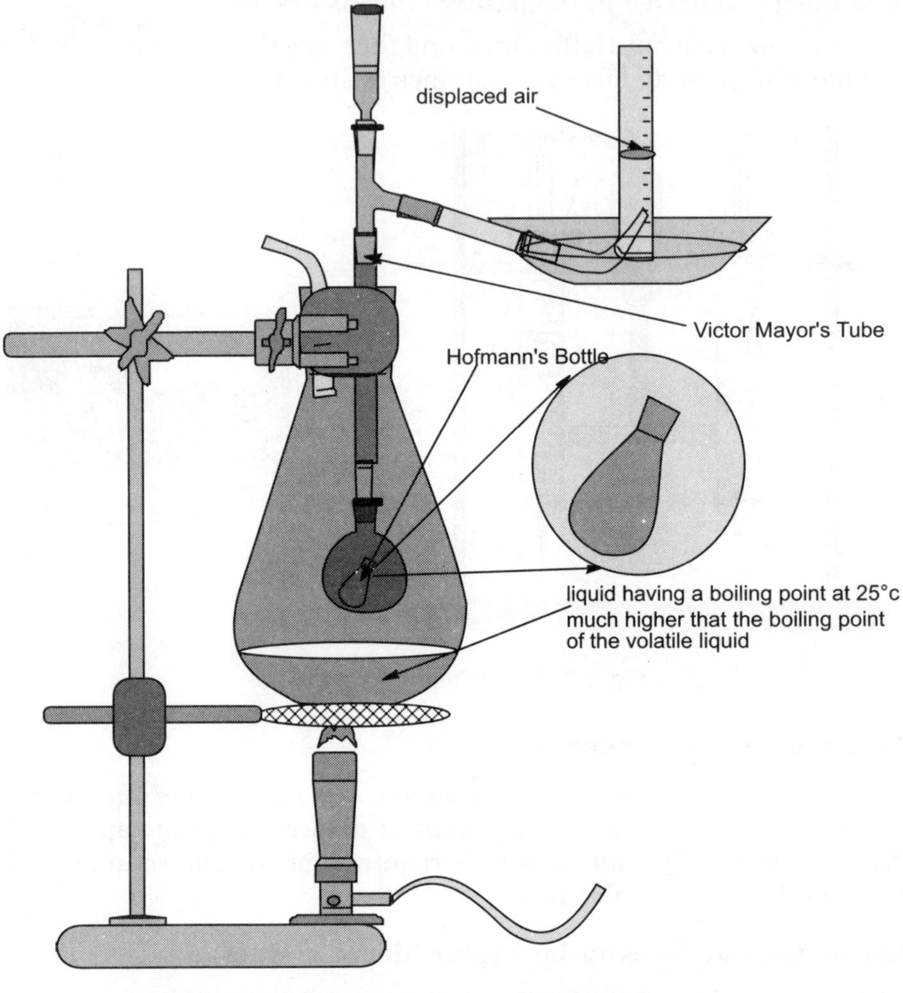

Fig. 9.8. The Apparatus Victor Mayor's Experiment

Experiment : The apparatus set up in the figure consists of an inner tube known as Victor Mayor's tube with a bulb and a side tube near the top, side tube ends under a beehive self in a trough of water. Victor Mayor (1848-1897) designed this apparatus the tube is tightly corked and is suspended in an outer tube Gacket) in, which is boiling liquid, with boiling point at 20°C above the boiling point of those substances of which the vapour density is to be determined.

Working: Liquid in the jacket is kept boiling till the temperature in the inner tube becomes constant, i.e. when no more air bubbles can escape from it, now a graduated tube filled with water is placed inverted over the beehive self. Hofmann's bottle, small phial loosely stoppered containing known weight of volatile substance, is dropped in the tube bursts open and the substance changes into vapour. Vapour remains inside the tube and the air inside the tube is displaced out and bubbles in the graduated tube. The volume of the displaced air is noted after levelling.

Calculation of Vapour Density by Dumas' Method

J.B Duma's designed the apparatus in the year 1800-1884. This apparatus consists of a bulb better known as Dumas' Bulb, which has a long narrow neck (see fig. 9.9). It is cleaned, dried and weighted accurately.

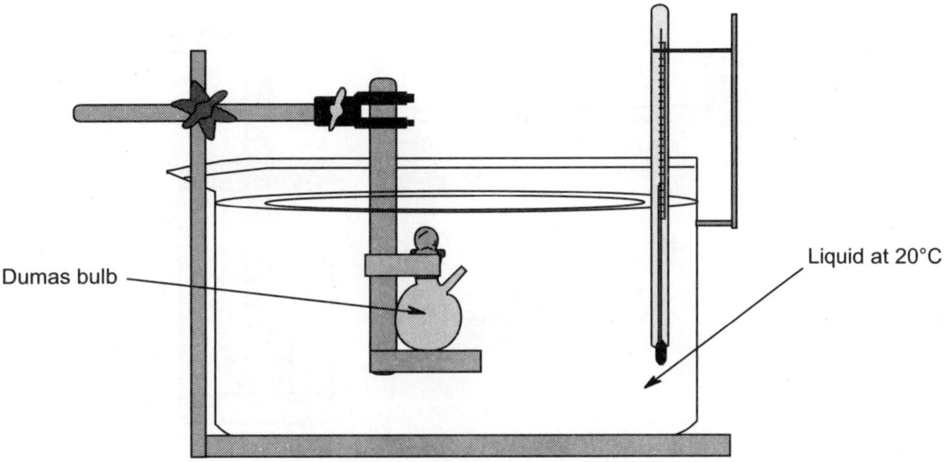

Fig. 9.9. The Dumas Apparatus

A few of liquid, under examination is drowned into the bulb, by warming and cooling it, while keeping its mouth dipped under the liquid. It is immersed in a bath of boiling liquid (constant temperature), of which; boiling point is necessary 20°C higher than the boiling point of the liquid under examination. The open end projects outside. The thermometer indicates constant temperature when whole of the liquid is vapourized and has displaced all air in the bulb; the open end of the bulb is sealed in a flame. Bulb is full of vapour at a temperature equal to the temperature of the boiling point of the liquid and at an atmospheric pressure, at the moment. Bulb is cooled and weighted, the difference of the weight to give the weight of the vapour. Volume of the bulb is determined by unsealing it under water, which fills the bulb. It is weighted to get the weight of the water filling the bulb from where volume of water (or volume of vapour) is calculated.

Weight of the bulb full of Air $= W$ g

Weight of the bulb full of vapour $= w$ g

Temperature at commencement $= t$°C and pressure, p mm $= w_1 - Wcc$

$$\text{Corrected volume} = \frac{273p(w_1 - W)}{760(t + 273)\text{cc}}$$

Hence, the density of the vapour pressure at N.T.P. $= \dfrac{w - W + 0.00129V}{V_1}$ g/cc.

Modification of Victor Mayor's Method (Lumsden's Method)

A number of methods are found which are modifications of Victor Mayor's method. The best among them is the Lumsden's method.

Here, the theory behind is that, instead of measuring the volume of vapour obtainable from a given weight of liquid under atmospheric pressure, the volume is kept constant and the increase of pressure found. The apparatus is shown in Fig. 9.10.

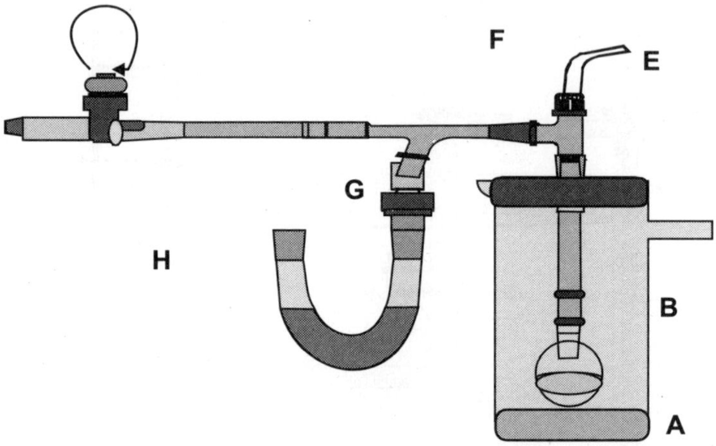

Fig. 9.10. The Lumsden's Apparatus and Experiment

The bulb A is a large flask of boiling type fitted with an outlet at B through which an air condenser passes. The inner tube C is the vapourizing tube at D and two other openings at E and F, which are closed when the apparatus is in use. The capillary tube D is connected to a manometer H–G, by means of which the apparatus can be introduced to the atmospheric pressure. There is a fixed mark on the vertical limb of the manometer G. For the heating, the boiling flask A may be placed on an asbestos box, together with the heating liquid and this is boiled it is also advisable to protect the top of the inner flask from the heat of the bummer by placing the asbestos board on the top of the stopper of the outer tube. Enough mercury is place into the manometer so that when the liquid is at fixed mark G, there is a little over the graduate position H. The preliminary heating is continued until when the stopcock I is closed, the mercury in G remains at the mark. This indicates that constant temperature has been reached, when this the case, Hofmann's Bottle filled with the liquid, which has been weighted is dropped in (asbestos must be placed inside the inner tube to arrest its fall). As the liquid vapourizes, the pressure increases and when the steady pressure is obtained the tube H is raised and the mercury is brought back to G mark.

It is essential to keep the mercury near the mark gradually raising H as the pressure increases. The difference between the two gives increase of pressure. It is very much sure that before removed, the easiest way of doing this is to blow a current of air through the apparatus from a foot-blower. The methods of calculation the vapour density is used for the calculation of atomic weight of different substances.

Composition of the seawater and the water of the stream do not only consist of hydrogen and oxygen but there are many elements dissolved in them so, these elements are listed in form of a table in a chart given below:

Element	Sea water (*PPB)	Stream (PPB)
Hydrogen	1.21×10^8	1.01×10^8
Helium	.00069	---
Lithium	180	2.89
Beryllium	.005	---
Boron	4304	17.99
Carbon (inorganic and dissolved organic)	28050	11622
Nitrogen (dissolved, N_2, NO_3^-, NO_2^-, NO_4^-)	15, 12	955
and also dissolved organic.	669	225
Oxygen (dissolved)	6025	8.9×10^4
Fluorine	1320	104
Neon	0.110	---
Sodium	1.07×10^7	7205
Magnesium	1.30×10^6	3645
Aluminium	1	51
Silicon	2895	4875
Phosphorous	61	80
Sulphur	9.06×10^5	3800
Chlorine	1.89×10^7	8248
Argon	433	---
Potassium	3.78×10^5	1410
Calcium	4.12×10^6	14698
Scandium	0.00052	.00039
Titanium	1.01	1.01
Canadium	2.45	0.0089
Chromium	0.31	.9
Manganese	.2	8.21
Iron	2.0	41
Cobalt	0.45	.21
Nickel	1.689	2.12

*PPB: -Parts Per Billion, 1 part in one billion, 1 part in 1,000,000,000 parts

Element	Sea water (*PPB)	Stream (PPB)
Copper	0.489	9.9
Zinc	4.98	28.99
Gallium	.00289	.1
Germanium	0.0045	---
Arsenic	3.7	1.99
Selenium	.2	.21
Bromine	67000	20
Krypton	.212	---
Rubidium	121	1.49
Strontium	8101	59
Yttrium	0.0132	0.07
Zirconium	0.254	---
Niobium	0.015	---
Technetium	---	---
Ruthenium	.00069	---
Rhodium	---	---
Palladium	---	---
Silver	.032	.22
Cadmium	.11	.22
Indium	0.0001	---
Tin	.01	---
Antimony	.24	---
Tellurium	---	1
Iodine	63.9	6.99
Xenon	0.045	---
Cesium	.40	0.0345
Praseodymium	0.00065	0.00699
Neodymium	0.00278	0.039
Promethium	---	---
Samarium	0.00065	0.0079
Europium	0.000129	0.001
Gadolinium	0.00069	0.0079
Terbium	0.00143	0.0012
Dysprosium	0.0009	0.044
Holmium	0.00022	0.0011
Erbium	0.000867	0.0033

Element	Sea water (*PPB)	Stream (PPB)
Thulium	0.000166	0.0011
Ytterbium	0.000811	0.334
Lutetium	0.000144	0.0011
Molybdenum	9.99	0.49
Barium	14.98	60
Lanthanum	0.0033	0.045
Cerium	0.0041	0.79
Hafnium	0.0069	---
Tantalum	0.00245	---
Tungsten	0.1	0.022
Rhenium	0.00784	---
Osmium	---	---
Iridium	---	---
Platinium	---	---
Gold	0.00034	0.0021
Mercury	0.022	0.0698
Thallium	0.011	---
Lead	0.023	---
Bismuth (thorium, and uranium decay series, Polonium, Astatine, Radon, Francium etc.	0.022	---
Thorium	0.01	.1
Uranium	3.22	0.34

HYDROGEN PEROXIDE

Introduction

In the year 1818 a French chemist Louis Jacues Themard first prepared hydrogen peroxide from the reaction of barium peroxide and dilute sulphuric and he named it as *eau oxegenee* (oxygen producer).

Occurrence

Hydrogen peroxide occurs in traces in nature, mostly in rain and snow it has not yet been detected in space. Hydrogen peroxide mainly occurs in the following reaction sequences they are as follows:

(*a*) When the hydrogen-oxygen flame explodes the liquid air, the hydrogen and the oxygen present inside the molecule react to form hydrogen peroxide.

(*b*) When water is irritated by an ultra-violet radiation of wavelength below 180 nm, hydrogen peroxide can be generated, but it is better to keep the water in a vapour state.

(*c*) Hydrogen peroxide can also be prepared by making a silent electrical sparking (discharge) through the water vapour under a pressure of .01014 Pa, this then cooled to get hydrogen peroxide.

Laboratory Preparation of Hydrogen Peroxide

Hydrogen peroxide is prepared in the laboratory by the reaction of sodium peroxide and dilute sulphuric acid.

Experiment: In a flask sodium peroxide is kept and dilute sulphuric acid is allowed to drop from a funnel the reaction starts immediately and the liquid thus obtained is hydrogen peroxide, which is 28% to 33% pure (apparatus set up is shown in Fig. 9.11).

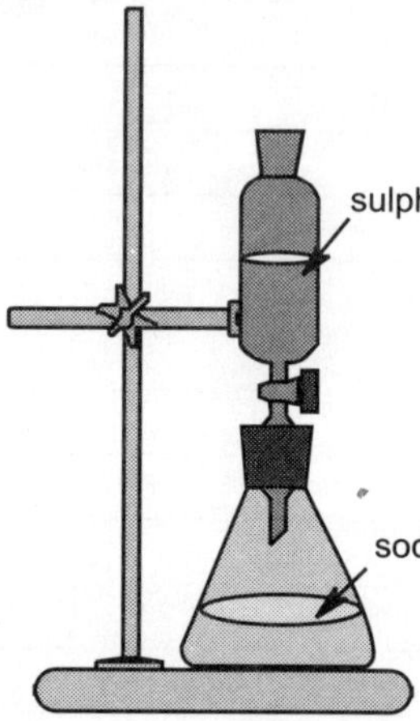

Fig. 9.11. The Laboratory preparation of Hydrogen Peroxide

Reaction Equation:

Sodium Peroxide + Dil Sulphuric Acid ⟶
Hydrogen Peeoxide + Sodium Sulphate

$$Na_2O_2 + H_2SO_4 \longrightarrow H_2O_2 + Na_2SO_4 \Downarrow$$

Other Methods of Preparation

From Barium Peroxide: Hydrogen peroxide can be prepared also from barium peroxide paste. This paste made in cold water is introduced to a thin stream of dilute sulphuric acid where barium sulphate and hydrogen peroxide will be formed:

Barium Peroxide + Dil Sulphuric Acid ⟶ Hydrogen Peroxide + Barium Sulphate

$$BaO_2 + H_2SO_4 \longrightarrow H_2O_2 + BaSO_4 \Downarrow$$

From Carbon Dioxide and Barium Peroxide: If barium peroxide paste is introduced to a steady stream of carbon dioxide then also hydrogen peroxide is formed where barium carbonate is precipitated.

From Electrolysis of Potassium Bisulphate: When Potassium Bisulphate is electrolyzed in an aqueous form hydrogen is collected at the negative electrode of the apparatus and at the negative electrode of the apparatus and at the positive end potassium persulphate is produced.

$$\text{Potassium Sulphate} \xrightarrow{\text{electrolysis}} \text{Potassium Persulphate}$$

This potassium persulphate reacts with steam to form hydrogen peroxide (steam should be passed through out the experiment).

Potassium Persulphate	+	Steam	⟶	Hydrogen Peroxide	+	Potassium Sulphate
$K_2S_2O_8$	+	$2H_2O$	⟶	H_2O_2	+	$2KHSO_4 \Downarrow$

Hydrogen peroxide is carried out with steam when condensed yields liquid hydrogen peroxide.

Purification of Hydrogen Proxide

Hydrogen peroxide obtained from above processes mentioned is not pure, therefore it need purification. For this following processes are carried out:

Evaporation over water bath: By this process hydrogen peroxide is evaporated on a water bath where the water vapour and the other dissolved gases are vapourized and leaves behind a 50% concentrated hydrogen peroxide solution.

Dehydration in Vacuum Desiccator: In this process hydrogen peroxide is dehydrated under reduced pressure inside a desiccator (Fig. 9.12).

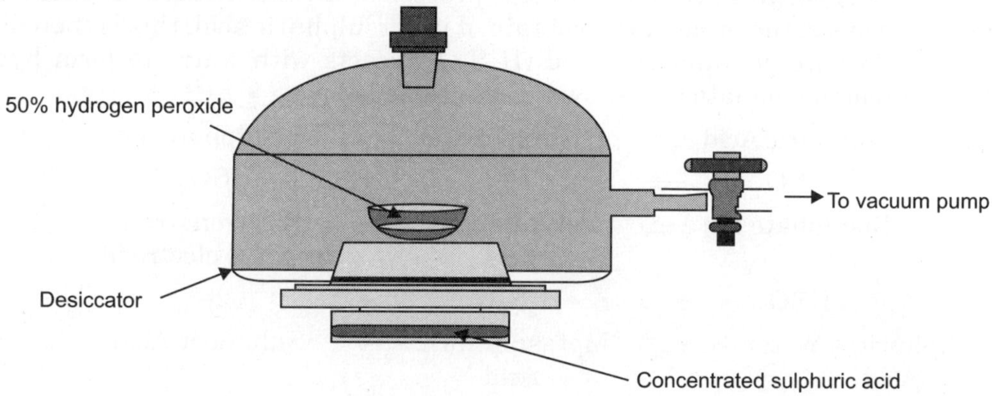

Fig. 9.12. The Dehydration of Hydrogen Peroxide in Vacuum Desiccator

Vacuum Distillation : Fifty per cent solution is distilled under reduced pressure. When almost pure hydrogen peroxide is obtained as second fraction at 70°C, first fraction is water at 50°C. Last traces of water are removed by chilling hydrogen peroxide with carbon dioxide snow and separating 100% crystalline hydrogen peroxide which are remelted, the experimental set up is shown in Fig. 9.13.

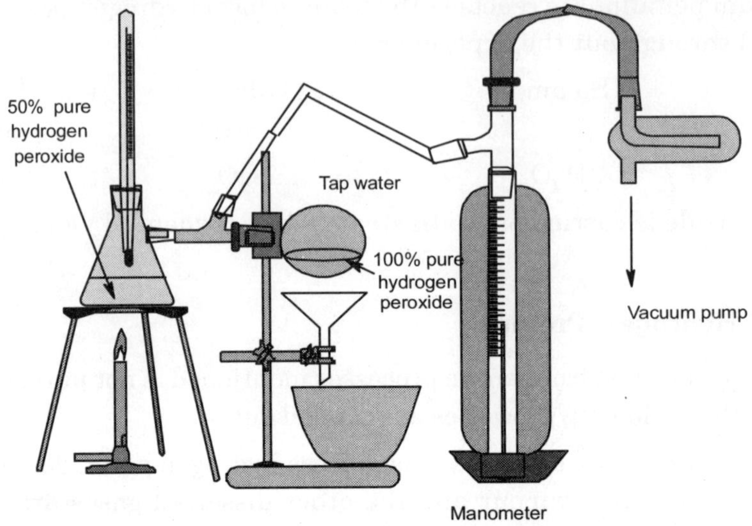

Fig. 9.13. The Vacuum Distillation of Hydrogen Peroxide

Industrial Process of Hydrogen Peroxide Preparation

By Electrolysis of Sulphuric Acid : In this process 50% concentrated sulphuric acid is set under electrolysis, the product thus obtained is persulphuric acid, this is then distilled under vacuum, where persulphuric acid ($H_2S_2O_8$) reacts with water to form hydrogen peroxide. The reaction equation is shown in the table below :

Sulphuric Acid	$\Longrightarrow$	Hydrogen Ion	+	Bisulphate ion
H_2SO_4	$\Longrightarrow$	H^+	+	HSO_4^-
Bisulphate ion	$\Longrightarrow$	Persulphuric Acid	+	Electron (from positive electrode)
$2HSO_4^-$	$\Longrightarrow$	$H_2S_2O_8$	+	$2e^-$
Persulphuric Acid + Water	$\Longrightarrow$	Metasulphuric Acid	+	Sulphuric Acid
$H_2S_2O_8 + H_2O$	$\Longrightarrow$	H_2SO_5	+	H_2SO_4
Metasulphuric Acid + Water	$\Longrightarrow$	Hydrogen Peroxide	+	Sulphuric Acid
$H_2SO_5 + H_2O$	$\Longrightarrow$	H_2O_2	+	H_2SO_4

(Reaction at	Hydrogen Ion	+	Electron	$\Longrightarrow$	Hydrogen
cathode)	H^+	+	$2e^-$	$\Longrightarrow$	H_2

2. By the method of auto Oxidation : This process is fully consisting of organic reactions, employing a method of reducing a derivative of antroquinone to anthraquinol by dissolving it in an organic solvent and passing hydrogen gas through this mixture in the presence of a catalyst palladium. When the anthraquinol comes in contact with air 20% solution of hydrogen peroxide is obtained and the antroquinone derivative is rejected. (See Fig. 9.14).

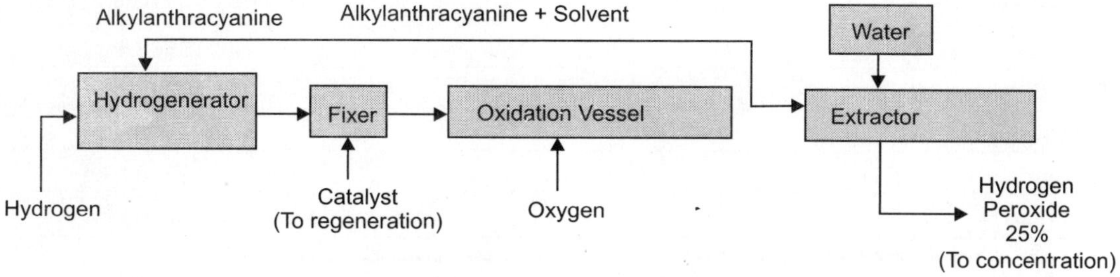

Fig. 9.14. The Industrial Process of Hydrogen Peroxide from Alkylanthracyanine

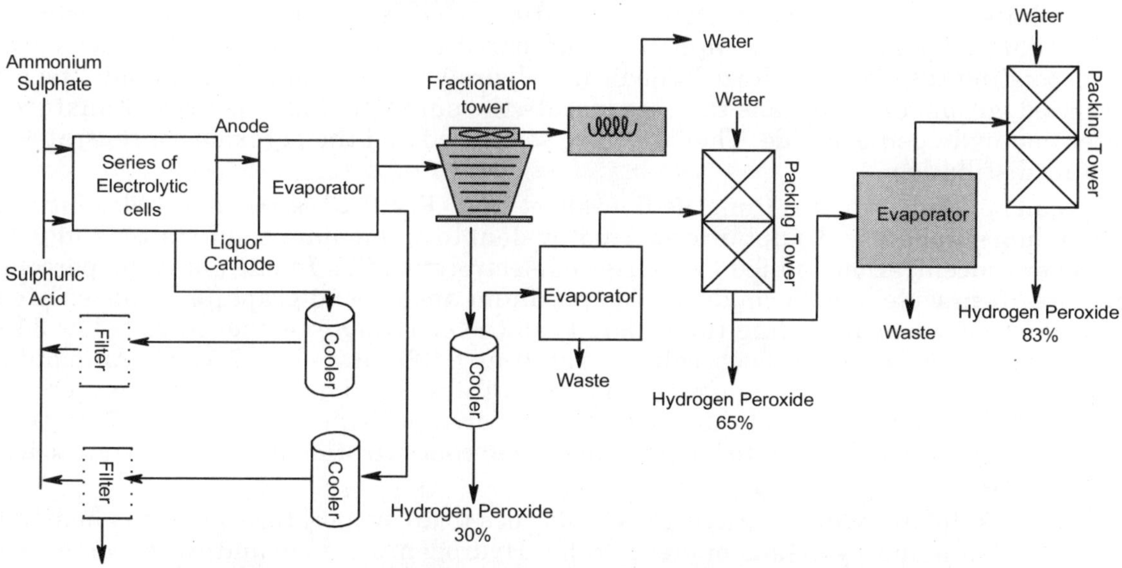

Fig. 9.15. The Flow Diagram of the Industrial Preparation of Hydrogen Peroxide from Ammonium Sulphate by Electrolysis.

Reaction

$$\text{Ammonium Bisulphate} \xrightarrow{\text{Direct Current}} \text{Ammonium Persulphate} + \text{Hydrogen}$$

$$2NH_4 2HSO_4 \longrightarrow (NH_4)_2S_2O_8 + H_2$$

$$\text{Ammonium Persulphate} + \text{Water} \longrightarrow \text{Ammonium Bisulphate} + \text{Hydrogen Peroxide}$$

$$(NH_4)_2O_2O_8 + 2H_2O \longrightarrow 2NH_4HSO_4 + H_2O_2$$

(80 to 85% current efficiency)

Material and Energy Requirement

Basis of 1 ton 30% hydrogen Peroxide

Ammonia	22.679715 kg
Sulphuric acid	27.215568 kg
Demineralized Water	977.40935 litre
Electricity	5700 kWhr
Steam	7620.35904 kg
Platinum	Very Little

Process

Hydrogen Peroxide is manufactured by the hydrolysis of a solution containing the persulphate (S_2O) ion. The particular solution used (persulphuric acid ammonium persulphate or potassium persulphate) is prepared in an electrolytic cell in which the corresponding sulphate is oxidized anodically Hydrolysis of persulphate (outside the cell) is carried out under conditions that regenerate the sulphate and vapourize a mixture of water and hydrogen peroxide. The latter are separated and the peroxide concentrated by vacuum distillation.

It can be easily realized from the flow sheet that (Fig. 9.15) a regenerated solution of ammonium sulphate and sulphuric acid (equivalent to ammonium bisulphate) is adjusted to proper concentration and fed to a series of electrolytic cells. In one plant the porcelain cells are 70cm wide and 95 cm deep. The platinum anodes and graphite cathodes are so arranged that water in cooling tubes may keep the electrolyte temperature below 35°C. The potential drop among each cell, operated at 5400 amps is 5.7 volts. Ammonium thiocyanate (0.1 9 per litre) is usually added to the electrolyte to increase oxygen over voltage. Anode current density is 0.4 to 1.1 amperes per sq cm. A current of air swept continuously through the cell to keep the hydrogen concentration below 5 per cent so as to avoid explosion.

The persulphate solution leaving the cells goes to an evaporator where it is heated by both live steam and by means of steam cells. Hydrogen peroxide and water vapours go overhead. The ammonium bisulphate solution is cooled, filtered to remove any sediment, adjusted in concentration, and returned to the cells. The cathode liquor is similarly cooled, filtered, and reused. The hydrogen peroxide-water vapour mixture is sent to Stone were distillation towers which are operated under vacuum (40 mm of Hg). A 30 to 40 per cent hydrogen peroxide solution is removed from the bottom of the tower. This solution may be concentrated to 80-85% or 90-98% H_2O_2 in two stages.

The 30 to 40% Hydrogen peroxide solution recovered from the cell liquors are adjusted with sulphuric acid (about 0.5 g per litre) and fed to the stone were of porcelain evaporator where the solution is heated to 65°C at 40 to 50 mm of Hg. The vapours evolved are scrubbed in a packed tower with distilled or demineralized water in such a ratio to the peroxide vapours that a 65% of hydrogen peroxide solution leaves the bottom of the tower.

This product may be fed to a second evaporator (at 75°C and the same pressure) and a second packed tower to produce an 80 to 85% solution of hydrogen peroxide. The product is cooled, Stabilized with acids or an in organic inhibitor and sent to storage.

Physical Properties of Hydrogen Peroxide

Melting Point Boiling Point	-0.461°C - 155.5°C
Density	1.441 g/cc (at 20°C)
Surface Tension	75.94 dynes/cm (at 18°C)
Viscosity	1307 centipoises
Heat of Fusion	2516 cal/g mole 14.1 kcal/g mole
Heat of Sublimation	14.1 kcal/g mole

Chemical Properties

Concentrated and pure hydrogen peroxide is a clear, colourless liquid have a same viscosity and appearance like water, but has a greater density, hydrogen peroxide also does have a greater bond strength.

EXOTHERMIC

Hydrogen peroxide is very stable compound at dilute state also. It has a remarkable property of breaking up itself to water and oxygen very easily, when hydrogen peroxide decomposes it gives up considerable amount of energy and the reaction is termed as exothermic. The reaction is shown below:

Hydrogen peroxide like ozone is very good oxidizing agent because it looses its oxygen atom very easily; some reactions with equations will prove the above system.

A: Hydrogen Peroxide reacts with black lead sulphide to white lead sulphide

Lead Sulphide + Hydrogen Peroxide ⟶ Lead Sulphate + Water

$$PbS + 4H_2O_2 \longrightarrow PbSO_4 + 4H_2O$$

B: Hydrogen Peroxide converts Ferrous salts to ferric salts

Ferrous Sulphate + Sulphuric Acid + Hydrogen Peroxide ⟶ Ferric Sulphate + Water

$$2FeSO_4 + H_2SO_4 + H_2O_2 \longrightarrow Fe_2(SO_4)_3 + 2H_2O$$

C: Hydrogen Peroxide converts Sulphurous Acid to Sulphuric Acid

Sulphurous Acid + Hydrogen Peroxide ⟶ Sulphuric Acid + Water

$$H_2SO_3 + H_2O_2 \longrightarrow H_2SO_4 + H_2O$$

D: Hydrogen Peroxide converts Sodium Nitrite to Sodium Nitrate

Sodium Nitrite + Hydrogen Peroxide ⟶ Sodium Nitrate + Water

$$NaNO_2 + H_2O_2 \longrightarrow NaNO_3 + H_2O$$

E: - Hydrogen Peroxide liberates iodine from potassium iodide

Potassium Iodide + Hydrogen Peroxide ⟶ Potassium Hydroxide + Iodide

$$2KI + H_2O_2 \longrightarrow 2KOH + I_2$$

Hydrogen peroxide has also a property of bleaching action and is also used for bleaching delicate materials. Hydrogen peroxide is applied to human hair the carotene dye is bleached.

Bhaviour Towards Other Oxidising Agents

Hydrogen peroxide reacts with oxidizing agents in such way that one oxygen atom liberated by the reaction and the oxidizing agent forms an oxygen molecule. This can be termed as the reducing property of hydrogen peroxide also. This can by the following reactions:

$$
\begin{array}{rcl}
H_2O_2 & \longrightarrow & H_2O + O \\
O_3 & \longrightarrow & O_2 + O \\
O + O_2 & \longrightarrow & O_2 \\
\hline
H_2O_2 + O_3 & \longrightarrow & H_2O + 2O_2
\end{array}
$$

Hydrogen peroxide also reacts with silver peroxide to produce metallic silver water and oxygen.

Silver Peroxide + Hydrogen Peroxide ⟶ Silver + Water + Oxygen

$$Ag_2O + H_2O_2 \longrightarrow 2Ag + H_2O + O_2$$

Chlorine is transformed into hydrogen chloride when it reacts with hydrogen peroxide.

Chlorine + Hydrogen Peroxide ⟶ Silver + Oxygen

$$Cl_2 + H_2O_2 \longrightarrow 2HCl + O_2$$

Hydrogen peroxide can decompose acidified potassium permanganate solution

Hydrogen Peroxide + Potassium Permanganate + Sulphuric Acid ⟶ Potassium Sulphate + Manganese Sulphate + Water + Oxygen

$$5H_2O_2 + 2KMnO_4 + 3H_2SO_4 \longrightarrow K_2SO_4 + 2MnSO_4 + 8H_2O + 5O_2$$

Reduction Property

Hydrogen peroxide acts as a reducing agent, which can be proved in the following reaction:

$H_2O_2 \rightarrow 2H + (O - O)\ 2O^-$ (oxide ion)

(Oxidation Process) $(O - O) \xrightarrow{\text{Gains electron 2e}} 2O^-$

(Reduction Process) $(O - O) \xrightarrow{\text{Looses electron 2e}} O_2 + 2\,(-)$ (oxygen molecule)

Tests of Hydrogen Peroxide

1. Hydrogen peroxide turns starch iodide paper blue, by liberating iodine from potassium iodide.
2. Hydrogen peroxide and potassium dichromate is shaken with dilute sulphuric acid and eather the eather layer turns blue due the formation of perchromic acid.
3. Pertitanic acid is formed when hydrogen peroxide reacts with titanium dioxide (TiO_2)

We shall now discuss about the differences between the water and the hydrogen peroxide, have been discussed in the table given below:

Water	Hydrogen Peroxide
1. Water is known as the monoxide of hydrogen.	1. It is the peroxide of hydrogen.
2. Water is termed as the natural oxide of hydrogen.	2. Hydrogen peroxide is the oxide of water $2H_2O \rightarrow 2H_2 + O_2$
3. Water has a boiling point of 100°C without nay explosive decomposition.	3. Boiling point of hydrogen peroxide is 150°C and it boils with explosive decomposities.
4. Water is neutral in nature (neither acidic nor alkaline)	4. Hydrogen peroxide behaves acid.
5. Water is an excellent solvent	5. Hydrogen peroxide does not appear as good solvent.
6. Distil water or pure water does not act as a bleaching agent.	6. Hydrogen peroxide acts as a good bleaching agent.
7. Water forms $CuSO_45H_2O$ $FeSO_4.7H_2O$	7. Hydrogen Peroxide forms $(NH_4)_2$, $SO_4.H_2O_2$.
8. Water liberates hydrogen and oxygen on electrolysis.	8. Hydrogen peroxide liberates oxygen and water on decomposition.
9. Water not a germicide or disinfectant.	9. Hydrogen peroxide is both germicide and as well as disinfectant.
10. Water is produced from hydrogen and oxygen.	10. Hydrogen peroxide is produced from sodium peroxide and sulphuric acid.
11. As water is neutral it cannot be used as oxidizing or reducing purpose.	11. Hydrogen peroxide can be used both as a reducing and oxidizing agent.

Determination of Strength of Hydrogen Peroxide

It is very important to determine the strength of hydrogen peroxide. The strength of hydrogen peroxide is expressed in the terms of volumes oxygen at S.T.P. (standard temperature and pressure that one volume of hydrogen peroxide gives out on heating). The following process can do the quantitative measurement, where the volume of hydrogen peroxide is given.

Example: Calculate the strength of the hydrogen peroxide in percentage marked by volumes by definition 50 volumes of hydrogen peroxide means that 1 volume of this grade hydrogen peroxide gives 50 volumes of oxygen at S.T.P. or litre of 50 volumes of hydrogen peroxide gives 50 litres of oxygen at S.T.P. or 50/22.4 mol of oxygen.

Form the equation

$$2H_2O_2 \longrightarrow 2H_2O + O_2$$

We find that 1 mode of oxygen = 2 mole of hydrogen peroxide

$$\therefore \frac{50}{22.4} \text{ mole of oxygen} \equiv \frac{2\times 50}{22.4} = 4.46 \text{ mole of hydrogen peroxide}$$

Thus 50 volumes of hydrogen peroxide contain 4.46 mole per litre in the solution or it is 4.46 mole per litre in the solution or it is 4.46M = (4.46 × 2) that is 8.92N.

Thus strength = molarity × molar mass.

$151.64g^{L-1}$ – 4.46 × 34 (Molar mass is molecular mass) or 15.164% is the actual strength of the hydrogen peroxide.

Structure of Hydrogen Peroxide

Hydrogen peroxide has the skid type configuration; the molecules are not situated on the same plane (see Fig. 9.16). In the figure two states of hydrogen peroxide is shown (a) the gaseous state and (b) the crystalline state. In the crystalline state the dihedral angle is reduced to 91.3° and the angle between the oxygen bonds is much increased at 100°.

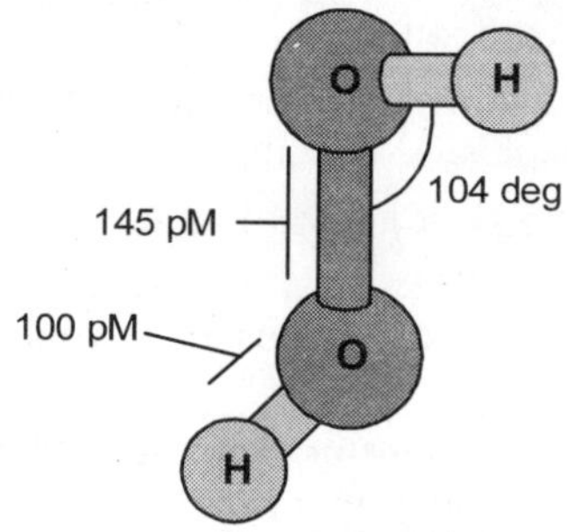

Fig. 9.16. The structure of Hydrogen Peroxide

The O – O bond linkage has derived from a p - p overlap of the p-orbitals of the two bonding oxygen atoms (O – O). Each of the two bonding of hydrogen and oxygen that is (H-O) is cause (by a s-p overlap between s-p overlap between a s-orbital of hydrogen and p-orbital of oxygen atom. This will be clear from Fig. 9.17.

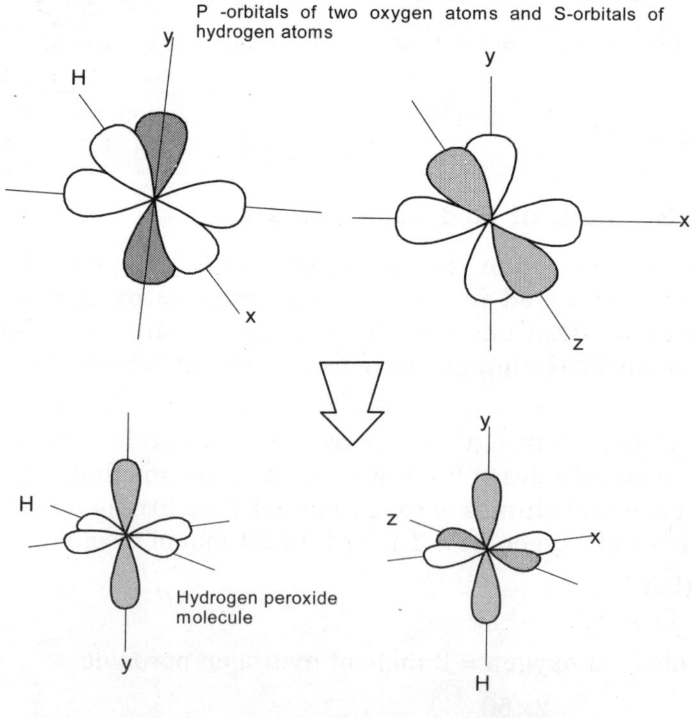

Fig. 9.17. The Hydrogen peroxide orbitals

CHAPTER

10

Nitrogen and Its Compounds

Introduction

Lavoisier first isolated nitrogen in the years (1743-1794). He named it as azote, which was not a supporter of life. Daniel Rutherford (1772) named it *nephatic air* meaning poisonous air, Chaptal in 1822 named it nitrogen the producer of nitrates.

Occurrence

Nitrogen occurs in atmosphere 78%by volume, which according to Lavoisier were 4/5th of the total atmosphere. Nitrogen is the most essential element for the nutrition for the plants. The plants absorb nitrogen through the roots (nodules) with the help of notrosomonas bacteria. Besides nature nitrogen also occurs in many different chemical fertilizers, chemical compounds and also in animal proteins. In this chapter we shall be discussing the different properties of nitrogen its lab preparation and also industrial preparation, the nitrogen cycle etc.

Nitrogen Cycle

It is the repetitive process in nature by mans of which nitrogen is synthesized from its parent constituents and these parent constituents are recovered from nitrogen. This is an organic, inorganic biological process governed by both plant another living organisms.

Nitrogen exists in nature in different forms both element and compounds also ionic and radical forms.

The exchange-conversion taking place in the nitrogen cycle is rather complex reactions for elementary chemistry, because biological reactions are involved. So an outline diagram in the form of a cycle chart is shown in the Fig. 10.1.

Nitrogen cycle also occurs in the oceans. But the estimation of proper information about this is not much; still we see that 20% of the nitrogen fixed on earth occurs in sea. This nitrogen seems to be primarily fixed by the ocean-bacteria. It has been estimated that a single cyano- bacterium known broadly as trichodesmium could fix as much as 1050 tons of nitrogen per day. This nitrogen fixation is the major function of the nitrogen cycle; the other processes are the ammonification (fixation of nitrogen in form of ammonia) nitrification (extraction of nitrogen from nitrates). De-nitrification (conversion of nitrogen to nitrates).

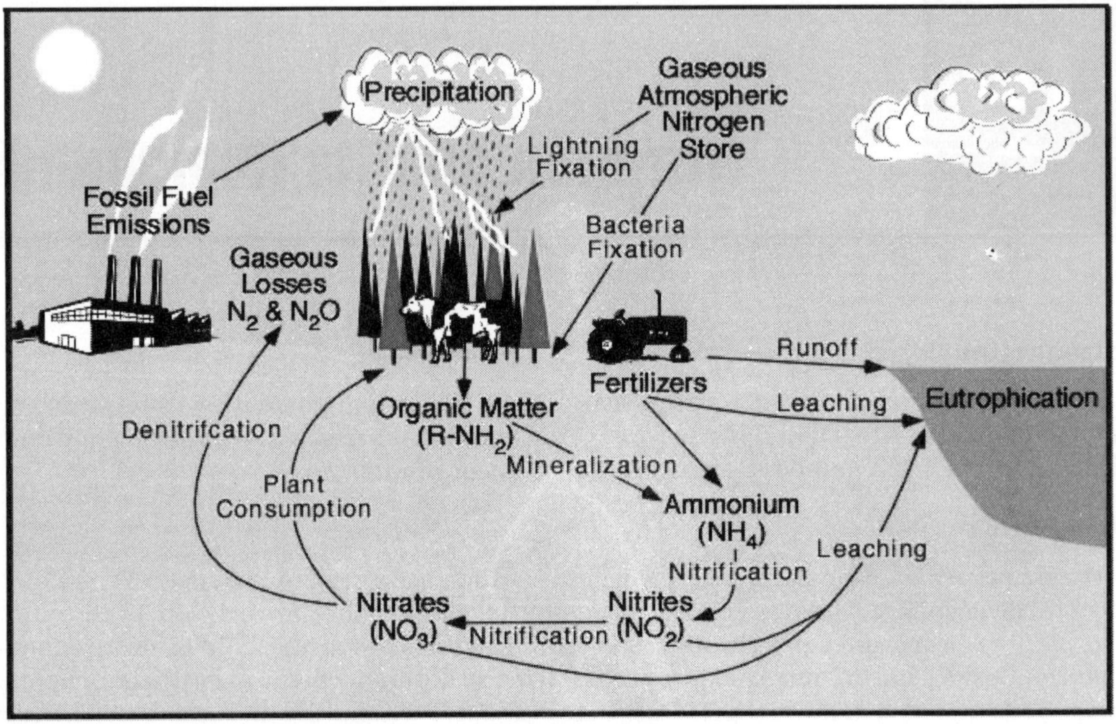

Fig. 10.1. The Nitrogen Cycle

Laboratory Preparation of Nitrogen

Theory: Nitrogen in the laboratory is prepared from the reaction of sodium nitrite and ammonium chloride, in the form of an equivalent mixture.

Experiment: An equivalent mixture (ratio 1: 1) is taken in a round bottom flask fitted with a cork and a delivery tube. The mixture is kept in a strong aqueous state with water; the flask is heated gently, as soon as the nitrogen starts liberating the heat is gradually reduced. The gas is collected by the downward displacement of water. (See Fig. 10.2 for the apparatus set up).

Reaction Equation

Ammonium Chloride + Sodium Nitrate ⟶ Ammonium Nitrate + Sodium Chlorde

$$NH_4Cl + NaNO_2 \longrightarrow NH_4NO_2 \Downarrow + NaCl \Downarrow$$

Ammonium Nitrate	$\longrightarrow$	Nitrogen	+	Water
$2NH_4NO_2$	$\longrightarrow$	$N_2 \Uparrow$	+	$4H_2O[NH_4OH]$

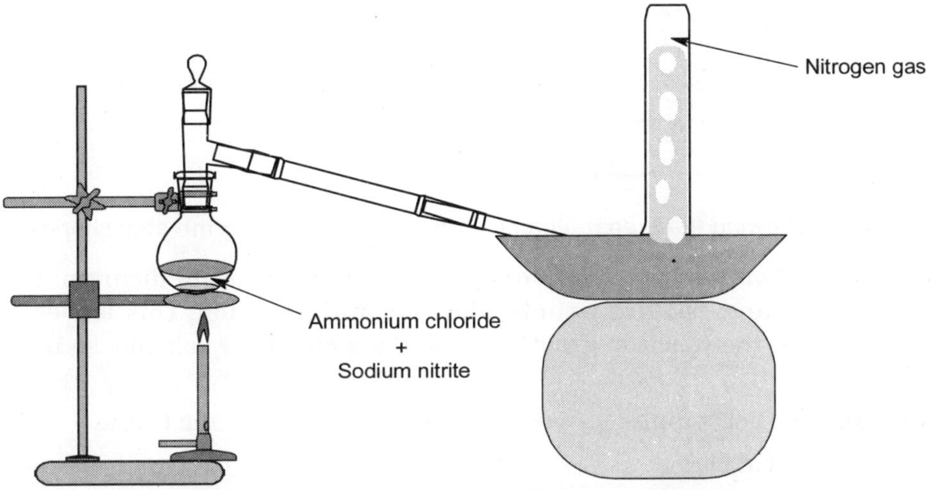

Fig. 10.2. Showing the lab preparation of Nitrogen

Apparatus	Reagents	Temperature	Pressure
Round Bottomed flask, clamp stand, delivery tube, gas jar beehive self. Corks etc.	Ammonium Chloride Sodium Nitrate.	100°C to 110°C	755mm of Hg.

Precaution: 1. The flask should not be heated very strongly and for long time because the evolved nitrogen will start reacting with the hydrogen present in the water producing ammonia, a compound of nitrogen and hydrogen.

2. The delivery tube must be removed first before removing the burner, because the sudden removal of the burner will cause a partial vacuum, which will cause the water to rush in and break the flask.

Other Methods

1. By removal of Oxygen from Air: In a small crucible as shown in the Fig. 10.3, floating in water a piece of white phosphorous is placed and is ignited by touching it with a red hot iron rod, immediately it is covered with a gas jar. The phosphorous reacts with oxygen present in the air to form phosphorous pentoxide, which is dissolved in water to form phosphoric acid and the left over gas is nitrogen.

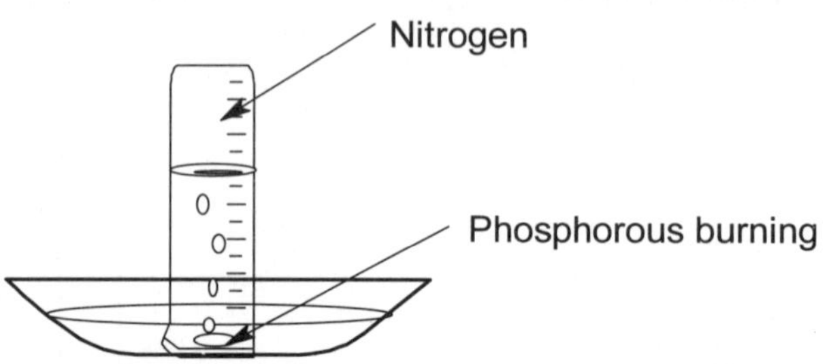

Fig.10.3. Showing the preparation of Nitrogen by consuming the atmosheric oxygen

2. By heating Ammonium Dichromate : When ammonium dichromate is ignited, violent flash occurs and red fire irrupts like volcanic irruption, this leaves behind the green chromic oxide, the reaction exactly looks like a chemical volcano, so it is named as volcano experiment.

$$\text{Ammonium Dichromate} \longrightarrow \text{Nitrogen} + \text{Chromic Oxide} + \text{Water}$$
$$(NH_4)_2Cr_2O_7 \qquad N_2 + Cr_2O_3 + 4H_2O$$

3. Nitrogen obtained from Air : Oxygen present in the air can be also removed by passing the air over hot iron fillings, which combines with oxygen and *forms* iron oxide and the nitrogen is separated (the apparatus shown in the Fig. 10.3).

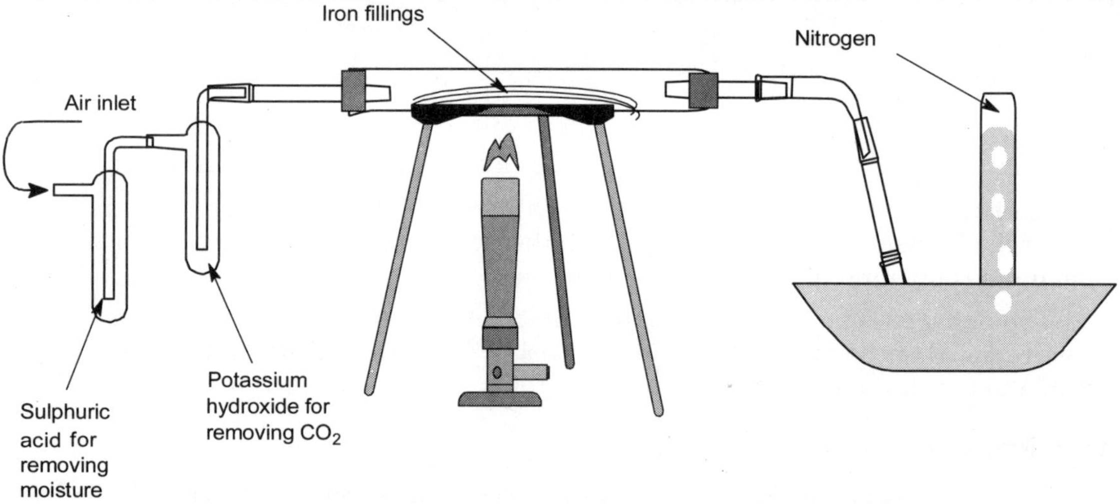

Fig. 10.4. Showing the Preparation of Nitrogen from Air

Industrial Process of Nitrogen Preparation

Nitrogen in industry is prepared by the fractional distillation of liquid air, by the use of rectifying column. Liquid air usually 55% to 63% of oxygen. The boiling point of liquid oxygen is –183°C and that of nitrogen is –195°C so the oxygen is liberated (separated) first and then the nitrogen is separated. It is not very much difficult to separate nitrogen from oxygen by fractional distillation. Nitrogen from liquid air however contains rare or inert gases as impurities, which are successively separated according to their boiling.

The apparatus here used is known as the *Coffey's Still* (see Fig. 10.5) consists of two columns made up of wooden walls with an inner steal lining, the two columns are provided with copper plates for collecting the condensate. The two columns are placed side by side the column B is the analyzer column and the column to which an outlet tube E is fitted. This conveys the vapour to the rectifier column C; the plates of the rectifier and the analyzer are both perforated (small perforations) a safety valve. A zigzag large pipe F is fitted inside the rectifier. The rectifying column has an outlet tube, which drains out the final condensate to the hot feint for vaporization.

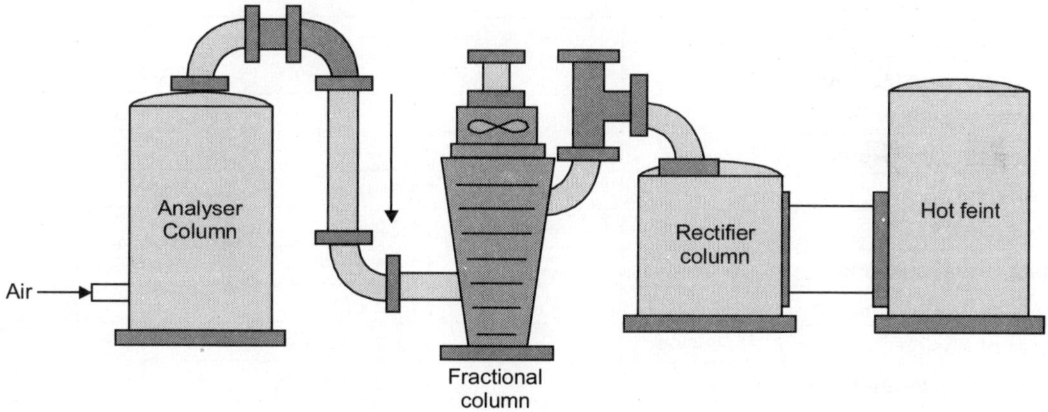

Fig. 10.5. Showing Coffey's still

At the beginning of the distillation air is introduced into the analyzer column in this column the water vapours are absorbed and the air is send to the fractionation column where the oxygen is separated from the liquid air (air is liquefied in this column). The rectification column separates the oxygen from the nitrogen in the liquid State, and the liquid nitrogen is send to the hot feint for vaporization. The total operation is very much complicated, which is not easy to discuss in this book.

Properties of Niterogen

Elementary Nitrogen: Nitrogen is chemically inert at normal temperature (27°C) to all elements and reagents except lithium. But when heated it directly combines with oxygen, boron, silicon and calcium.

Active Nitrogen: Under reduced pressure inside a discharge tube, when electrical discharge done the nitrogen comes to an unstable but active state from which it again returns back to its original molecular state with a yellowish after glow. Active nitrogen is much reactive than ordinary molecular nitrogen, which has been proved below:

Properties	Value
Melting Point	–210°C
Boiling Point	–197°C
Density	0.96737(g), 0.804(L), 1.0265(s)
Ionization Potential	
I	14.52eV
II	30.0
III	46.2
IV	77.0
V	97.04
Atomic Radius	70Å
Ionic Radius	1.72Å
Pauling's Crystal Radius	1.1Å
Electronegativity	3.0
Stater's Atomic Number	.57
(Zeff/mol)	3.65
(Zeff/mol/r2)	11.32
Heat of vapourization	133.0 cal/mol
Heat of Transformation	54.72 cal/mol
Heat of Fusion	172.3 cal/mol
Critical Temperature	126.26 ± 04°K
Critical Pressure	33.4 ± 0.2atm

1. In this state the nitrogen reacts with phosphorous, when nitrogen and phosphorous are subjected to electrical discharge under reduced pressure, phosphorous nitride is formed.

$$\text{Phosphorous} + \text{Nitrogen} \xrightarrow{\text{electrical discharge}} \text{Phosphorous Nitride}$$

$$2P_2 + N_2 \longrightarrow 2P_2N$$

POISON

2. At high temperature under same condition active nitrogen also reacts with sodium to form sodium Nitride

$$\text{Sodium} + \text{Nitrogen} \xrightarrow{\text{electrical discharge}} \text{Sodium Nitride}$$

$$6Na + N_2 \longrightarrow 2Na_3N$$

3. Nitrogen also combines with its own oxide and forms molecular nitrogen

$$\text{Nitrogen (atomic)} + \text{Nitric Oxide} \longrightarrow \text{Nitrogen (molecular)} + \text{Oxygen}$$

$$N_2 + 2NO \longrightarrow 2N_2 + O_2$$

Hydrocarbons (compounds of carbon and hydrogen) react with active nitrogen forming hydrocyanic acid at a considerable high temperature.

Molecular Properties

At standard temperature and pressure (S.T.P.) elemental nitrogen exists as gas, which has density of 0.96737g./litre. This value indicates that the molecular formula is N_2. The molecule N_2 at its ground state has a magnetic susceptibility of -0.430×10^{-6} at 25°C and therefore has no resultant electronic angular momentum either the orbital or the spin var The electronic formula :N: :N: indicates a triple covalent bond N – N. The molecule has electronic structure according to the molecular orbital theory and Valence Bond Theory complete electronic structure is

$$1s^2, 1s^2, \delta s^2, \delta^2, \delta 2^2 px, \pi 2^2 py, \pi 2^2 pz$$

The inner atomic forces in the nitrogen molecule are very high. The energy of dissociation the nitrogen molecule is 225.8Kcal/mol. Spectroscopical studies indicate that at ordinary temperature molecular nitrogen consists of molecules with symmetrical and asymmetric nuclear spins in the ratio 2 : 1 because the nitrogen molecule is both very stable and highly symmetrical having small intermolecular forces. The phase change of nitrogen is shown below:

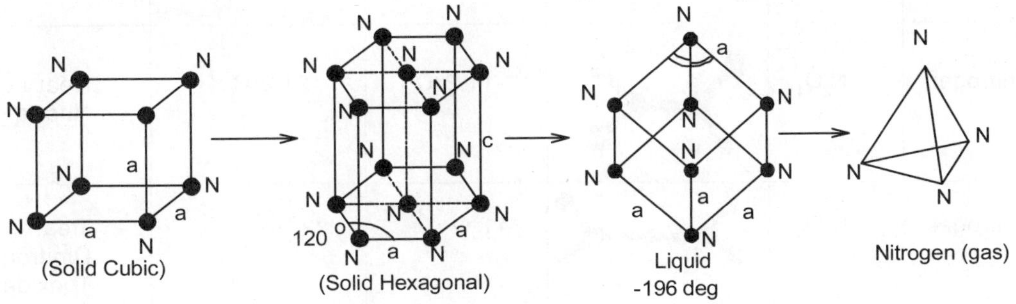

Fig. 10.6.

Nuclear Properties

Nitrogen as it occurs in nature has two isotopes ^{14}N and ^{15}N in the abundance ratio of 99.635 to 0.365. In addition the radioactive isotope ^{12}N and ^{13}N ^{16}N ^{17}N have first been made 1 variety of nuclear reactions. The first two are positron emitters and half-lives of 0.125 seconds and 9.932 mines respectively. ^{16}N and ^{17}N are electron emitters and half lives of 7.35 sec 4,14 sec respectively. Unfortunately none of these have sufficient long life for conveniently used as tracer ^{15}N has been used as a tracer due to its extended life, by using nitrogen in which ^{15}N has been concentrated by mass spectroscopic technique.

Compounds of Nitrogen

There are wide range of compounds of nitrogen at high temperature or under reduced pressure at electrical discharge, nitrogen forms many compounds, which we are going to discuss.

Compounds of Nitrogen and Oxygen

The chart below shows the list of compounds of nitrogen and oxygen with their physical properties in brief.

Name	Formula	Structure	Melting Point	Boiling Point	Method of Preparation
Nitrous Oxide	N_2O		–90.8°C	–88.5°C	Heat ammonium sulphate
Nitric Oxide	NO		–163.6°C	–151.7°C	Nitric acid & Copper
Dinitrogen Trioxide	N_2O_3		–103°C	3.5°C	Condense NO and NO_2
Dinitrogen	N_2O_4	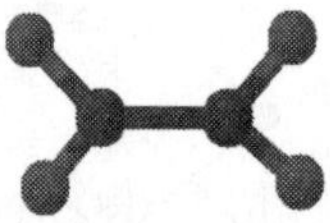	–11.2°C	21.2°C	Heat Lead Nitrate
Dinitrogen	N_2O_5		41°C	------	Treat Dinitrogen Trioxide with ozone

We shall now discuss the Compounds of Nitrogen in detail.

Nitric Oxide (NO): This compound is generally obtained from the heated electrical discharge of nitrogen and oxygen, which yields lots of energy

$$\text{Nitrogen} + \text{Oxygen} \longrightarrow \text{Nitric Oxide} + \text{Heat} \Uparrow$$

$$N_2 + O_2 \longrightarrow 2NO + \text{Heat}$$

The Gas formed in the process needs much energy or heat. The process often in the lightning:

The process below is a very common process often used in the labs for detecting the nitrite compounds; involve the synthesis of Nitric Oxide.

$$\text{Ferrous Sulphate} + \text{Sodium Nitrite} + \text{Sulphuric Acid} \longrightarrow \text{Nitric Oxide} + \text{Ferric Sulphate} + \text{Sodium Sulphate} + \text{Water}$$

$$2FeSO_4 + 2NaNO_2 + 2H_2SO_4 \longrightarrow 2NO + Fe_2(SO_4)_3 + Na_2SO_4 + H_2O$$

Very pure Nitric oxide can be prepared by the action of sulphuric acid and potassium nitrate.

$$\underset{\text{Nitrate}}{\text{Potassium}} + \text{Mercury} + \underset{\text{Acid}}{\text{Sulphuric}} \longrightarrow \underset{\text{Oxide}}{\text{Nitric}} + \underset{\text{Sulphate}}{\text{Potassium}} + \underset{\text{Sulphate}}{\text{Mercuric}} + \text{Water}$$

$$2KNO_3 + 6Hg + 4H_2SO_4 \longrightarrow 2NO + K_2SO_4 + 3Hg_2SO_4 + 4H_2O$$

The above reaction is used for the estimation of Nitric oxide in the Lunge Nitro meter.

Nitric oxide is a colourless gas slightly heavier than air and sparingly soluble in water. The gas is sparingly soluble in water. The gas is freely soluble in cold ferrous sulphate solution forming a dark brown liquid of ferric nitrososulphate ($FeNOSO_4$). It also dissolves readily in an alkaline sulphide (NaOH + FeS) forming nitrososulphate of sodium $Na_2SO_3(NO)_2$. Nitric oxide is stable compound unless heated up to 1200°C. It breaks into nitrogen and oxygen.

$$\text{Nitric Oxide} \longrightarrow \text{Nitrogen} + \text{Oxygen} + \text{energy 43000 cal}$$

$$2NO \longrightarrow N_2 + O_2$$

Nitric Oxide is reduced to nitrogen when passes over red-hot copper nickel or iron.

Composition: The volumetric composition of nitric oxide can be demonstrated by showing that volume of nitrogen, which remains after the removal of oxygen from a known volume of gas by a heated iron wire, is equal to one half of original volume of nitric oxide. Gravimetrically, the composition of nitric oxide can be demonstrated by heating a weighted quantity of finely divided nickel in the gas and then weighing the residual nitrogen and the molecular weight of the gas calculated from its relative density, the molecular formula is readily deduced. The molecular formula from the contribution of the gas is shown below:

The molecule contains an *odd* electron, which, accounts *for* its chemical activity.

Nitrogen Dioxide (Nitrogen Peroxide) (NO_2) It is prepared by the action of nitric oxide and oxygen under heated electrical discharge,

$$\text{Nitric Oxide} + \text{Oxygen} \longrightarrow \text{Nitrogen Dioxide}$$

$$2NO + O_2 \longrightarrow 2NO_2$$

Nitrogen dioxide can also be prepared in the laboratory by heating lead nitrate (Fig. 10.7).

$$\text{Lead Nitrate} \longrightarrow \text{Nitrogen Dioxide} + \text{Oxygen} + \text{Lead Oxide}$$

$$2Pb(NO_3)_2 \longrightarrow 4NO_2 + O_2 + 2PbO$$

The gaseous product is passed through the freezing mixture kept surrounding the reservoir, the gas condenses as yellow liquid and the oxygen passes out. The nitrogen dioxide in room temperature is a brown gas. Which, when liquefied becomes yellow and has a boiling point of 21.2°C. If solidified to a colourless crystal at 10°C. The density of nitrogen dioxide vapour decreases with the temperature, which indicates that the gas consists of a mixture of N_2O_4 molecule and NO_2 molecules. This is supported by the fact that the colour of the vapour darkens with the rise of the temperature and this is due to decrease of density.

Nitrogen Tetraoxide ⟶ Nitrogen Dioxide

$$N_2O_4 \longrightarrow 2NO_2$$

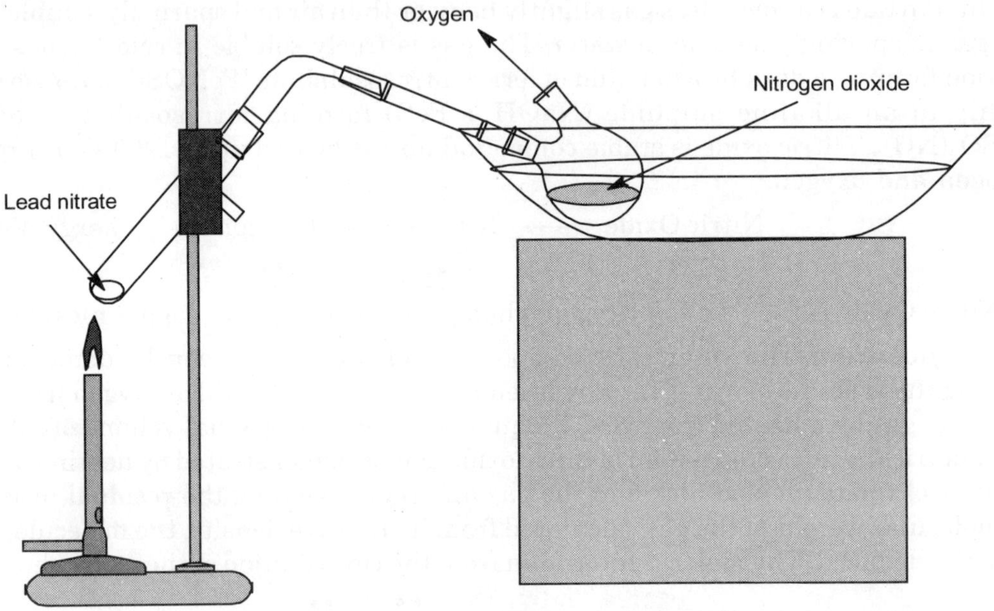

Fig. 10.7. The Lab Preparation of Nitrogen Dioxide

Below 140°C nitrogen becomes paler and its density continues to decrease further due to dissociation* into nitric oxide and oxygen. The vapour of the dioxide supports the combustion of a taper and lights a glowing splinter. Phosphorous and carbon already burning vigorously continues to burn in the gas. Hot sodium burns in the gas and potassium bursts into flame, nitrogen dioxide acts as oxidizing agent (donor of oxygen). Copper and iron, when heated strongly in the gas from their oxides setting free nitrogen, hydrogen sulphide is also set free from sulphur.

Nitrogen Dioxide + Hydrogen Sulphide ⟶ Nitrogen Oxide + Water + Sulphur

$$NO_2 + H_2S \longrightarrow NO + H_2O + S$$

Finely divided copper reacts with nitrogen dioxide to form cuprous oxide at room temperature

Copper + Nitrogen Dioxide ⟶ Cupric Oxide + Nitric Oxide

$$2Cu + NO_2 \longrightarrow Cu_2O + NO$$

Dinitrogen Trioxide (N_2O_3): It is stable only in liquid state. It dissolves immediately into a mixture of nitric oxide and nitrogen dioxide.

*Dissociation : It is the process of breaking up substances on heating and recombines them on cooling.

Carbon Monoxide + Nitrogen Dioxide ⟶ Carbon Dioxide + Nitric Oxide

$$CO + NO_2 \longrightarrow CO_2 + NO$$

At room temperature the gas contains only 10% of the undissociated molecules of N_2O_3. A mixture of nitric oxide and nitrogen dioxide is proper proportions is required to produce dinitrogen trioxide also known as *nitrogen sequioxide and nitrous anhydride.* This can be prepared by the reaction of dilute (60%) nitric acid with arsenious oxide or starch.

Arsenious Oxide + Nitric Acid + Water ⟶ Dinitrogen Trioxide + Arsenic Acid

$$As_2O_3 + 2HNO_3 + 2H_2 \longrightarrow ON_2O_3 + 2H_3AsO_4$$

The gas set free when cooled in a freezing mixture condenses into a deep blue liquid. The gas gets rapidly absorbed in potassium or sodium hydroxide solution forming nitrite and with sulphuric acid forming nitrososulphuric acid.

Potassium Hydroxide + Dinitrogen Trioxide ⟶ Potassium Nitrite + Water

$$KOH + N_2O_3 \longrightarrow KNO_2 + H_2O$$

Dinitrogen Trioxide + Sulphuric Acid ⟶ Nitrososulphuric Acid + Water

$$N_2O_3 + 2H_2SO_4 \longrightarrow 2SO_2(OH).O.NO + H_2O$$

Nitrogen Pentoxide (N_2O_5) : This can be prepared by the action of chlorine on silver nitrate.

Silver Nitrate + Chlorine ⟶ Silver Chloride + Nitrogen Pentoxide + Oxygen

$$4AgNO_3 + 2Cl_2 \longrightarrow 4AgCl + 2N_2O_5 + O_2$$

When ozone passes through nitrogen dioxide under low temperature nitrogen pentoxide is produced.

Nitrogen Dioxide + Ozone ⟶ Nitrogen Pentoxide + Oxygen

$$NO_2 + O_3 \longrightarrow 2N_2O_5 + O_2$$

Nitrogen pentoxide is also prepared from the careful dehydration of a very pure nitric acid with phosphorous pentoxide, when phosphorous pentoxide is carefully added to the acid. The reaction vessel is cooled in freezing mixture, a thick ropy mass is developed which is distilled in retort, and freezing mixture collects nitrogen pentoxide crystals.

The nitrogen pentoxide vapourizes with the rising of temperature, which is 31°C, with some decomposition of course. It decomposes into nitrogen dioxide and oxygen.

Nitrogen Pentroxide ⟶ Nitrogen Dioxide + Oxygen

$$2N_2O_5 \longrightarrow 4NO_2 + O_2$$

It is the anhydride of nitric acid, when dissolved in water it produces nitric acid with evolution of heat. It is very powerful oxidizing agent phosphorous, sodium and potassium if wormed burns in liquid nitrogen pentoxide. Charcoal is not attached by the liquid nitrogen pentoxide but burns brilliantly in the vapour.

The crystalline nitrogen pentoxide as seen by the X-ray analysis has a structure, which is familiar to nitronium nitrate $(NO_2)^+ (NO_3)^-$. In the gaseous state the nitrogen pentoxide has a chemical formula of

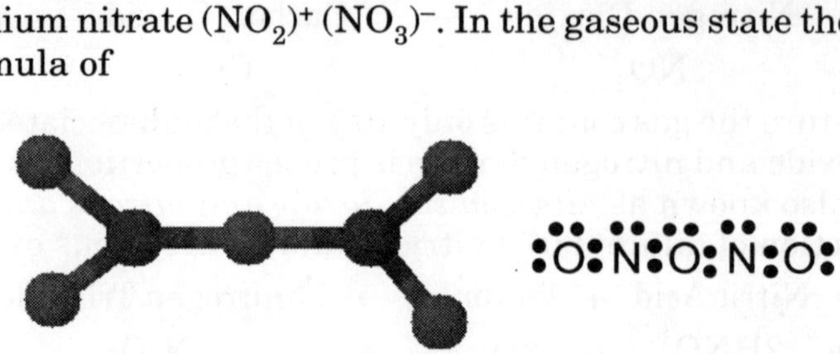

Nitrogen Trioxide (NO_3): This is the compound of nitrogen, which is produced by the reaction of nitrogen dioxide and ozone, it is however stable only under a temperature of –140°C and decomposes at higher temperature into nitrogen dioxide and oxygen.

Nitrogen Pentoxide + Ozone $\xrightarrow{-140°C}$ Nitrogen Trioxide + Oxygen

$$2N_2O_5 + 2O_3 \longrightarrow 4NO_3 + 2O_2$$

The reaction becomes reverse when the temperature becomes above –140°C

Nitrogen Trioxide $\xrightarrow{\text{-Above -140°C}}$ Nitrogen Dioxide + Oxgyen

$$2NO_3 \longrightarrow 2NO_2 + O_2$$

Nitrous Oxide (N_2O): *Joseph Priestely* first invented Nitrous oxide in the year 1772, he made it by removing oxygen from nitric oxide by moist iron fillings.

Nitric Oxide + Water + Iron ⟶ Nitrous Oxide + Ferric Hydroxide

$$2NO + H_2O + Fe \longrightarrow N_2O + Fe(OH)_2$$

But in the ear 1799 *Sir Humphrey Davy* prepared mostly pure nitrous oxide by heating ammonium nitrate. This gas causes a hysteric effect when inhaled in large amount can cause death, it is used as anesthetic when used with oxygen.

Laboratory Preparation of Nitrous Oxide

Theory: Nitrous Oxide is prepared in the laboratory by heating a mixture of Sodium Nitrate and Ammonium Sulphate in a test tube, the gas is collected by the down ward displacement of worm water or mercury.

(The apparatus set up is shown in Fig. 10.8).

Ammonia Sulphate + Sodium Nitrate ⟶ Sodium Sulphate + Ammonium Nitrate

$$(NH_4)_2SO_4 + 2NaNO_3 \longrightarrow Na_2SO_4 \Downarrow + 2NH_4NO_3 \Downarrow$$

Ammonia Nitrate ⟶ Nitrous Oxide + Water

$$2NH_4NO_3 \longrightarrow N_2O \Uparrow + 2H_2O \Uparrow$$

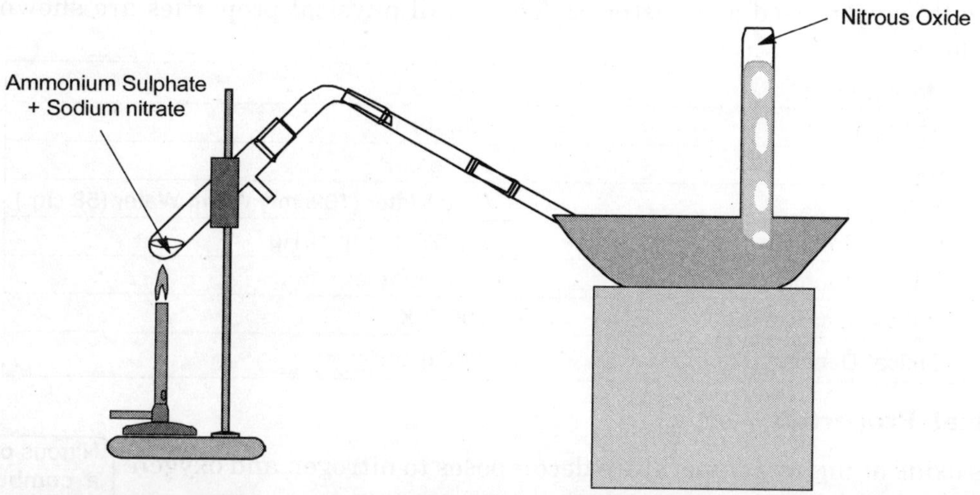

Apparatus	Reagents	Temperature	Pressure
Test tube, delivery tube, gas jar, burner, etc.	Sodium Nitrate, Ammonia Sulphate.	185°C to 190°C	Normal

Fig. 10.8. The Lab Preparation of Nitrous Oxide

Precaution

1. The gas should be collected by the drowned displacement of warm water, or mercury.
2. The heating should be done very carefully below 200°C temperature.
3. The delivery tube should be removed first before removing the burner otherwise the water will be sucked in the tube, which will cause crack.
4. Ammonium Chloride instead of ammonium sulphate cannot be used because the chlorine evolved will violently react with nitrous oxide and may cause damage.

Other Methods of Preparation of Nitrous Oxide

1. *By removing oxygen from Nitric Acid:* Nitrous oxide can be prepared by removing oxygen from nitric acid by means of stannous chloride and hydrochloric acid.
2. *By creating an Electrical Discharge between Nitrogen and Oxygen:* When an electrical discharge is carried out between a mixture of nitrogen and oxygen at very low temperature and pressure nitrous oxide is produced.

Physical Properties

Nitrous oxide is a colourless gas with a sharp sweet odour and taste. It is slightly heavier than air, soluble in cold water and in alcohol. The gas when inhaled causes a hysteric laughter so named as laughing gas, but long time inhaling can cause death, this gas mixed with oxygen used as anesthetic. The detail physical properties are shown in the table below:

Melting Point	–90.86°C
Boiling Point	–88.48°C
Density	.198g/cm^3
Solubility	Cold Water (109cm^3) Worm Water (58 cm^3)
Critical Pressure	2.137972 mm of Hg
Critical Volume	96.7 cm^3/mol^{-1}
Critical Temperature	309.6°K
Critical Density	.272 g/cm^3

Chemical Properties

Nitrous oxide at higher temperature decomposes to nitrogen and oxygen.

Nitrous oxide is a combustible gas

$$\text{Nitrous Oxide} \longrightarrow \text{Nitrogen} + \text{Oxygen}$$
$$2N_2O \longrightarrow 2N_2 + O_2$$

For this reason nitrous oxide is a supporter of combustion, when a glowing splinter is introduced it bursts into a flame. Also charcoal phosphorous etc. burns vigorously when introduced into a jar of nitrous oxide and liberates nitrogen.

Nitrous oxide reacts with phosphorous to produce phosphorous pentoxide:

$$\text{Nitrous Oxide} + \text{Phosphorous} \longrightarrow \text{Phosphorous Pentoxide} + \text{Nitrogen}$$
$$5N_2O + 2P \longrightarrow P_2O_5 + 5N_2$$

With carbon it forms carbon dioxide

$$\text{Nitrous Oxide} + \text{Carbon} \longrightarrow \text{Carbon Dioxide} + \text{Nitrogen}$$
$$N_2O + C \longrightarrow CO_2 + 2N_2$$

With magnesium it produces magnesium oxide:

$$\text{Magnesium} + \text{Nitrous Oxide} \longrightarrow \text{Magnesium Oxide} + \text{Nitrogen}$$
$$Mg + N_2O \longrightarrow MgO + N_2$$

Uses: Nitrous oxide is used as a mild anesthetic in minor operations.

Reaction with Halogen: Here we are going to discuss about the reactions of nitrogen compounds (mainly oxides) with halogens. Nitrogen trioxide when reacts with chlorine forms chlorine monoxide and nitrogen dioxide.

$$\text{Chlorine} + \text{Nitrogen Trioxide} \xrightarrow{25^\circ C} \text{Chlorine Monoxide} + \text{Nitrogen Dioxide}$$
$$Cl + NO_3 \longrightarrow ClO + NO_2$$

When atomic fluorine reacts with nitrogen dioxide, fluorine nitrite is produced as vapours.

$$\text{Fluorine} + \text{Nitrogen Dioxide} \longrightarrow \text{Fluorine Nitrite}$$
$$F + NO_2 \longrightarrow FNO_2$$

Similarly iodine and bromine reacts with nitrogen dioxide forming their respective halides.

Nitrogen Dioxide	+	Iodine	$\longrightarrow$	Iodine Nitrite
NO_2	+	I_2	$\longrightarrow$	$2INO_2$
Bromine	+	Nitrogen Dioxide	$\longrightarrow$	Bromine Nitrite
Br_2	+	NO_2	$\longrightarrow$	$2BrNO_2$

Nitrogen pentoxide reacts with sodium fluoride to form sodium nitrate and fluorine nitrite.

Nitrogen Pentoxide	+	Sodium Fluoride	$\longrightarrow$	Sodium Nitrate	+	Fluorine Nitrite
N_2O_5	+	NaF	$\longrightarrow$	$NaNO_3$	+	FNO_2

Nitrogen pentoxide reacts with iodine and liberates nitrogen with the formation iodine pentoxide.

Nitrogen Pentoxide	+	Iodine	$\longrightarrow$	Iodine Pentoxide	+	Nitrogen
N_2O_5	+	I_2	$\longrightarrow$	I_2O_5	+	N_2

Some metallic chloride like zinc chloride reacts with nitrogen dioxide to form zinc chloride and nitrosyl chloride.

Zinc Chloride	+	Nitrogen Dioxde	$\longrightarrow$	Zinc Nitrate	+	Nitrosyl Chloride
$ZnCl_2$	+	$4NO_2$	$\longrightarrow$	$Zn(NO_3)_2$	+	2NOCl

Iodine adds to nitric oxide to form nitric acid.

Iodine	+	Water	+	Nitric Oxide	$\longrightarrow$	Nitric Acid	+	Hydrogen Iodide
$3I_2$	+	$4H_2O$	+	2NO	$\longrightarrow$	$2HNO_3$	+	6HI

When Potassium Iodide reacts with nitrogen dioxide iodine is liberated

Potassium Iodide	+	Nitrogen Dioxide	+	Water	$\longrightarrow$	Potassium Hydroxide	+	Iodine	+	Nitric Oxide
2Kl	+	NO_2	+	H_2O	$\longrightarrow$	2KOH	+	I_2	+	NO

When nitric oxide reacts with bromine it forms nitrosyl bromide.

Bromine	+	Nitric Oxide	$\longrightarrow$	Nitrosyl Bromide
Br_2	+	2NO	$\longrightarrow$	2NOBr

Fluorine also reacts with nitric oxide forming nitrosyl fluoride.

Oxyacids of Nitrogen

The list of oxy-acids of nitrogen in the form of a table is given below with their chemical formula for student's convenience.

Hyponitrous Acid	$H_2N_2O_2$
Nitramide	H_2NO_2
Hyponitric Acid	$H_2N_2O_3$
Nitrous Acid	HNO_2
Pernitric Acid	HNO_4
Nitric Acid	HNO_3
Nitroxyl	HNO

Hyponitrous Acid ($H_2N_2O_2$): This acid is prepared by the reaction of hydrogen chloride and silver hyponitrite ($Ag_2N_2O_2$). It forms colourless crystals.

Hydrogen Chloride + Silver Hypochlorite ⟶ Hyponitrous Acid + Silver Chloride

$$2HCl + Ag_2N_2O_2 \longrightarrow H_2N_2O_2 + 2AgCl$$

Hyponitrous acid decomposes readily with explosion when gently heat, into nitrous oxide and water.

Hyponitrous Acid ⟶ Nitrous Oxide + Water

$$H_2N_2O_2 \longrightarrow N_2O + H_2O$$

EXPLOSIVE

Hyponitrous acid as known in chemistry is a weak acid, its detail molecular formula is not known but resents studies show that structure of the acid is HON = NOH. The molecular mass of the free acid corresponds with the formula $H_2N_2O_2$. When sulphuric acid reacts with hyponitrous acid it is dehydrated and nitrous oxide is liberated.

Hyponitrous Acid reacts with different metals like sodium to form sodium hyponitrite with the evolution of hydrogen.

Sodium + Hyponitrous Acid ⟶ Sodium Hyponitrite + Hydrogen

$$2Na + H_2N_2O_2 \longrightarrow Na_2N_2O_2 + H_2$$

Sodium Hyponitrite also decomposes to give nitrous oxide and sodium hydroxide when mixed with water.

Sodium Hyponitrite + Water ⟶ Nitrous Oxide + Sodium Hydroxide

$$Na_2N_2O_2 + H_2O \longrightarrow N_2O + 2NaOH$$

Sodium hyponitrite when heated in dry state is liberated as it decomposes.

Sodium Hyponitrite ⟶ Sodium Nitrite + Nitrogen + Sodium Oxide

$$3Na_2N_2O_2 \longrightarrow 2NaNO_2 + N_2\Uparrow + 2Na_2O$$

All the compounds of hyponitrous decompose on heating in dry state.

Nitrous Acid, HNO_2: Chemically it is unstable, it is prepared from the reaction of nitrogen trioxide and water nitrous acid is also produced

Nitrogen Trioxide + Water ⟶ Nitrous Acid

$$2N_2O_3 + 2H_2O \longrightarrow 4HNO_2$$

Laboratory Preparation on Nitrous Acid

Theory : In laboratory nitrous acid is prepared from the action of sulphuric acid and sodium nitrite. The sulphuric acid is ice cooled (See Fig. 10.9).

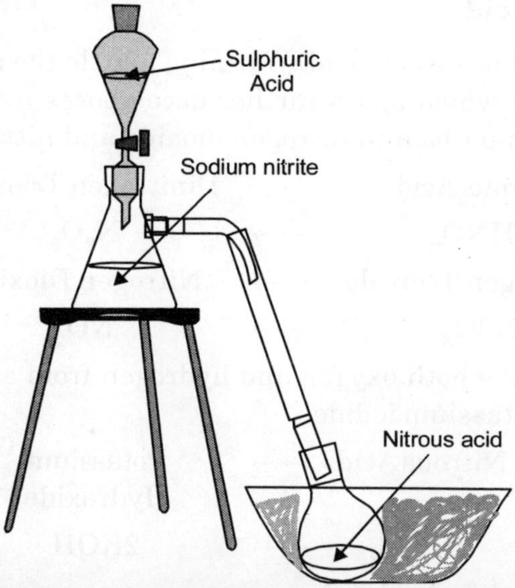

Apparatus	Reagents	Temperature	Pressure
Flask, delivery tube, thistle, funnel, gas jar etc.	Sodium Nitrite, Ice cold sulphuric acid	Below 0°C	Normal

Fig. 10.9. The Lab Preparation of Nitrous acid

Precaution : The reaction must be carried out under a temperature of below zero, because the nitrous acid produced will be decomposed at high temperature, so ice cooled sulphuric acid has been used.

Nitrous acid can also be prepared by the action of hydrochloric acid and sodium nitrite

$$\text{Sodium Nitrite} + \text{Hydrochloric Acid} \longrightarrow \text{Nitrous Acid} + \text{Sodium Chloride}$$

$$NaNO_2 + HCl \longrightarrow HNO_2 + NaCl$$

Nitrous acid is prepared from the reaction of potassium nitrite and sulphuric acid.

$$\text{Potassium Nitrite} + \text{Sulphuric Acid} \longrightarrow \text{Potassium Bi-Sulphate} + \text{Nitrous Acid}$$

$$KNO_2 + H_2SO_4 \longrightarrow KHSO_4 + HNO_2$$

Nitrous Acid is also prepared by the action of barium nitrite and sulphuric acid.

Here the Barium and the potassium sulphate is filtered off and the remaining clear solution initrous acid.

$$\text{Barium Nitrite} + \text{Sulphuric Acid} \longrightarrow \text{Barium Sulphate} + \text{Nitrous Acid}$$

$$Ba(NO_2)_2 + H_2SO_4 \longrightarrow BaSO_4 + 2HNO_2$$

Properties of Nitrous Acid

As known the nitrous acid is a weak acid. Warming a little the acid then decomposes into water and dinitrogen and which again further decomposes into dinitrogen trioxide and which again further decomposes into nitrogen dioxide and nitrogen dioxide.

Nitrous Acid ⟶ Dinitrogen Trioxide + Water

$$2HNO_2 \longrightarrow N_2O_3 + H_2O$$

Dinitrogen Trioxide ⟶ Nitrogen Dioxide + Nitric Oxide

$$N2O_3 \longrightarrow NO_2 + NO$$

Nitrous acid can remove both oxygen and hydrogen from any compound for example iodine is liberated from potassium iodide.

Potassium Iodide + Nitrous Acid ⟶ Potassium Hydroxide + Nitric Oxide + Iodine

$$2Kl + 2HNO_2 \longrightarrow 2KOH + 2NO + I_2$$

It also decolourizes acidified potassium permanganate

Potassium Permanganate + Sulphuric Acid + Nitrous Acid ⟶ Potassium Sulphate + Manganese Sulphate + Nitric Acid + Water

$$2KMnO_4 + 3H_2SO_4 + 5HNO_2 \longrightarrow K_2SO_4 + 2MnSO_4 + 5HNO_3 + 3H_2O$$

It also turns acidified potassium dichromate to green chromic sulphate

Potassium Dichromate + Sulphuric Acid + Nitrious Acid ⟶ Potassium Sulphate + Chromic Sulphate + Nitric Acid + Water

$$K_2Cr_2O_7 + 4H_2SO_4 + 3HNO_2 \longrightarrow K_2SO_4 + Cr_2(SO_4)_3 + 5HNO_3 + 3H_2O$$

Turns Stannous chloride to stannic chloride

Stannous Chloride + Hydrochloric Acid + Nitrous Acid ⟶ Nitric Oxide + Stannic Chloride + Water

$$SnCl_2 + 2HCl + 2HNO_2 \longrightarrow 2NO + SNCl_4 + 2H_2O$$

The nitrous acid also precipitates sulphur from hydrogen sulphide (being an oxyacid)

Nitrous Acid + Hydrogen Sulphide ⟶ Nitric Oxide + Sulphur + Water

$$2HNO_2 + H_2S \longrightarrow 2NO + SP + 2H_2O$$

The compounds of nitrous acid are called nitrites. It reacts with different metals to form nitrates e.g. sodium, potassium and also heavy metals like zinc, iron etc. nitrates are stable and they are best prepared removing oxygen from their respective nitrates, by heating them with lead.

Sodium Nitrate + Lead ⟶ Lead Oxide + Sodium Nitrite

$$NaNO_3 + Pb \longrightarrow PbO + NaNO_2$$

Test for Nitrites

1. All the metallic nitrites are soluble in cold water except silver nitrite.
2. When heated produces nitrogen (except those of light metals).
3. When reacts with acid produces nitrogen dioxide.
4. Turns ferrous sulphate into ferrous nitrososulphate (Ring Test).
5. To the solution of a nitrite few drops of acetic acid, $HC_2H_3O_2^*$, then add few drops of silver nitrate, until a precipitate occurs, now add a little sodium chloride solution, then add and centrifuge the solution discard the precipitate, add thiourea $CS(NH_2)_2$ and agitate the mixture if NO_2^- ions are present there will an effervescence, now add a few drops of ferric chloride, the solution will become red $Fe(SCN)^{2+}$.

Nitric Acid (HNO_3): It was known as aqua frotis meaning brave water, discovered by Geber (779A.D) by distilling copper as (ferrous sulphate) with potassium nitrate and potassium aluminium sulphate. But in the year 1650 Glauber prepared nitric acid by heating potassium nitrate with sulphuric acid. (Which is the lab preparation of nitric acid).

Cavendish in the year 1785 obtained nitric acid by passing an electrical discharge through a mixture of nitrogen and moist oxygen. In the rain water portions of nitric acid are obtained due to lightning. The atmospheric nitrogen and the oxygen combine during lightning and stay in the clouds during condensation and when the rain drops these nitrogen dioxide gets dissolved in the water.

Laboratory Preparation of Nitric Acid

Theory: In lab nitric acid is prepared by the action of potassium nitrate and sulphuric acid (Glauber's Process). The gas is collected in the reservoir flask and then cooled under tap water (see Fig. 10.10).

$$\text{Sulphuric Acid} + \text{Potassium Nitrate} \longrightarrow \text{Nitric Acid} + \text{Potassium Bisulphate}$$

$$H_2SO_4 + KNO_3 \longrightarrow HNO_3 + KHSO_4 \Downarrow$$

Also a second reaction takes place at much higher temperature (800°C), which yields more acid

$$\underset{\text{Nitrate}}{\text{Potassium}} + \underset{\text{Bisulphate}}{\text{Potassium}} \xrightarrow{800°C} \underset{\text{Sulphate}}{\text{Potassium}} + \underset{\text{Acid}}{\text{Nitric}}$$

$$KNO_3 + KHSO_4 \longrightarrow K_2SO_4 + HNO_3$$

---

*Since acetic acid is an carboxylic acid, any such unusual configurations can be made, as because of covalent bonding, the detail cause is not the syllabs of this book but the other possible configurations are CH_3COOH, $C_2\text{~}O_2$, $C_2H_3HO_2$.

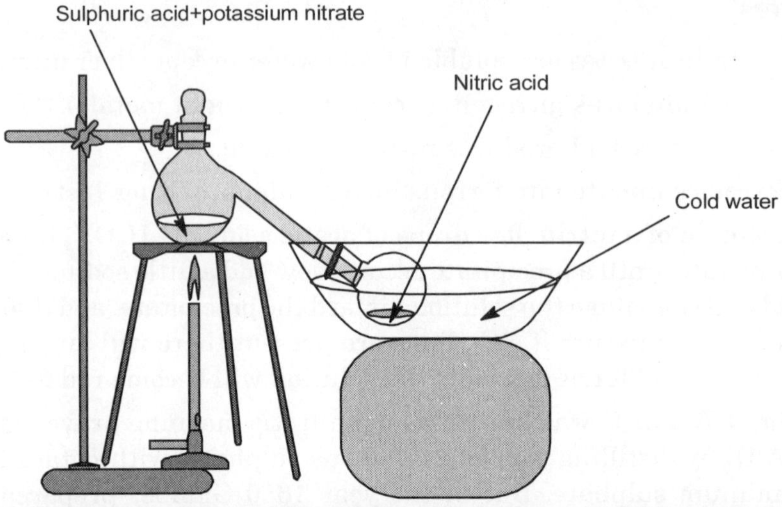

Apparatus	Reagents	Temperature	Pressure
All glass retort, burner, reservoir flask, trough etc.	Potassium Nitrate, Sulphuric Acid	Below 200°C	Normal

Fig. 10.10. The Lab Preparation of Nitric Acid

Precaution : (1) the second reaction is not allowed to carry out because the higher temperature can cause damage to the retort and secondly the nitric acid at such a high temperature will decompose, into nitrogen dioxide water and oxygen.

(2) The potassium sulphate formed gets deposited inside the wall of the glass retort is so much hard that is irremoveable.

Other Methods of Preparation of Nitric Acid

Nitric Acid is also prepared from the reaction of sodium nitrate and sulphuric acid:

Sodium Nitrate + Sulphuric Acid ⟶ Sodium Bisulphate + Nitric Acid

$$NaNO_3 + H_2SO_4 \longrightarrow NaHSO_4 + HNO_3$$

When nitric oxide reacts with ozone and water nitric acid is produced.

Nitric Oxide + Water + Ozone ⟶ Nitric Acid

$$2NO + H_2O + O_3 \longrightarrow 2HNO_3$$

Nitrogen Pentoxide (anhydride of nitric acid) when reacts with water nitric acid is produced

Nitrogen Pentoxide + Water ⟶ Nitric Acid

$$N_2O_5 + H_2O \longrightarrow 2HNO_3$$

Lead nitrate is heated with sulphuric acid nitric acid is produced

Lead Nitrate + Sulphuric Acid ⟶ Nitric Acid + Lead Sulphate

$$Pb(NO_3)_2 + H_2SO_4 \longrightarrow 2HNO_3 + PbSO_4$$

Anhydrous Nitric Acid (NO_5): Nitric Acid also can be obtained from *Deville's Method,* in which it is found in perfect transparent brilliant colourless crystals, having a melting point of 85°C and boiling point of 113°C.

Sometimes the crystals undergo decomposition when kept under sealed glass tubes at ordinary temperatures and the tube bursts with a dangerous explosion from the pressure exerted by the gases. The crystals dissolve rapidly in water, emitting much heat and producing ordinary nitric acid.

The process consists of a uniform current of perfectly dry chlorine gas very slowly over crystals of dry silver nitrate. The slat is heated to 200°C at first till the decomposition has taken place and then the temperature is lowered to 150°C. The operation is very delicate and requires a good precaution. The experimental set up is shown in Fig. 10.11.

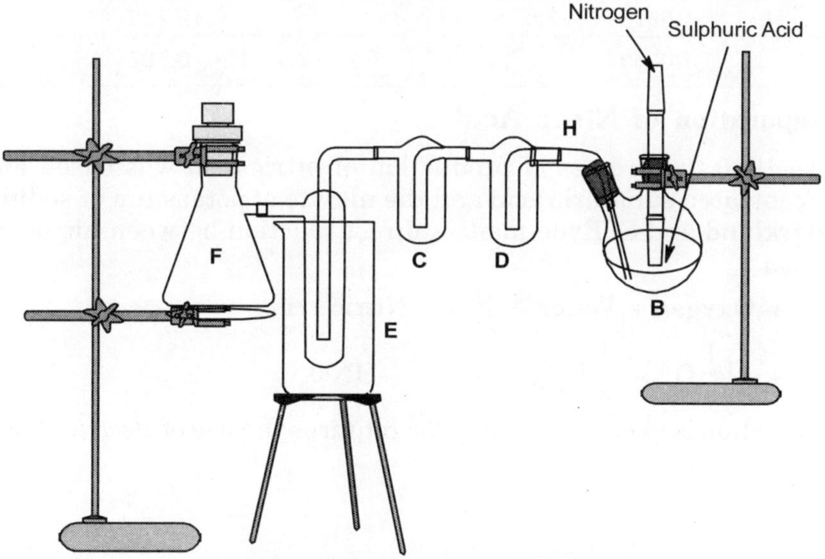

Fig. 10.11. The Preparation of Nitric Acid by Deville's Methods

Through the funnel sulphuric acid is poured in the vessel B. Pure chlorine gas is blown in the vessel B through the tube H. The bulb C is filled with dry calcium chloride, which acts as a dehydrating agent. The second bulb is filled with sulphuric acid (D) and pumice stone, the E and F are made of continuous glass because the acid immediately attacks and destroys the corks and Para-rubber joints. A portion of the lower oxides of nitrogen is liquefied and collected in the tube F and is then separated. The chlorine displaces the nitric acid and nitrate from silver nitrate and silver chloride is formed and the nitric acid with oxygen escapes if the reservoir F is surrounded with freezing mixture the nitric acid is converted to crystals. The reaction is stated below:

Silver Nitrate + Chlorine $\longrightarrow$ Silver Chloride + Oxygen +Nitric Acid (anhydrous)

$$2AgNO_3 + Cl \longrightarrow 2AgCl + O + 2NO_5$$

The anhydrous nitric acid is very much soluble in water and forms ordinary nitric acid, and also very much concentrated; the table below shows the concentration with specific gravity:

Specific gravity	NO_3 in 1/100 part
1,5000	79.700
1.4670	70.136
1.4107	58.181
1.3056	41.444
1.2523	34.271
1.1958	27.098
1.0878	12.752
1.0053	0.797

Industrial Preparation of Nitric Acid

Theory: Before 1900 the process of production of nitric acid was based entirely on the reaction of concentrated sulphuric acid and the nitrate of potassium or sodium. But in the year 1903 E.Berkland and S. Eyde made a direct reaction between nitrogen and oxygen to form nitric acid

$$\text{Nitrogen} + \text{Oxygen} + \text{Water} \longrightarrow \text{Nitric Acid}$$

$$\frac{1}{2}N + 1\frac{1}{4}O_2 + \frac{1}{2}H_2O \longrightarrow HNO_3$$

The whole reaction is exothermic and also requires the use of *electrical arc furnace* (see Fig. 10.12)

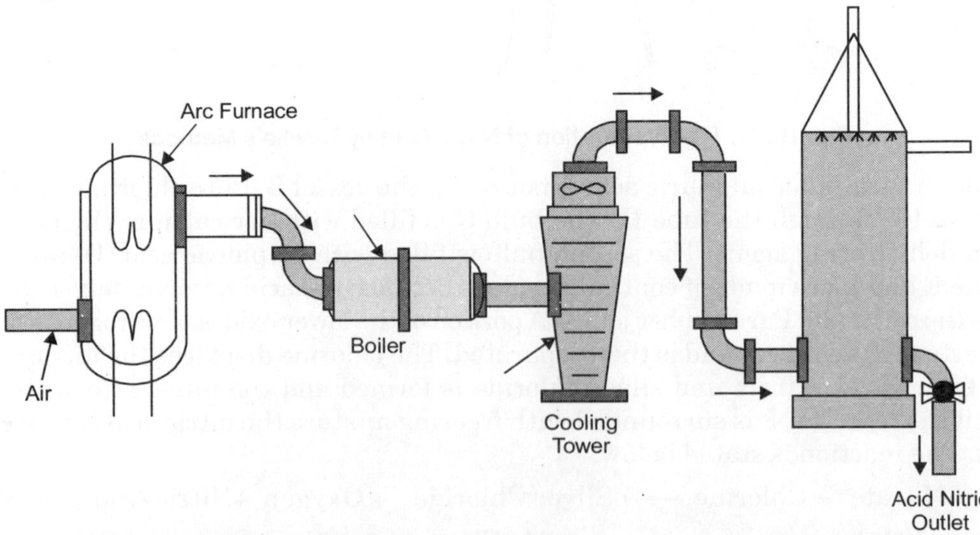

Fig. 10.12. Brikland-Eyde Process for Industrial Nitric Acid Preparation

Besides all these the reaction is pollution creator. So a modern method of manufacture of nitric acid is done by the catalytic oxidation of ammonia (a compound of nitrogen and hydrogen) over platinum, to give nitric oxide. *W. Ostwald* first studied this reaction in the year of 1901. The production of nitric acid according to his process named as Ostwald's Process (see Fig. 10.13).

The reaction taking place in the Ostwald's Process shown:

$$\text{Ammonia} + \text{Oxygen} \longrightarrow \text{Nitric Oxide} + \text{Water}$$
$$NH_3 + 1^1/_4O_2 \longrightarrow NO + 1^1/_4O_2$$
$$\text{Nitric Oxide} + \text{Oxygen} \longrightarrow \text{Nitrogen Dioxide}$$
$$NO + {}^1/_2O_2 \longrightarrow NO_2$$
$$\text{Nitrogen Dioxide} + \text{Water} \longrightarrow \text{Nitric Acid}$$
$$NO_2 + {}^1/_2H_2O \longrightarrow {}^2/_3\ HNO_3$$

In a typical industrial unit of manufacturing nitric acid a mixture of air with 10% ammonia by volume is passed very rapidly over a series of platinum wire gauge at the temperature of 885°C and 5 atmospheric pressure. Contact time with catalyst is minimized to avoid unnecessary side reactions. Conversion efficiency here is about 98%. The flowing gases are passed through an absorption column to yield 60% aqueous nitric acid at about 40°C temperature.

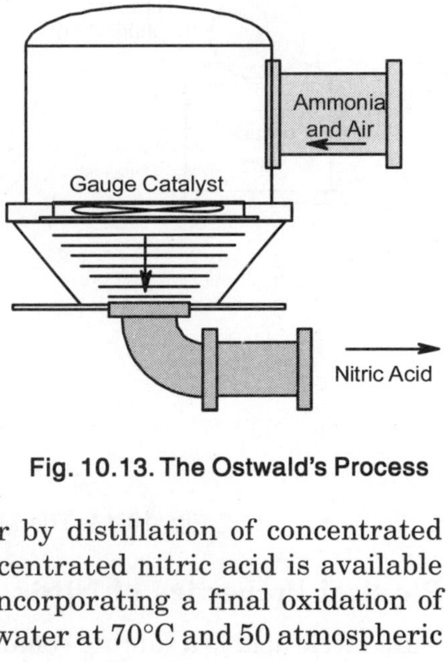

Fig. 10.13. The Ostwald's Process

Then the process of concentration and distillation of 60% aqueous nitric acid produced in most ammonia burning plants is limited by the formation of a maximum boiling azeotrope at (122°C) 68.59% by weight. Further, concentration to 98% can be effected by countercurrent dehydration using sulphuric acid or by distillation of concentrated magnesium nitrate solution. Alternatively the 99% concentrated nitric acid is available directly from the process of oxidation of ammonia by incorporating a final oxidation of nitrogen dioxide with the theoretical amounts of air and water at 70°C and 50 atmospheric pressure over a period of four hour.

$$\text{Nitrogen Dioxide} + \text{Oxygen} + \text{Water} \longrightarrow \text{Nitric Acid}$$
$$2NO_2 + 1/2O_2 + 1/2H_2O \longrightarrow 2HNO_3$$

The largest use of nitric acid is in the manufacture of NH_4NO_3 (ammonium nitrate) for the fertilizers and these accounts for some 80% of the production. Many plants have a capacity of 2000 tonnes per day or more and great care must be taken to produce the ammonium nitrate in a ready usable form (e.g. pills of about 3 mm diameter). About one per cent of a conditioner II is unstable at high temperatures (above 300°C) nitrous oxide breaks into nitrogen and oxygen and water, this reaction is also explosive:

$$\text{Ammonium Nitrate} \xrightarrow{200°C-300°C} \text{Nitrous Oxide} + \text{Water}$$
$$NH_4NO_3 \longrightarrow N_2O + 2H_2O$$

For this reason a temperature limit of 140°C is maintained in the reaction unit of ammonium nitrate.

Many inorganic substances catalyze the decomposition.

Nitric Acid from Nitre and Sulphuric Acid : This is the oldest productive process of nitric acid, which was used during the world wars. The mixture of potassium nitrate and sulphuric acid is kept in an iron retort and distilled. The process is same as the laboratory process. The vapours of nitric acid are condensed through the silica condensers, when condensed acid flows into the reservoir (see Fig. 10.14).

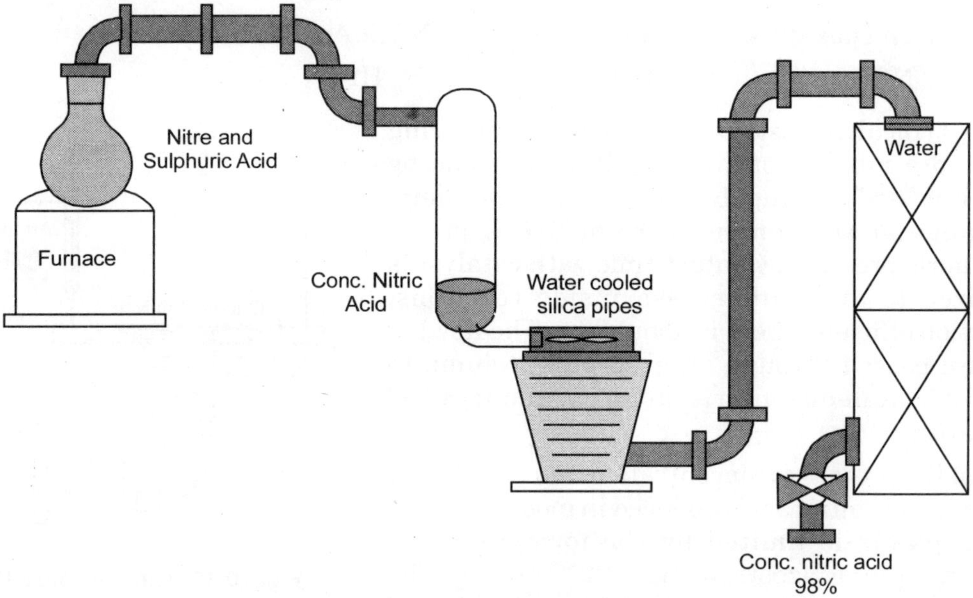

Fig. 10.14. The Manufacture of Nitric Acid from Nitre Process

Physical Properties of Nitric Acid

Pure nitric acid is a fuming colourless liquid with a specific gravity 1.53. The boiling point of which is 86.5°C and freezing point is at – 41°C into a transparent crystals. The 98% pure nitric acid is very much hydroscopic and also absorbs water from water vapour. Nitric acid mixes by all means with water, 68% solution is available in market, which is known as commercial nitric acid and boils at 121°C it can be distilled to pure nitric acid (98%) the pure nitric acid is called fuming nitric acid. Because it consists of nitrogen dioxide gas dissolved in it also the colour of it is slightly brown. Some of the physical properties are shown in the table on next page :

Vapour Pressure	56.99
Density	1.5043g/cm^3
Viscosity	7.46 centipoise
Specific Conductance	3.72×10^{-2}
Di-electric Constant	50 ± 10%
Heat of Formation	–174.1 kJ/mol^{-1}
Heat of Formation (Gibb's free Entropy)	–80.79 kJ/mol^{-1}
Standard Entropy of Heat	155.59 kJ/mol^{-1}

Structure of Nitric Acid

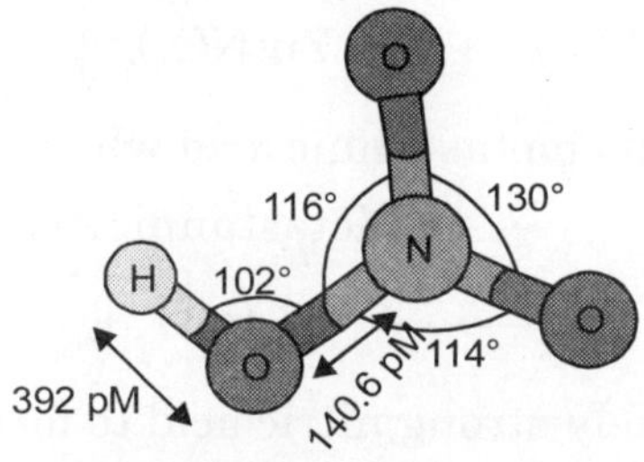

The Structure of Nitric Acid

Chemical Properties

It is one of the strongest acids used in the chemical laboratory

$$\text{Nitric Acid} \longrightarrow \text{Nitrogen Dioxide} + \text{Water} + \text{Oxygen}$$
$$4NHO_3 \longrightarrow 4NO_2 + 2H_2O + O_2$$

From the above reaction it is proved that nitric acid contains oxygen. Nitric acid as known well quite miscible with water. It is a strong monobasic (as it contains only one hydrogen molecule). It dissociated completely when diluted, into positive hydrogen ions and negative nitrate ions and H^+ and NO_3^-respectively.

$$\text{Nitirc Acid} \quad \text{Hydrogen Ion} + \text{Nitrate Ion}$$
$$HNO_3 \rightleftharpoons H^+ + NO_3^-$$

It reacts with different metals, e.g. copper, zinc, iron and lighter metals like potassium, sodium etc. Copper reacts with nitric acid in many ways, the detail reactions with equation are shown below:

1. The action of nitric acid (dilute) on copper yields nitric oxide and copper oxide

$$\text{Copper} + \text{Dillute Nitric Acid} \longrightarrow \text{Copper Oxide} + \text{Nitric Oxide} + \text{Water}$$
$$3Cu + 2HNO_3 \longrightarrow 3CuO + 2NO + H_2O$$

2. Concentrated nitric acid react with copper in some other way here nitrogen dioxide is formed.

$$\text{Copper} + \text{Nitric Acid} \longrightarrow \text{Copper Nitrate} + \text{Nitrogen Dioxide} + \text{Water}$$
$$Cu + 4HNO_3 \longrightarrow Cu(NO_3)_2 \Downarrow + 2NO_2 \Uparrow + 2H_2O$$

3. With mercury, mercuric nitrite and nitrogen dioxide is formed.

Mercury + Nitric Acid ⟶ Mercuric Nitrite + Nitrogen Dioxide + Water

$$Hg + 4HNO_3 \longrightarrow Hg(NO_3)_2 \Downarrow + 2NO_2 \Uparrow + 2H_2O$$

4. When dilute nitric acid reacts with mercury mercurous nitrite is formed

Mercury + Nitric Acid ⟶ Mercurous Nitrite + Nitric Oxide + Water

$$6Hg + 8HNO_3 \longrightarrow 3Hg_2(NO_3)_2 \Uparrow + 2NO \Uparrow + 4H_2O$$

5. When zinc reacts with dilute nitric acid nitrous oxide and zinc nitrate is formed.

Zinc + Nitric Acid ⟶ Zinc Nitric + Nitric Oxide + Water

$$Zn + 4HNO_3 \longrightarrow Zn(NO_3)_2 \Downarrow + 2NO_2 \Uparrow + 2H_2O$$

6. Tin in the same way forms metastannic acid when reacts with nitric acid.

Tin + Nitric Acid ⟶ Metastannic Acid + Nitric Oxide + Water

$$Sn + 4HNO_3 \longrightarrow H_2SnO_3 \Downarrow + 4NO_2 \Uparrow + H_2O$$

7. Iron reacts with moderately strong nitric acid to form ferric nitrate and nitrogen dioxide.

Iron + Nitric Acid ⟶ Ferric Nitrate + Nitric Oxide + Water

$$Fe + 6HNO_3 \longrightarrow Fe(NO_3)_3 \Downarrow + 3NO_2 \Uparrow + 3H_2O$$

Nitric acid in very dilute stage reacts with lighter metals like sodium, potassium and magnesium.

8. When nitric acid reacts with sodium, sodium nitrate and hydrogen is evolved.

Sodium + Nitric Acid ⟶ Sodium Nitrate + Hydrogen

$$2Na + 2HNO_3 \longrightarrow 2NaNO_3 + H_2$$

9. With potassium the same reaction occurs, but potassium reacts in oxide form.

Potassium Oxide + Nitric Acid ⟶ Sodium Nitrate + Water

$$K_2O + 2HNO_3 \longrightarrow 2KNO_3 + H_2O$$

10. With manganese nitric acid reacts to form manganese nitrate.

Manganese + Cold Dil Nitric Acid ⟶ Manganese Nitrate + Hydrogen

$$Mn + 2HNO_3 \longrightarrow Mn(NO_3)_2 + H_2$$

11. When Magnesium reacts with nitric acid magnesium nitrite is produced, with hydrogen gas.

Magnesium + Cold Dil Nitrate Acid ⟶ Magnesium Nitrate + Hydrogen

$$Mg + 2HNO_3 \longrightarrow Mg(NO_3)_2 + H_2$$

Nitric acid also reacts with non-metallic acid when reacts with carbon, iodine, phosphorous and sulphur.

Sulphur + Nitric Acid ⟶ Sulphuric Acid + Nitrogen Dioxide + Water

$$S + 6HNO_3 \longrightarrow H_2SO_4 + 6NO_2 + 2H_2O$$

13. Carbon reacts with Nitric Acid to form carbonic Acid.

Carbon + Nitric Acid ⟶ Carbonic Acid + Nitrogen Dioxide + Water

$$C + 4HNO_3 \longrightarrow H_2CO_3 + 4NO_2 + H_2O$$

14. Iodine reacts with nitric acid to *form* iodic acid

Iodine + Nitric Acid ⟶ Iodic Acid + Nitrogen Dioxide + Water

$$I_2 + 10HNO_3 \longrightarrow 2HIO_3 + 10NO_2 + 4H_2O$$

15. Phosphorous also reacts with nitric acid producing phosphoric acid.

Phosphorous + Nitric Acid ⟶ Phosphoric Acid + Nitrogen Dioxide + Water

$$2P_2 + 20HNO_3 \longrightarrow 4H_3PO_4 + 20NO_2 + 4H_2O$$

Nitric Acid also reacts with compounds like it reacts with elements.

16. With Hydrogen Sulphide it forms Sulphur and Nitric Oxide

Hydrogen Sulphide + Nitric Acid ⟶ Sulphur + Nitric Oxide + Water

$$3H_2S + 2HNO_3 \longrightarrow 3S + 2NO + 4H_2O$$

17. Sulphuric Acid is formed when nitric acid reacts with sulphur dioxide.

Sulphur Dioxide + Nitric Acid + Water ⟶ Sulphuric Acid + Nitric Oxide

$$3SO_2 + 2HNO_3 + 2H_2O \longrightarrow 3H_2SO_4 + 2NO$$

18. Ferrous sulphate with Sulphuric Acid forms browns ring when reacts with nitric acid.

(a) Ferrous Sulphate + Nitric Acid + Sulphuric Acid ⟶ Ferric Nitrate + Nitric Oxide + Water

$$6FeSO_4 + 2HNO_3 + 3H_2SO_4 \longrightarrow 3Fe_2(SO_4)_3 + 2NO + 4H_2O^*$$

(b) Nitrogen Dioxde ⟶ Nitric Oxide + Oxygen

$$NO_2 \longrightarrow 4NO + O_2$$

(c) Ferrous Sulphate + Nitric Oxide ⟶ Ferrous Nitrososulphate

$$4FeSO_4 + 4NO \longrightarrow Fe(NO)SO_4$$

Formation of Ammonium Nitrate

Ammonium Nitrate and Zinc Nitrate are produced when zinc reacts with very dilute Nitric Acid

Zinc + Dil Nitric Acid ⟶ Ammonium Nitrate + Zinc Nitrate + Water

$$4Zn + 10HNO_3 \longrightarrow NH_4NO_3 + 4Zn(NO_3)_2 + 3H_2O$$

---

*These reactions are included in the Rign Test, which is the confirmatory test for nitrates and nitric acid, students should try these tests.

Similarly tin also forms ammonium nitrate and stannous nitrate.

$$\text{Tin} + \text{Dil Nitric Acid} \longrightarrow \text{Ammonium Nitrate} + \text{Stannous Nitrate} + \text{Water}$$

$$4Sn + 10HNO_3 \longrightarrow NH_4NO_3 + 4Sn(NO_3)_2 + 3H_2O$$

With iron it forms ferrous nitrate and ammonium nitrate

$$\text{Iron} + \text{Dil Nitric Acid} \longrightarrow \text{Ammonium Nitrate} + \text{Stannous Nitrate} + \text{Water}$$

$$4Fe + 10HNO_3 \longrightarrow NH_4NO_3 + 4Fe(NO_3)_2 + 3H_2O$$

Oxidation (removal of hydrogen)

It oxidizes glucose to carbon dioxide and water

$$\text{Nitric Acid} \longrightarrow \text{Nitrogen Dioxide} + \text{Water} + \text{Oxygen}$$

$$2HNO_3 \longrightarrow 2NO_2 + H_2O + O$$

$$\text{Glucose} + \text{Oxygen} \longrightarrow \text{Carbon Dioxide} + \text{Water}$$

$$C_6H_{12}O_6 + 3O_2 \longrightarrow 6CO_2 + 6H_2O$$

Nitration

The process of replacing one or two hydrogen atom with nitrogen atom. Nitric acid replaced hydrogen atoms with nitrate group in organic compounds like benzene.

$$\text{Benzene} + \text{Nitric Acid} \longrightarrow \text{Nitrobenzene} + \text{Water}$$

$$C_6H_6 + HNO_3 \longrightarrow C_6H_5NO_2 + H_2O$$

Experiment for liberation of oxygen from nitric acid.

When nitric acid is heated with saw-dust, it catches fire, due to liberation of oxygen from nitric acid, and high temperature (see Fig. 10.15).

Special Reactions of Nitric Acid

It is a powerful oxidizing agent and reacts with organic matter such as with proteins in human skin to produce trinitrophenol or picric acid with the liberation of nitrogen monoxide. This is known as *xanthoproteic reaction*. Organic matter is thus turned yellow.

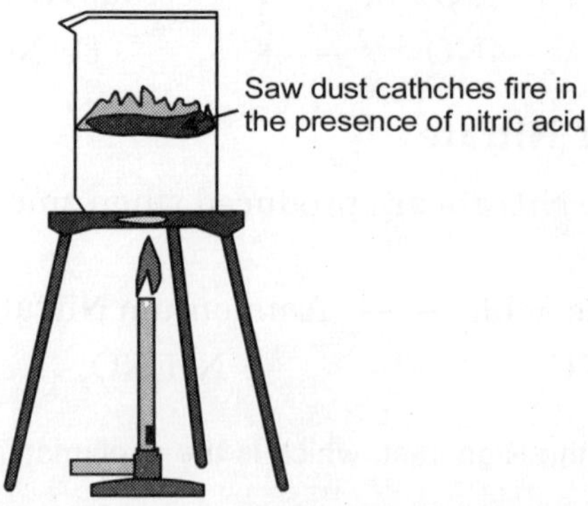

Fig. 10.15. The Expermimental Set up for the Liberation of Oxygen from Nitric Acid

Here comes the end of the chemistry of nitric acid, but it is not the real end in the chemistry, much is left, which is undone.

Hyponitrous Acid ($H_2N_2O_2$): This acid is chemically unstable, but it is found that sodium hyponitrate is more stable this hyponitrate is prepared by the action of hydroxylamine in menthol with ethyl nitrate with the catalysis of sodium ethoxide:

(Menthol) hydroxylamine + Ethyl Nitrate $\xrightarrow{\text{Sodium ethoxide}}$ Sodium Hyponitrate + Water + Ethyl Alcohol

$$NH_2OH + C_2H_5NO_3 \xrightarrow{C_2H_5Na} Na_2N_2O_3 + H_2O + C_2H_5OH$$

The chemical structure is not known

Nitramide, or Hydronitrous Acid (H_2NO_2): This acid is also chemically unstable and sodium salt of this is prepared which is stable. The reaction takes place between sodium nitrite and metallic sodium in presence of Ammonia.

Sodium Nitrate + Sodium (metal) $\xrightarrow{\text{Ammonia}}$ Sodium Hydronitrite

$$NaNO_2 + Na \xrightarrow{NH_3} Na_2NO_2$$

Chemical structure is also unknown*

Pernitric Acid, or Peroxynitric Acid (HNO_4): This is prepared by the action of hydrogen peroxide on dinitrogen pentoxide.

Dinitrogen Pentoxide + Hydrogen Peroxide ⟶ Nitric Acid + Pernitirc Acid

$$N_2O_5 + H_2O_2 \longrightarrow HNO_3 + HNO_4$$

This Pernitric Acid explodes when heated and is also very unstable it is changed to nitric acid by liberating oxygen when exploding.

Pernitric Acid ⟶ Nitric Acid + Oxygen

$$HNO_4 \longrightarrow HNO_3 + O$$

The structure of Pernitric acid is shown in Fig. 10.16 :

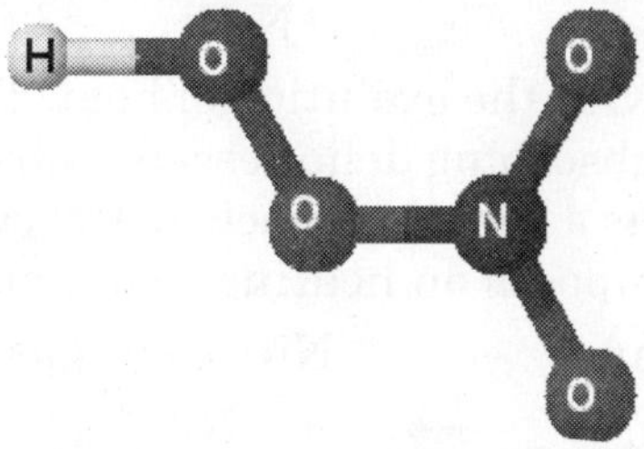

Fig. 10.16. The structure of pernitric acid

*This not that it is completely not known, because absence of evidence is not the evidence of absence.

Compounds of Nitrogen and Hydrogen

Under this head we shall learn about the compounds, which does form when nitrogen reacts with hydrogen, thus they are commonly called as nitrogen hydrides.

The list of different nitrogen hydrides is shown below:

Compound	Formula
Ammonia	NH_3
Hydrazine	N_2H_4
Diamine	N_2H_2
Hydrozoic Acid	HN_3
Hydrooxylamine	NH_2OH

Hydrazide (N_2H_4): Hydrazide has derived from the reaction of ammonia and sodium hypochlorite.

Hydrazine is a weak base, bi-functional in nature. Hydrazine can also be prepared by the direct action of nitrogen and hydrogen but it requires high pressure.

Ammonia + Sodium Hypochlorite ⟶ Hydrazine + Sodium Chloride + Water

$$2NH_3 + NaOCl \longrightarrow N_2H_4 + NaCl + H_2O$$

Hydrazine is a weak base, bi-functional in nature. Hydrazine can also be prepared by the direct action of nitrogen and hydrogen but it requires high pressure.

Nitrogen (gas) + Hydrogen(gas) ⟶ Hydrazine

$$N_2 + 2H_2 \longrightarrow N_2H_4$$

Hydrazine (anhydride) has a melting point of 2.1°C and a boiling point of 115°C. It is a colourless liquid with fuming nature. Hydrazine is highly stable. Hydrazine reacts with water due to hydrolysis.

Hydrazine + Water ⟶ Nitrogen Pentahydrate Ion + Hydroxyl Ion

$$N_2H_4 + H_2O \longrightarrow N_2H_5 + OH^-$$

Hydrazine also burns in air with the evolution of heat. That is why it is used in rockets as fuels. Hydrazine is moisture absorbing deliquescent substance and also absorbs carbon dioxide from air. Hydrazine acts as a base and reacts with acid and gets neutralized forming salt and water. Hydrazine decomposes on heating to ammonia and nitrogen

Hydrazine ⟶ Nitrogen (gas) + Ammonia

$$3N_2H_4 \longrightarrow N_2 + 4NH_3$$

Hydrazine has some amazing character which are described below:

1. Hydrazine reacts with iodine to produce hydrogen iodide in the presence of sodium bi-carbonate.

Hydrazine + Iodine $\longrightarrow$ Hydrogen Iodide + Nitrogen

$$N_2H_4 + 2I_2 \longrightarrow 4HI + N_2$$

2. Potassium Permanganate gets decolourized when hydrazine reacts with it.

Potassium Permanganate + Hydrazine + Sulphuric Acid $\longrightarrow$ Potassium Sulphate + Manganese Sulphate + Water + Nitrogen

$$4KMnO_4 + 5N_2H_4 + 6H_2SO_4 \longrightarrow 2K_2SO_4 + 4MnSO_2 + 16H_2O + 5N_2$$

3. Hydrazine reacts with copper oxide and forms cuprous oxide

Hydrazine + Copper Oxide $\longrightarrow$ Cuprous Oxide + Water + Nitrogen

$$N_2H_4 + 4CuO \longrightarrow 2Cu_2O\Downarrow + 2H_2O + N_2\Uparrow$$

4. Ferric salts are reduced (oxygen is removed) to ferrous salts by hydrazine

Hydrazine + Ferric Chloride $\longrightarrow$ Ferrous Chloride + Hydrogen Chloride + Nitrogen

$$N_2H_4 + 4FeCl_3 \longrightarrow 4FeCl_2\Downarrow + 4HCl + N_2\Uparrow$$

5. Iodates reacts with hydrazine to form iodides.

Potassium Iodate + Hydrazine $\longrightarrow$ Potassium Iodide + Water + Nitrogen

$$2KIO_3 + 3N_2H_4 \longrightarrow 2KI + 6H_2O + 3N_2$$

6. Noble metals like gold silver and platinum reacts with hydrazine in form of their respective metals to produce their respective metals.

Silver Nitrate + Hydrazine $\longrightarrow$ Silver + Nitrogen + Nitric Acid

$$4AgNO_3 + N_2H_4 \longrightarrow 4Ag + N_2 + 4HNO_3$$

Aurum Chloride + Hydrazine $\longrightarrow$ Gold + Nitrogen + Hydrogen Chloride

$$3AuCl_4 + 3N_2H_4 \longrightarrow 3Au + 3N_2 + 12HCl$$

Platinum Chloride + Hydrazine $\longrightarrow$ Platinum + Nitrogen + Hydrogen Chloride

$$PtCl_4 + N_2H_4 \longrightarrow Pt + N_2 + 4HCl$$

Structure of Hydrazine: The structure of hydrazine is much similar to the structure of hydrogen peroxide. The structure of hydrazine is shown in Fig. 10.17.

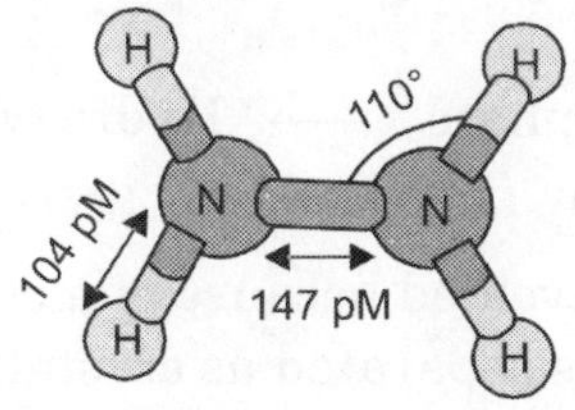

Fig. 10.17. Stracture of Hydrazine.

Hydroxylamine, (NH_2OH) : This compound has derived from ammonia, in which one of the hydrogen atom of ammonia has replaced by one hydroxyl $(OH)^-$ ion. So it is named as hydroxylamine. Like ammonia and hydrazine it is also a base and gets neutralized when reacts with acid and forms salt and water.

Hydroxylamine has many different processes of preparation.

1. It can be prepared directly from the nitric oxide and hydrogen (nascent) when oxygen is removed from nitric oxide, hydroxylamine is produced.

$$\text{Nitric Oxide} + \text{Hydrogen} \longrightarrow \text{Hydroxylamine}$$
$$NO + 3H \longrightarrow NH_2OH$$

2. Secondly by passing an electric current through nitric acid, here nitric acid is mixed with sulphuric acid and the whole solution is kept in compartments called *cells*. Each cell contains 50% sulphuric acid and 50% nitric acid.

$$\text{Nitric Acid} + \text{Hydrogen} \longrightarrow \text{Hydroxylamine}$$
$$HNO_3 + 2H \longrightarrow NH_2OH$$

3. Hydroxylamine can also be prepared by product of acetic acid manufacture from petrol. The process consists of the vapour phase nitration of ethane or propane to nitroethane ($CH_3CH_2NO_2$). On treatment with strong sulphuric acid it is rearranged to the hydroxamic acid from which acetic acid and hydroxylamine are formed.

$$\text{Nitroethane} \xrightarrow{\text{Sulphuric Acid}} \text{Hydroxamic Acid} \longrightarrow \text{Acetic Acid} + \text{Hydrooxylamine}$$
$$CH_3CH_2NO_2 \xrightarrow{H_2SO_4} H_3C\text{–}\underset{\underset{OH}{|}}{C}\text{=N–OH} \longrightarrow CH_3COOH + NH_2OH$$

4. Hydroxylamine in anhydrous state is prepared by the addition of sodium methoxide in methyl alcohol in which Hydroxylamine hydrochloride is present.

$$\underset{\text{Methoxide}}{\text{Sodium}} + \underset{\text{Hydrochloride}}{\text{Hydroxylamine}} \longrightarrow \underset{\text{Alcohol}}{\text{Methyl}} + \underset{\text{Chloride}}{\text{Sodium}} + \text{Hydroxylamine}$$
$$NaOCH_3 + NH_2OH.HCl \longrightarrow CH_3OH + NaCl + NH_2OH$$

Sodium chloride precipitated is filtered off and the filtrate is distilled under pressure at first the methyl alcohol distils over and then the hydroxylamine comes out.

5. It is also prepared by the distillation of phosphate of hydroxylamine under reduced pressure.

$$\text{Hyroxylamine Phosphate} \longrightarrow \text{Hydroxylamine} + \text{Phosphoric Acid}$$
$$(NH_2OH)_3H_3PO_4 \longrightarrow 3NH_2OH + H_3PO_4$$

Generally the distillation is avoided because of a chance of explosion so the filtrate is cooled when the hydroxylamine is separated as crystals.

Hydroxylamine forms colourless crystals, which are very unstable in anhydrous state but quite stable in aqueous state as hydroxide the hydroxylamine decomposes like ammonia in the anhydrous state as below equation shows:

$$\text{Hydroxylamine} \longrightarrow \text{Ammonia} + \text{Nitrous Oxide} + \text{Water}$$
$$4NH_2OH \longrightarrow 2NH_3 \Uparrow + N_2O \Downarrow + 3H_2O$$

Though unstable, hydroxylamine is a very reactive substance, when brought in contact with chlorine it bursts into flame producing hydroxylamine chloride.

Hydroxylamine + Chlorine ⟶ Hydroxylamine Chloride

$$2NH_2OH + Cl_2 \longrightarrow 2NH_2OH.Cl$$

Hydroxylamine solution can act both as the remover of oxygen and hydrogen atoms.

(a) It reacts with copper oxide to form cuprous oxide

Copper Oxide + Hydroxylamine ⟶ Cuprous Oxide + Nitrous Oxide + Ferric Hydroxide

$$4CuO + 2NH_2OH \longrightarrow 2Cu_2O \Downarrow + N_2O \Uparrow + Fe(OH)_3 \Downarrow$$

(b) A solution of hydroxylamine and sodium hydroxide removes hydrogen atom from ferrous hydroxide, producing ferric hydroxide.

Hydroxylamine + Ferrous Hydroxide + Water ⟶ Ammonia + Ferric Hydroxide

$$2NH_2OH + 2Fe(OH_2) + H_2O \longrightarrow NH_3 + 2Fe(OH)_3$$

(c) With hydrogen peroxide hydroxylamine forms ammonia

Hydroxylamine + Hydrogen Peroxide ⟶ Ammonia + Water + Oxygen

$$NH_2OH + H_2O_2 \longrightarrow NH_3 + H_2O + O_2$$

Hydroxylamine is used in organic chemistry for synthesis and has a more or less pyramidal structure which is shown in Fig. 10.18.

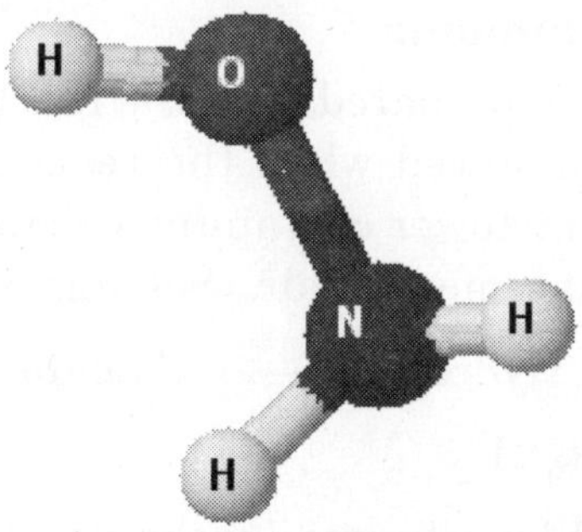

Fig. 10.18. The Structure of Hydroxylamine

AMMONIA

Ammonia Spirit or Hartshon (NH_3) Ammonia as named by Priestely as *Alkaline Air.* It is very much volatile compound of nitrogen and hydrogen and also known as *nitrogen hydride.* Gaseous ammonia was probably known from the ancient times. In the year 1756 Dr Joseph Black first isolated it and proved its distinction from the nitrogen sesquicarbonate.

Priestely afterwards with greater accuracy and by more investigation prepared it in a pure state and discovered more of its properties.

Ammonia though not very scattered in the nature, but found in many vegetable products, the decomposition of animal matter, either by fermentation or by distillation yields it in moderate abundance, it is also found with acids in many strata of mineral kingdom. In the atmosphere a very minute proportions of ammonia is present, which exert a beneficial effect upon the vegetation. Volcanic districts are highly productive of ammonia, standing second in this respect only the Guano Islands of Africa. Seawater contains a low percentage of ammonium salts. Fresh water also holds it in combination, but in an extremely small amount. A perceptible quantity of this compound generally combined with acids, is found in mineral spring soils almost universally contain it more specially, however, these, ferruginous and argillaceous classes. In many substances where ammonia is apparently absent, a careful and patient investigation can prove its presence and often where it is detected unexpectedly; it is definitely judged that its formation may be of atmosphere. Whereas a calm and considerable review of the material and of the circumstances attending such preparations, may lead to a more simple and conclusive demonstration of its presence, thus it occurs in atmosphere.

Apparently no ammonia exists in rainwater when performing analysis, but on examining rainwater recently collected in town or country an expert chemist carl detect its minute appearance. Animal waste especially those reptiles are rich in ammoniac compounds.

We shall now study the laboratory process of ammonia.

Laboratory Preparation of Ammonia

Theory: Ammonia in laboratory is prepared from the reaction of ammonium chloride and calcium hydroxide; the gas is produced when the reaction mixture is heated. The gas produced is passed through a lime tower containing calcium oxide, for drying. The gas is collected by the downward displacement of air. (See Fig. 10.19).

Ammonium Chloride + Calcium Hydroxide ⟶ Calcium Chloride + Ammonia + Water

$$2NH_4Cl + Ca(OH)_2 \longrightarrow CaCl_2 + 2NH_3 + 2H_2O$$

Precaution: *(a)* Gas should be collected by the downward displacement (of air and not water because water reacts with ammonia to produce its hydroxide (ammonium hydroxide).

(b) Only calcium oxide should be used for drying the gas and not sulphuric acid or phosphorous pentoxide because ammonia will react with them forming ammonium sulphate and ammonium phosphate.

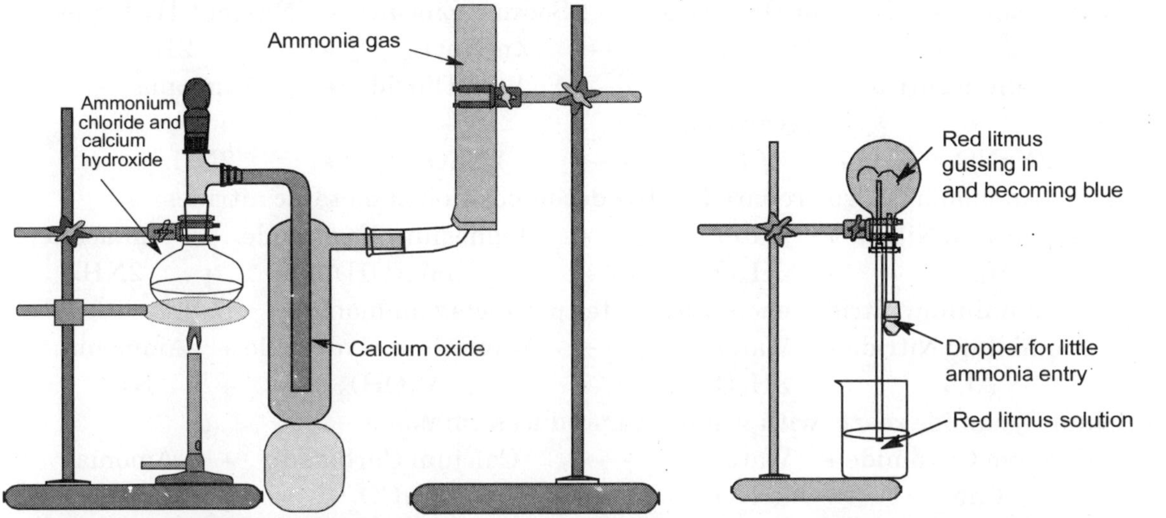

Laboratory Preparation of Ammonia Gas

Apparatus	Reagents	Temperature	Pressure
Round bottom flask, gas jar, burner, lime tower	Ammonium Chloride, Calcium oxide and Calcium hydroxide	Nealry 120°C	Normal

Fig. 10.19. Fountain experiment

Drying of Amnıonia

Ammonia gas cannot be dried with the help of sulphuric acid and phosphorous pentoxide because both will react with ammonia as stated below:

Ammonia + Sulphuric Acid ⟶ Ammonium Bisulphate

$$NH_3 + H_2SO_4 \longrightarrow (NH_4)HSO_4$$

Water + Phosphorous Pentoxide + Ammonia ⟶ Ammonium Phosphate

$$3H_2O + P_2O_5 + 6NH_3 \longrightarrow 2(NH_4)3PO_4$$

Other Methods of Ammonia Preparation

I. Ammonia can also be prepared by the reaction of nitric oxide or nitrogen dioxide with hydrogen in the presence in of platinum wire as catalyst but this requires a high temperature.

Nitric oxide + Hydrogen ⟶ Ammonia + Water

$$2NO + 5H_2 \longrightarrow 2NH_3 + 2H_2O$$

II. Ammonia in the pure form be prepared by the action of zinc and sodium nitrate also sodium hydroxide here nascent hydrogen is produced, which reacts with sodium nitrate to form ammonia.

(a) Zinc + Sodium Hydroxide ⟶ Sodium Zincate + Nascent Hydrogen

$$Zn + 2NaOH \longrightarrow Zn(NaO)_2 + 2H$$

(b) Sodium Nitrite + Nascent Hydrogen ⟶ Sodium Dioxide + Ammonia

$$2NaNO_2 + 6H \longrightarrow 2NaO_2 + 2NH_3$$

III. Ammonia is also prepared by the decomposition of metallic nitrides.

Magnesium Nitride + Water ⟶ Magnesium Hydroxide + Ammonia

$$Mg_3N_2 + 6H_2O \longrightarrow 3Mg(OH)_3 + 2NH_3$$

Also aluminium nitride reacts with water producing ammonia.

Aluminium Nitride + Water ⟶ Aluminium Hydroxide + Ammonia

$$AIN + 3H_2O \longrightarrow Al(OH)_3 + NH_3$$

Calcium Cynamide reacts with water and produces ammonia

Calcium Cynamide + Water ⟶ Calcium Carbonate + Amonia

$$CaN_2 + 3H_2O \longrightarrow CaCO_3 + 2NH_3$$

When Hydroxylamine reacts with hydrogen peroxide ammonia is produced.

Hydroxylamine + Hydrogen Peroxide ⟶ Ammonia + Oxygen + Water

$$NH_2OH + H_2O_2 \longrightarrow NH_3 + O_2 + H_2O$$

Physical Properties

The physical properties of ammonia are also similar to those of water and hydrogen fluorides, but dissimilar to other hydrogen compounds in the point of view off freezing point heat etc. The detail physical properties are given in the table below :

Melting Point	–77.74°C
Boiling Point	–33.35°C
Heat of Fusion at melting Point	5657J $mole^{-1}$
Heat of Vapourization at boiling point	23350J $mole^{-1}$
Critical Temperature	132.4°C
Critical Pressure	112.5 atmopheric
Diaelectric Constant at –77.7°C	25
Density at –70°C	0.7253g/cm^3
Density at –30°C	0.6778g/cm^3
Enthalpy of Formation	46.19kJ $mole^{-1}$
Enthalpy of Formation (Gibb's)	16.64 $deg^{-1}mole^{-1}$
Molar Heat Capacity	35.66 deg^{-1} mol^{-1}
Viscosity	0.01350 Pascal
Vapour Pressure at –20°C	1426.8 torr (190.23 Kpa)
Vapour Pressure at 0°C	3221.0 torr (429.43 Kpa)
Solubility in Water	
1 atmosphere at 0°C	42.8% by weight
1 atmosphere at 20°C	33% by weight
1 atmosphere at 40°C	23.4 by weight

Experiment of the Solubility of Ammonia

Ammonia is very much soluble in water and the solution is alkaline this amount of solubility can create a fountain which can be proved by the *ammonia fountain experiment.*

In this experiment ammonia is filled in the flask having connected with a delivery tube and a dropper full of water it is then kept inverted and the open end of the tube is dipped in the water solution containing phenolphthalein solution. Some of the water from the dropper is introduced in the flask containing ammonia, immediately the ammonia gets dissolved in the water creating a partial vacuum thus causing more and more phenolphthalein solution to rush in like a fountain. When the transparent phenolphthalein solution comes in contact with the alkaline ammonia solution it changes its colour from colourless to pink (Fig. 10.20 showing the fountain experiment) .

Molecular Structure

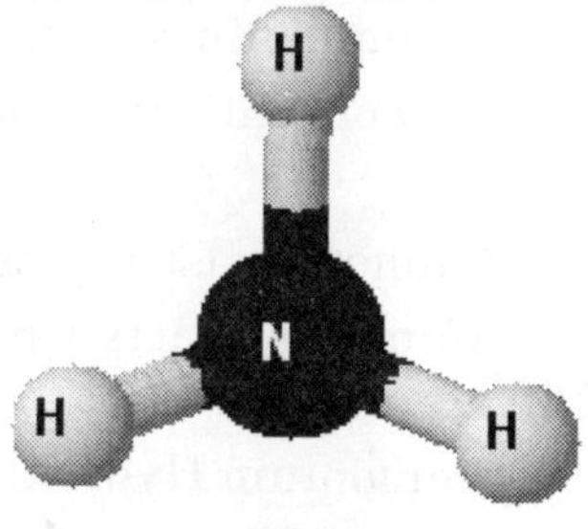

Fig. 10.20

The ammonia molecule has a pyramidal structure of the type as shown in Fig. 10.20, in which the nitrogen atom has achieved a stable electronic configuration forming three electron pair bonds with the three hydrogen atoms. The HNH bond angle in the pyramid is 106.75° which is the best explained as resulting from the use of Sp^3 hybrid bonding orbitals by the nitrogen atom. This should yield tetrahedral bond angles with the result that one Sp^3 orbital is occupied by the unshared pair which is connected repulsively near to the nitrogen nucleus on the shared pairs of relatively near to the nitrogen nucleus on the shared pairs of the electrons forming the N-h bonds. Produces a slight compression of the HNH bond angles, thus according for the fact that they are slightly less than tetrahedral (109.5°). The dipole moment of the ammonia molecule is 1.5 debyes, is a resultant of the combined polarities of these N-H bonds and of the unsaturated electron pair in the highly directional Sp^3 orbital. The pyramidal ammonia molecule turns inside out readily and it starts oscillating between the two extreme positions at the precisely determined.

Use of ammonia clocks

Chemical Properties

The chemical properties of ammonia can be classified under three groups

(1) Addition reaction commonly called ammonation

(2) Amonolysis or subtraction reaction

(3) Oxidation-Reduction reaction.

(1) Ammonation : Reactions of this kind includes those in which ammonia molecule and to other molecules or ions either through the mechanism of covalent bond formation using the unshared electrons on the nitrogen atom, or through hydrogen bonding. Most familiar of the ammonation reaction is the reaction of ammonia and water.

Ammonia + Water $\longrightarrow$ Ammonium Hydroxide

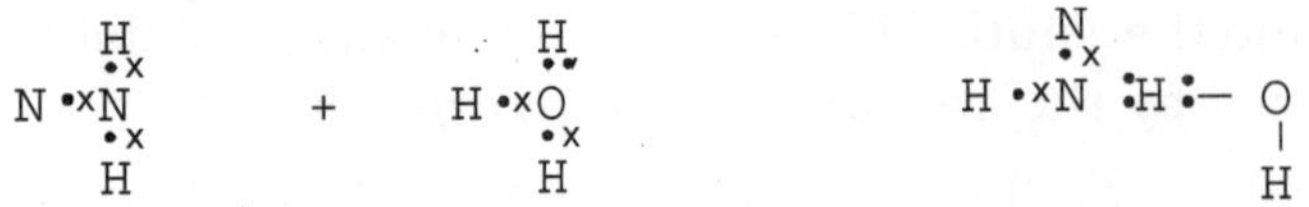

> Best example of ammonation is the reaction of ammonia and water.

The strong tendency of water and ammonia to combine is proved by the high stability of ammonia in water (in fountain experiment) about 700 volumes in one volume of water at 20°C and one atmosphere pressure. As indicated by the ionization constant 1.77×10^{-5} at 25°C. The phase diagram for the NH_3H_2O system indicates the existence of $2NH_3H_2O$ also exists. Ammonia reacts with acids in hydroxide form.

1. With sulphuric acid it forms ammonium sulphate

Ammonium Hydroxide + Sulphuric Acid $\longrightarrow$ Ammonium Bisulphate + Water

$$NH_4OH \quad H_2SO_4 \longrightarrow NH_4HSO_4 + H_2O$$

2. Ammonium Hydroxide reacts with hydrochloric acid to form ammonium chloride

Ammonium Hydroxide + Hydrogen Chloride $\longrightarrow$ Ammonium Chloride + Water

$$NH_4OH + HCl \longrightarrow NH_4Cl + H_2O$$

Ammonia also reacts with metals forming complexes commonly called *amines.*

Ammonia reacts with copper nitrate forming a soluble complex called Cupramonium Nitrate.

Ammonium Hydroxide + Copper Nitrate $\longrightarrow$ Cuprammonium Nitrate

$$NH_4OH + Cu(HO_3)_2 \longrightarrow Cu(NH_3)_4(NO_3)_2$$

2. Ammonia reacts with cobalt chloride to form ammonium cobahichloride a red coloured complex.

Ammonia + Cobalt Chloride $\longrightarrow$ Ammonium Cobaltichloride

$$6HN_3 + CoCl_2 \longrightarrow Co(NH_3)_6Cl_2$$

Ammonia also reacts with variety of molecules capable of acting as an electron acceptor (acids). The reaction of ammonia with such substances as sulphur trioxide, silicon tetrafluoride are typical examples, which have been illustrated below by means of equations.

(2) Amonolysis : Ammonolytic reactions include of ammonia in which an amide ($-NH_2$) and an imide group (=NH) or a nitride (=N) replaces one or more atoms or groups in the reaction. Examples have been shown in the reactions following.

(3) Oxidation-Reduction : These reactions may be subdivided into those which involve a change in the oxidation state of the nitrogen atom and these in which eliminated hydrogen is liberated. The example is catalytic oxidation of ammonium air to form nitric oxide, which is better known as *Kraut's Experiment.*

In this experiment, ammonia is heated up to 800°C in the presence of a catalyst wire gauge when ammonia decomposes to water and nitric oxide. Liquid ammonia (ammonium hydroxide) is heated in a flask and also a current of oxygen gas is also bubbled inside the flask and in the presence of hot platinum gauge the ammonia starts decomposing into oxide and water. (See Fig.10.21).

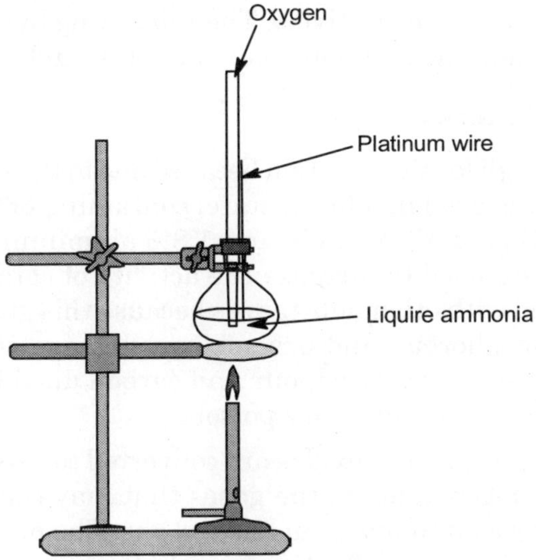

Fig. 10.21. The Kraut's Experiment

$$\text{Ammonia} + \text{Oxygen} \longrightarrow \text{Water} + \text{Nitric Oxide}$$
$$4NH_3 + 5O_2 \longrightarrow 6H_2O + 4NO$$

Ammonia also removes oxygen (reduces) from any metallic oxide examples copper oxide

$$\text{Copper Oxide} + \text{Ammonia} \longrightarrow \text{Copper} + \text{Nitrogen} + \text{Water}$$
$$3CuO + 2NH_3 \quad 3Cu + N_2 + 3H_2O$$

Ammonia also reacts with chlorine to form ammonium chloride and amine chloride.

$$\text{Chlorine} + \text{Ammonia} \longrightarrow \text{Ammonium Chloride} + \text{Amine Chloride}$$
$$Cl_2 + 2NH_3 \quad NH_4Cl + NH_2Cl$$

With active metals like sodium and magnesium ammonia forms amides and hydrogen is liberated.

$$\text{Sodium} + \text{Ammonia} \longrightarrow \text{Sodium Amide} + \text{Hydrogen}$$
$$2Na + 2NH_3 \quad 2NaNH_2 + H_2$$
$$\text{Magnesium} + \text{Ammonia} \longrightarrow \text{Magnesium Nitride} + \text{Hydrogen}$$
$$3Mg + 2NH_3 \quad Mg_3N_2 + 3H_2$$

INDUSTIRAL PROCESS OF AMMONIA PREPARATION

Haber Bosch Process

This process is the major source of industrial ammonia. In a typical process, water gas (a mixture of carbon monoxide, hydrogen and carbon dioxide) mixed with nitrogen is passed through a scrubber cooler as shown in the diagram (101) to remove dust and undecomposed material. The carbon dioxide and carbon monoxide are removed by a carbon dioxide purifier and ammoniac cuprous solution respectively. The remaining hydrogen and nitrogen gases are passed over a catalyst and the reaction occurs at 750°C and 1000 atmosphere pressure.

Catalyst Used in this Process

A single catalyst is not enough for this reaction because due to the strong state of equilibrium, here a mixture of catalyst is used in a fused state, containing 66% Iron Oxide (Fe_2O_3) and 31% Iron monoxide (FeO) potassium oxide and 1.8% aluminium oxide. The last two are called as promoters and are used to increase the activity of catalyst. The catalyst is kept pure and not contaminated with other substances because this gives result to the poisonous catalyst e.g. sulphur, phosphorous and arsenic can damage the catalyst permanently. Whereas gases such as oxygen, water vapour and carbon dioxide can reduce the activity of the catalyst; thus making it a temporary poison.

Before using the catalyst, the iron oxides are converted to iron; the catalyst is activated by this treatment and is not exposed to the gases that may contain poisons. Nitrogen is obtained from air and hydrogen comes from natural gas. The process involves preparation of the nitrogen-hydrogen mixture purification to exclude poison from catalysts, comparison to high pressure and circulation of synthesized gas through a closed loop, which contains two-synthesis converter after ammonia has been condensed from the partially reacted gas pressure.

$$\text{Hydrogen} + \text{Nitrogen} \xrightarrow{\text{1000 atmosphere}} \text{Ammonia}$$

$$3H_2(G) + N_2(G) \xrightarrow{750°C/\text{catalyst}} NH_3$$

The temperature of the reaction is carefully maintained and thus regulated because the higher temperatures can shorten the life of the catalyst and can reduce the conversion at equilibrium whereas the lower temperature reduce the rate of reaction. The reaction mixture entering the converter is relatively cool it is heated to the reaction temperature inside the reaction converter by the process of heat exchange. With the reacting gas and the converter is designed to remove some of the heat of reaction while the gas is passing through the catalyst.

The converters are large pressure vessels containing a heat exchanger and a catalytic basket. The converter for the 1000 atmosphere process contains 4.672272m^3 of catalyst in a converter for a 300 atmosphere of pressure, which produces 22.5kg of ammonia per hour, per cubic metre of catalyst. Molecules serve refractory compounds to reduce the volume and increase the life of the catalyst.

The highest reaction temperature is confounded to the catalyst basket inside the vessel and the thick walls of the vessel are held at a lower temperature.

Ammonia is also produced in industries by following this process. In this process gaseous nitrogen is passed over grinded calcium carbide, which is heated at 1100°C by carbon rods. Which are heated electronically inside the drums of calcium chloride or stirring the carbide mixed with some calcium chloride or calcium fluoride continuously through a furnace heated with electric arcs to produce Calcium Cynamide or nitrolim. (see Fig. 10.22).

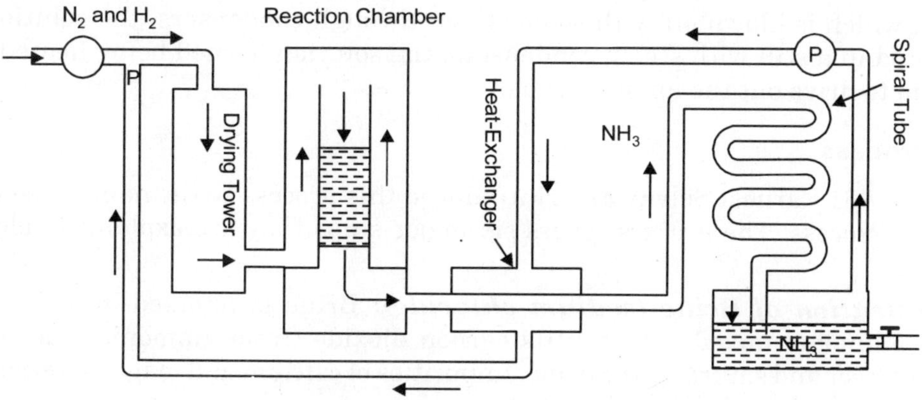

Fig. 10.22. Showing the Haber Bosch process

$$\text{Calcium Carbide} + \text{Nitrogen} \longrightarrow \text{Calcium Nitride} + \text{Carbon}$$

$$CaC_2 + N_2 \longrightarrow CaN_2 + 2C$$

Calcium Cynamide is a derivative of Cynamide the amide of the cyanic acid in which hydroxyl group is replaced by the amino group) (Fig. 10.23).

$$\text{Cyanic Acid} \xrightarrow{\text{Removal of Hydroxyl/}NH_2\text{-addition}} NH_2CN \xrightarrow{CaCl_2} CaN.CN$$

When Calcium Cynamide is heated with water under pressure calcium carbonate and ammonia is produced

$$\text{Calcium Cynamide} + \text{Water} \longrightarrow \text{Calcium Carbonate} + \text{Ammonia}$$

$$CaN_2 + 3H_2O \longrightarrow CaCO_3 + 2HN_3$$

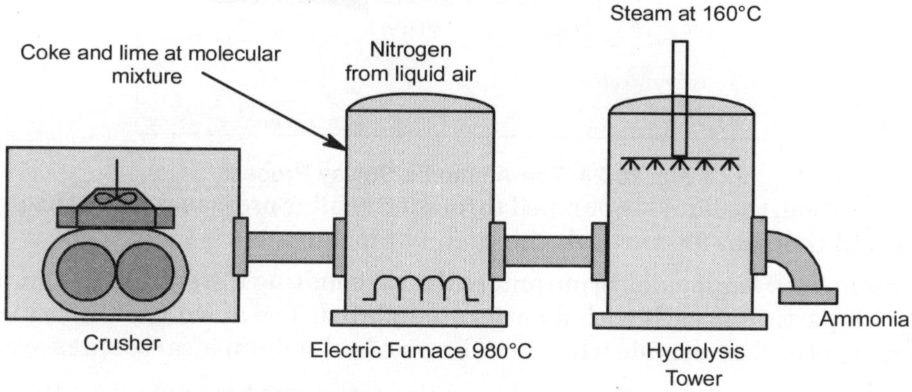

Fig. 10.23. The Cyanamide Process

In this process, the Cynamide is agitated with cold water to remove the unchanged carbide and then stirred with water and a little sodium carbonate in large iron autoclaves (pressure digesters) into which steam is blown until the pressure rises to 3-4 atmosphere. The pressure then rises automatically to 12-14 atmosphere owing to a production of ammonia, which is blown off with some steam through condensers, the solution formed being heated in a still with steam condensers, the solution formed being heated in a still with steam to drive out the ammonia gas.

Solvay Process

In the year 1861 Earnest Solvay First introduced this process so the name has been given as Solvay's Process. The process covers six major steps they are explained below all see Fig. 10.24.

1. Saturation of Brine (sodium chloride): Brine is pumped into the ammonia absorber; ammonia mixed with a little carbon dioxide (from ammonia recovery tower) enters this tower and saturates the brine. Impurities of calcium and magnesium salt present in brine are precipitated as bicarbonates

$$\text{Ammonia} + \text{Water} + \text{Carbon Dioxide} \longrightarrow \text{Ammonium Bicarbonate}$$

$$NH_3 + H_2O + CO_2 \longrightarrow NH_4HCO_3$$

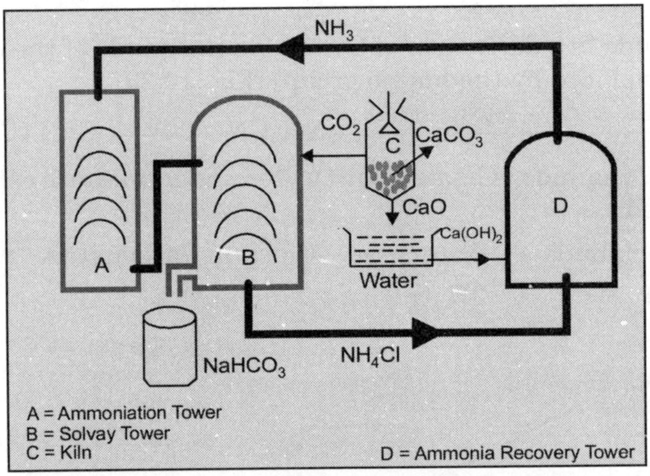

Fig. 10.24. The Ammonia Solvay Process

After saturation, the liquid is pumped through the filter press to remove the precipitated carbonates and then passed through the cooling pipes.

2. Carbonation: Ammoniac brine enters the carbonating tower from its top. This tower is packed with partitions each with a perforated plated. The down-coming brine meets an upward stream of carbon dioxide when small crystals of sodium bicarbonates are produced.

3. Filtration: The thick milky liquid from the carbonate tower is then filtered with the help of rotary vacuum filter. The solid bicarbonate, which is left on the filtre cloth, is scrapped periodically. The mother liquor or the filtrate is pumped to the top of ammonia recovery.

4. Calcination: Sodium bicarbonate obtained in filtration process heated strongly when sodium carbonate is formed.

Sodium Bicarbonate $\longrightarrow$ Sodium Carbonate + Water + Carbon Dioxide

$$2NaHCO_3 \longrightarrow Na_2CO_3 + H_2O + CO_2$$

5. Lime Kiln (Furnace): Here limestone is burnt to give carbon dioxide used in the carbonating tower. The quicklime CaO is converted into slaked lime, calcium hydroxide and pumped into ammonia recovery tower.

6. Recovery of Ammonia: The filtrate from vacuum filtre containing ammonium chloride with a little ammonium bicarbonate flows down the ammonia recovery tower while the slaked lime is admitted a little above the middle of the tower and steam is admitted from the bottom. Ammonium bicarbonate decomposes by the effect heat and ammonium chloride is decomposed by lime. Ammonia is set free along with a little carbon dioxide goes to the saturation tank the liquid flows out contains ammonia and ammonium chloride is the by-product.

Ammonium Chloride is the by product of Solvay process

MANUFACTURE OF AMMONIA (SOLVAY PROCESS)

Important Chemical Equations

1. Transformation of Ammonium Bicarbonate by passing Carbon Dioxide through ammonia and water:

Carbon Dioxide + Ammonia + Water $\longrightarrow$ Ammonia Bicarbonate

$$CO_2 + NH_3 + H_2O \longrightarrow NH_4HCO_3$$

2. Transformation of Ammonium Bicarbonate to Sodium Bicarbonate, by the reaction of Sodium Chloride and Ammonium Bicarbonate

Sodium Chloride + Ammonium Bicarbonate $\longrightarrow$ Sodium Bicarbonate + Ammonium Chloride

$$NaCl + NH_4HCO_3 \longrightarrow NaHO_3 + NH_4Cl$$

3. The Bicarbonate of Sodium is converted into Sodium Carbonate

Sodium Bicarbonate $\longrightarrow$ Sodium Carbonate + Water + Carbon Dioxide

$$2NaHCO_3 \longrightarrow Na_2CO_3 + H_2O + CO_2$$

The raw materials required for this process are sodium chloride, lime stone and ammonia. Sodium Chloride is easily available, however ammonia is expensive if not possible to recover from ammonium chloride and used again and again, the total process will be expensive.

Calcium Carbonate $\longrightarrow$ Carbon Dioxide + Calcium Oxide

$$CaCO_3 \longrightarrow CO_2 + CaO$$

The lime produced above is used to recover ammonia from ammonium chloride on heating.

Calcium Oxide + Ammonium Chloride $\longrightarrow$ Calcium Chloride + Ammonia + Water

$$CaO + 2NH_4Cl \longrightarrow CaCl_2 + 2NH_3 + H_2O$$

Thus, as most of the ammonia can be removed, sodium chloride and limestone are the only materials consumed in this process.

Chemical Composition of Ammonia

The composition and the molecular formula of ammonia can be determined in *two** ways:

1. Volumetric Method: In this method ammonia is subjected to an electrical sparking, which increases the volume of the gas to double because of decomposition of ammonia into nitrogen and hydrogen. This is then mixed with a known volume of oxygen and further sparking is continued. All the hydrogen combines with oxygen to form water, which gives the volume of hydrogen. The excess of oxygen is then removed by means of a solution of pyrogallol made alkaline and the volume of the residual nitrogen is measured.

It is found that 2 volumes of ammonia gas give 1 volume of nitrogen and 3 volumes of hydrogen. Hence, according to Avogadro's Hypothesis, two molecules of ammonia are made up of one molecule of nitrogen and three molecules of hydrogen. Therefore, the molecular formula is NH_3; this is confirmed by the determination of the density of ammonia, which gives the molecular weight of 17.

Ammonia reacts with chlorine to form hydrogen chloride and nitrogen. A long glass tube provided with a funnel and a stop-clock (see Fig. 10.25) the tube is divided into three equal divisions is filled with chlorine. Concentrated ammonia solution is introduced to chlorine through the stopcock drop by drop as each drop enters a vivid reaction occurs. More ammonia is added to the hydrochloric acid formed to ammonium chloride, now on adding acidified (with sulphuric acid) water to combine with the excess of ammonia, two-thirds of the tube will be filled with water. The residual nitrogen is, therefore, equal to one-third of the volume of chlorine originally taken. The three volume of chlorine has combined, with three volumes of hydrogen to form hydrogen chloride.

By Avogadro's Hypothesis, therefore, the ammonia is 1 : 3. The molecular weight of ammonia being 17 from density determination, it follows that a molecule and three atoms of hydrogen, the formula is therefore NH_3.

Testing and Estimation (1) Physical: Ammonia gas is easily identified by its strong pungent smell also its volatile nature*.

(2) Chemical: Ammonia turns moist red litmus paper to blue also produces white fumes of ammonium chloride when reacts with hydrogen chloride or chlorine. The best of all is the test of Nessler's reagent when ammonia is added it turns brown forming a precipitate of mercuroamine iodide. Ammonia is also detected by the action of Millon's reagent, when ammonia is added there is a black precipitate formed.

Fig. 10.25. The Volumetric Apparatus to determine the molecular formula of ammonia

Preparation of Nessler's Reagent

Nessler's reagent is prepared by dissolving 25 g of potassium iodide in 100ml of water and adding to the solution, except a 5ml of solution kept separate a cold saturated solution of mercuric chloride until a permanent faint red precipitate is produced. The 5ml are also added and a little more mercuric chloride is added to make the solution permanent. A solution of potassium hydroxide is added to this solution mixture a faint yellow coloured solution will be the result, the solution is degantated and kept in a black tint bottle.

Preparation of Millon's Reagent

This reagent also consists of two solutions (A) and (B) they are as follows:

Solution A: 12.5ml of sulphuric acid in 100ml of water after cooling then add 12.5 g of mercuric sulphate and the solution is stirred well.

Solution B: One gram of sodium nitrate is added to 100ml of water. The working solution is prepared by adding the solutions A and B.

Composition of Nitrogen

Using Dumas Method, which is gravimetric process, does this experiment and the other one is volumetric process. These processes are used to determine the composition of nitrogen in air.

Volumetric Process: In this process a sample of dry air is collected a sample of dry air is collected and freed from carbon dioxide, by passing the air through lime water and then through a solution of pyrogallol to make it free form oxygen. This decrease the volume of air so it is easy to calculate the composition of nitrogen from the differential ratio of the initial volume and the volume of air obtained after the absorption of carbon dioxide and oxygen.

Gravimetric Method : J.B Dumas and Boussingault first determined the composition of air by weight in 1841. In this method air freed from carbon dioxide and moisture, was passed over red-hot copper which removed the oxygen of air, the residual gas consisting collected in an evacuated glass globe and weighted. The increase in weight of copper gave the weight of oxygen (see Fig. 10.26). Metallic copper is placed in a long glass tube. The large glass globe is also evacuated and weighted the two are connected as in the figure at one end of the glass tube. The other end of the tube is connected with a bulb of potassium hydroxide solution and the other calcium chloride for the removal of carbon dioxide and moisture from air. A current of air thus purified and dried is made to pass slowly over the red-hot copper, which absorbs the oxygen forming the copper oxide. Nitrogen enters and collects in the globe, at the end of the experiment the stopcocks are closed and the globe is weighted after cooling the whole apparatus. The weight of the nitrogen is estimated.

---

*Volatile nature : Evaporates quickly.

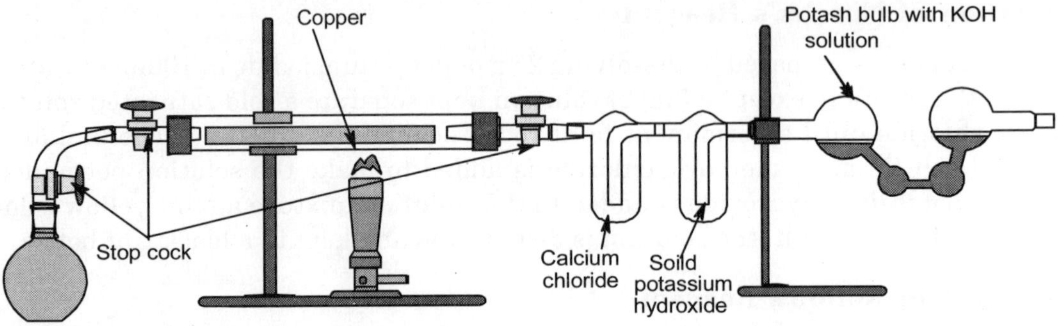

Fig. 10.26. The Experimental set for Measurement of Gravimetric Composition of Nitrogen

The tube is, therefore, evacuated by means of pump and weighted again. The difference of weight gives the weight of the nitrogen. This weight is added to the weight of nitrogen includes of course the weight of rare gases present in the air which cannot be removed by red-hot copper. Dumas and Boussingault found that 23 part by weight of oxygen and 77 part by weight of nitrogen in 100 part by weight of air.

To Prove that Elementary Gases likes Nitrogen are Di-Atomic

From Hyber's process we know that:

1 volume of Nitrogen + 3 volumes of hydrogen = 2 volumes of Ammonia

Applying Avogadro's hypothesis:

Let n be the number of molecule present in 1 volume of Gas.

∴ n be the number of nitrogen will combine with 3 volumes of hydrogen to produce 2n vol. of ammonia (Dividing by n).

1 molecule of hydrogen + 3 molecules of hydrogen = 2 molecule of ammonia.

∴ 1 molecule of nitrogen = 3 molecules of hydrogen = 1 molecule of ammonia. As ammonia is a compound so it must contain at least 1 atom of nitrogen.

1 atom of Nitrogen = 1 molecule of nitrogen.

Or, 1 molecule of nitrogen = 2 atoms of nitrogen.

∴ The elementary gases like nitrogen is di-atomic.

Compounds of Ammonia

Ammonium Amalgam: Ammonia readily combines with acids to form salts Lavoisier and later *Dumas* (1828) regarded these compounds of ammonia (1816) supposed that the salts contained the ammonium radical NH_4 which behaves like an alkali metal. Sal ammoniac is ammonium chloride, which is having a formula of NH_4Cl, is similar to KCl, potassium chloride. Berzzelius supported this view in 1820.

The ammonium theory was supported by the discovery of ammonium amalgam (the compound of mercury and other metal), in 1808 by Seeback in Jena and Berzzelius and **Protin** in Stockholm. If a solution of ammonium chloride is subjected to electrical current where mercury is used as the negative electrode (see Fig. 10.27). The mercury swells up and forms a soft pasty mass which rapidly decomposes evolving hydrogen and ammonia in the ratio 1 volume to 2

$$\text{Ammonium Amalgam} \longrightarrow \text{Ammonia} + \text{Hydrogen}$$

$$2HN_4 \longrightarrow 2NH_3 + H_2$$

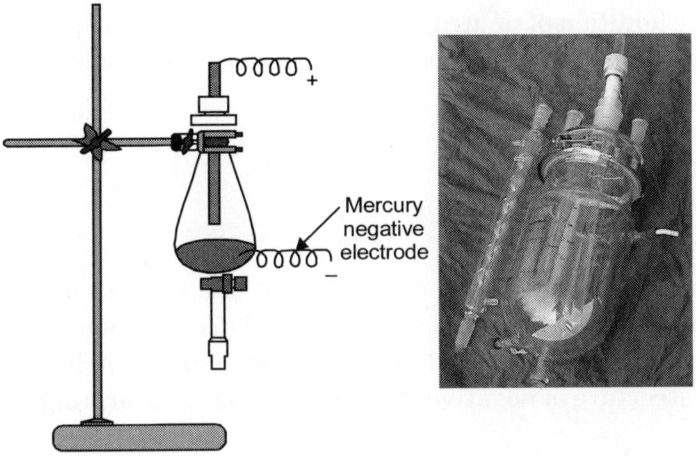

Fig. 10.27. The combustion pippette.

If a little sodium amalgam is added to a cold concentrated solution I = = = I of ammonium chloride, the amalgam swells up to a spongy mass. If this is put in water bubbles of hydrogen are evolved and the liquid smells of ammonia. Shelley (1870) found that ammonium amalgam compressed in a tube under piston obeys a form of ammonia and hydrogen in mercury. Ammonia is very much soluble in water and other experiment favour the existence of ammonium in the amalgam. Although the amalgam does not reduce solutions like ferric chloride or copper sulphate at the ordinary temperature, it reduces solutions of copper, cadmium, zinc and even barium salts at 0°C temperature. The voltage required to deposit sodium on a mercury negative electrode is similar to that required in the formation of ammonium amalgam.

Ammonium amalgam is also produced by the action of sodium amalgam on ammonium chloride, dissolved in anhydrous liquid ammonia. The deep-blue solution obtained by dissolving sodium or potassium in liquid ammonia, which may conduct electricity and may also contain metal ions. According to Schluback and Ballauf (1921), a colourless solution containing free ammonium is formed by adding blue solution of sodium.

In liquid ammonia and a solution of ammonium iodide, when this same blue solvent is added at –70°C.

$$\text{Sodium} + \text{Ammonium Iodide} \longrightarrow \text{Sodium Iodide}$$

$$Na + NH_4I \longrightarrow NaI$$

It decomposes at –40°C

Ammonium	⟶	Ammonia	+	Hydrogen
$2NH_4$	⟶	$2NH_3$	+	H_2

Ammonium Ion: The ammonium salts were once thought to contain 5 valent nitrogen e.g. $(H_4) \equiv N - Cl$ (ammonium chloride), but it is now considered that positive ammonium ion is formed by the addition of neutral ammonia molecule to the positive hydrogen ion (in the case of proton). Which means monovalent so that the whole ammonium ion is also monovalent

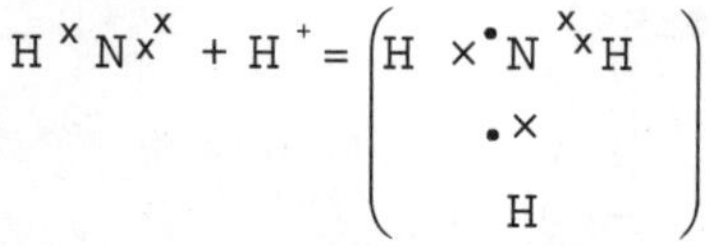

Four hydrogen are attached to the nitrogen at the corners of the tetrahedron with nitrogen at the centre (see Fig. 10.28). The tetrahedral arrangement of valences is necessary to form tetrahedron a dissymmetric structure. In ammonium salts the NH_4 ions forms a base of crystalline structure of negative ions e.g. Cl^-, OH^- etc) forming different compounds.

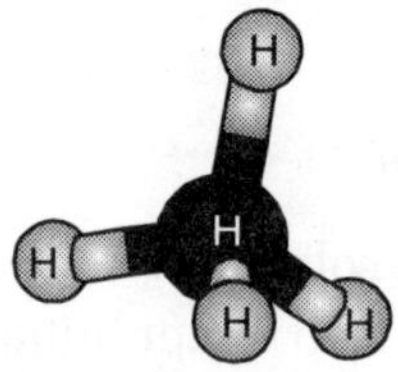

The 3-D structure of Ammonium Ion

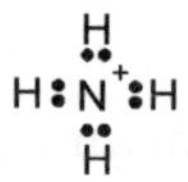

Lewis Structure

Fig. 10.28

Ammonium Chloride, (NH_4Cl): It is also called *Sal Ammoniac,* in nature it occurs in the volcanic gases. It is prepared by the action of ammonium hydroxide and hydrogen chloride and evaporating the residual solution.

Ammonium Hydroxide +	Hydrogen Chloride (Gas)	⟶	Ammonium Chloride +	Water
NH_4OH +	HCl	⟶	NH_4Cl +	H_2O

Ammonium Chloride is also prepared by the reaction of ammonium sulphate and sodium chloride (common salt)

Ammonium Sulphate + Sodium Chloride ⟶ Ammonium Chloride + Sodium Sulphate

$$(NH_4)_2SO_4 + 2NaCl \longrightarrow NH_4Cl + Na_2SO_4$$

Ammonia + Carbon Dioxide + Water ⟶ Ammonium Carbonate

$$2NH_3 + CO_2 + H_2O \longrightarrow (NH_4)_2CO_3$$

Ammonia Carbonate + Carbon Dioxide + Water ⟶ Ammonium Bicarbonate

$$(NH_4)_2CO_3 + CO_2 + H_2O \longrightarrow 2NH_4HCO_3$$

Sodium Chloride	+	Ammonium Bicarbonate	$\longrightarrow$	Sodium Bicarbonate	+	Ammonium Chloride
$NaCl$	+	$2NH_4HCO_3$	$\longrightarrow$	$NaHCO_3$	+	NH_4Cl

Freshnel first discovered this reaction in 1810.

In *ammonia-soda* process salt brine nearly saturated with common salt is taken and ammonia gas dissolved in it in an iron tower filled with bubblers* under pressure into the ammoniac brine solution in a tower called *Carbonating Tower* (see Fig. 10.29). Fitted inside with perforated bubblers, the lower part of the tower being cooled by iron pipes through which cold water circulates. The sodium bicarbonate precipitates and the liquid containing it in suspension is passed to rotating suction filter, on which the solid sodium bi carbonate is washed with a little water. The mother liquor contains undecomposed common salt and ammonium carbonate and ammonium chloride.

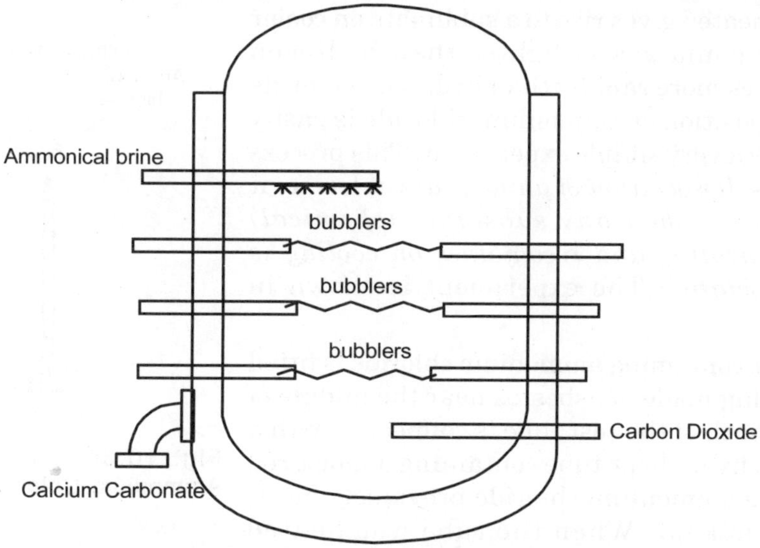

Fig. 10.29. The Carbonation Tower

Ammonium chloride is industrially prepared by the reaction of calcium chloride and ammonium hydroxide, for this purpose, the calcium chloride is obtained from the ammonia-soda process. Then the calcium chloride is kept in a tank provided with a stirrer to this tank ammonia is also introduced. The temperature is allowed to rise about 60°C, when calcium carbonate is produced in large quantities, as solid residue in the presence carbon dioxide.

Calcium Chloride	+	Ammonium Hydroxide	+	Carbon Dioxide	$\longrightarrow$	Calcium Carbonate	+	Ammonium Chloride	+	Water
$CaCl_2$	+	NH_4OH	+	CO_2	$\longrightarrow$	$CaCO_3$	+	$2NH_4Cl$	+	H_2O

*Bubblers—The carbonating tower consists of carbon dioxide gas under reduced pressure, thus the gas looks like bubbles.

This reaction is done in the carbonation tower (see. Fig. 10.29).

The filtrate liquid containing high concentration of ammonium chloride is concentrated and cooled in large crystalline pans where ammonium chloride separates out. The crystals are filtered in a centrifuge, slightly washed and dried by means of hot air at a low temperature. Ammonium chloride so made contains only 0.3- 0.4 per cent common salts as impurity. The ammonium chloride is a white crystalline solid freely soluble in water, when a dry sample of ammonium is vaporized; it suffers thermal decomposition into ammonia and hydrogen chloride gas.

$$\text{Ammonium Chloride} \longrightarrow \text{Ammonia} + \text{Hydrogen Chloride}$$
$$NH_4Cl \longrightarrow NH_3 + HCl$$

This reaction is reversible because the compound ammonium chloride recombines on cooling hence, the solid salt when heated gives rise to a sublimate on cooler surface. As ammonia gas is lighter than hydrogen chloride it diffuses more rapidly than hydrogen chloride thus the decomposition of ammonium chloride is easily demonstrated by a very simple experiment. This process is also known as *dissociation of ammonium chloride,* it can be defines as *when any substance (chemical) decomposes on heating and recombines on cooling is known as dissociation* The experiment is shown in Fig. 10.30.

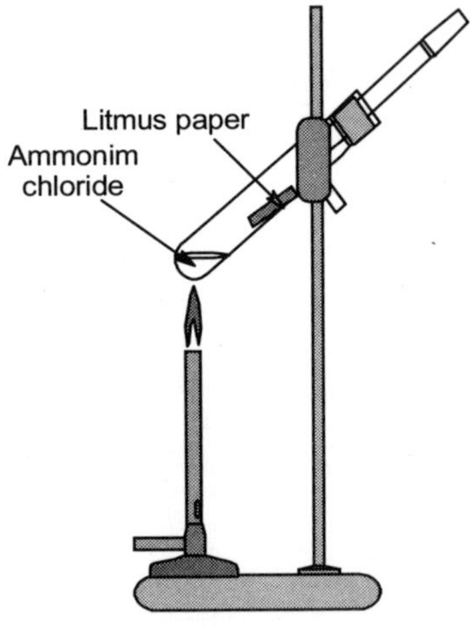

Fig. 10.30. The Dissociation of Ammonium Chloride

In a test tube containing ammonium chloride is fitted with a porous plug made of asbestos near the middle of the tube, the mouth of the test tube is connected with a stopper and a delivery bent tube containing a moist red litmus. With the ammonium chloride previously moist blue litmus was kept. When the tube was heated ammonia gas which was lighter came in contact with the red litmus changed the colour to blue. While the hydrogen chloride gas being much heavier and also evolved later came in contact with the blue litmus changed the colour to red, which was kept very near to the salt (See Fig. 10.30).

***Ammonium Carbonate, [$(NH_4)CO_3$]* :** It is also known as *Sal Volatile,* ammonium carbonate is prepared by heating a mixture of chalk and ammonium carbonate is obtained after the sublimation of the ammonium carbonate with little water.

$$\text{Ammonium Chloride} + \text{Calcium Carbonate} \longrightarrow \text{Calcium Chloride} + \text{Ammonium Carbonate}$$
$$2NH_4Cl + CaCO_3 \longrightarrow CaCl_2 + (NH_4)_2CO_3$$

Ammonium Carbonate is semi-transparent, white fibrous solid converted with a white opaque powder from outside. It is not a stable compound and continuously decomposes giving a strong odour of ammonia, so this is used as *'smelling salt'* in the cases of unconsciousness.

$$\text{Ammonium Bicarbonate} \longrightarrow \text{Ammonia} + \text{Carbon Dioxide} + \text{Water}$$
$$NH_4HCO_3 \longrightarrow NH_3 + CO_2 + H_2O$$

Ammonium Bicarbonate, ($NaHCO_3$) : The crystals of ammonium bicarbonate are prepared in the carbonation tower. The original liquid contains a much higher concentration of ammonia than the ammonium brine. The crystals and the ammonium bicarbonate crystals are filtered in a wooded box filter having a flat bottom. The crystals are dried on a system of belt conveyers, by means of hot air heat over a temperature of 51°C and packed in paper lined wooded higs or in small paper boxes.

Ammonium bicarbonate has a very strong volatile power so for this property ammonium bicarbonate is kept in proper sealed containers.

Ammonium Hydroxide, (NH_4OH) : Although crystalline solid $2NH_3H_2O$ and NH_3H_2O are known, it is doubled if they are ammonium oxide $(NH_2)_2O$ and ammonium hydroxide NH_4OH respectively. Sometimes it is assumed that aqueous solution of ammonium contains ammonium hydroxide, which is a base.

$$\text{Ammonia} + \text{Water} \rightleftharpoons \text{Ammonium Hydroxide}$$
$$NH_3 + H_2O \rightleftharpoons NH_4OH$$

But it is general properties can be explained by supposing that the ammonium ions, leaving the hydroxyl ions of the water which cause the alkaline reaction.

$$\text{Ammonia} + \text{Hydrogen} + \text{Hydroxyl Ion} \rightleftharpoons \text{Ammonium Ion} + \text{Hydroxy Ion}$$
$$NH_3 + H_2 + OH^- \longrightarrow NH_4^+ + OH^-$$

Ammonia is a weak base and this can be proved by the following equation:

$$H_3NH^+ + OH^- = H_3HN \longleftarrow OH$$

The aqueous solution of ammonia is prepared by passing the gas, distilled water; the flask is kept cool because this reaction produces heat. The liquid also explains considerably.

Sp Gr	Percentage of NH_3
0.875	36.90
0.880	35.20
0.890	31.85
0.990	28.50
0.910	25.15

Ammonium Sulphate [$(NH_4)_2SO_4$] : It is prepared by the reaction of ammonia on sulphuric acid, and evaporating the solution. (The industrial process is shown in Fig. 10.31).

$$\text{Ammonia} + \text{Sulphuric Acid} \longrightarrow \text{Ammonium Sulphate}$$
$$2NH_3 + H_2SO_4 \longrightarrow (NH_4)_2SO_4$$

In large scale it is prepared by the reaction of Calcium sulphate and carbon dioxide on ammonia where calcium carbonate and ammonium sulphate is produced. (See Fig. 10.31 for preparation).

Calcium Sulphate + Ammonia + Carbon Dioxde ⟶ Ammonium Sulphate + Calcium Cabonate

$$CaSO_4 + NH_3 + CO_2 \longrightarrow (NH_4)_2SO_4 + CaCO_3$$

Ammonium sulphate forms large crystals which is readily soluble in water when heated ammonium sulphate looses ammonia at 325°C it becomes ammonium hydrogen sulphide.

Ammonium Sulphate $\xrightarrow{325°C}$ Ammonium Bisulphate + Ammonia

$$(NH_4)_2SO_4 \longrightarrow NH_4HSO_4 + NH_3$$

On further heating it decomposes to nitrogen sulphur dioxide and ammonia

Ammonia Hydrogen Sulphate ⟶ Nitrogen + Sulphur Dioxide + Water + Ammonia

$$(NH_4)_2SO_4 \longrightarrow N_2 + 3SO_2 + 3H_2O + NH_3$$

Ammonium Sulphides, [$(NH_4)_2S$] : Ammonium sulphide is formed when an equal ammonia and hydrogen sulphide both gaseous state react with other. Bineau (1838) did this experiment.

Bineau saw that when ammonia and hydrogen sulphide gas reacted with each other at the ratio of 2 : 1 by volume at a temperature of -20°C gave crystals of ammonium mono sulphate $(NH_4)_2S$.

Ammonia + Hydrogen Sulphide ⟶ Ammonium Monosulphide

$$NH_3 + H_2S \longrightarrow (NH_4)_2S$$

But Bloxam found these crystals always contain ammonium hydrogen sulphide and although he obtained the mono sulphide by carefully adjusting their rates of flow, but the product still was contaminated by hydrosulphide. Pure ammonium hydrosulphide is prepared in fine needles on passing dry ammonia and dry hydrogen sulphide, alternately into eather. It dissociates very rapidly and for this cause it is kept in proper sealed bottles.

When hydrogen sulphide is passed into concentrated ammonia (0.880 Sp gr.) solution diluted with four times its volume of water, a solution of ammonia hydrogen sulphide is formed crystals of $(NH4)_2 2NH_4HS$ separate on cooling to 0°C. The fresh solution of ammonium hydrosulphide is colourless but becomes yellow owing to liberation of sulphur when reacts with atmospheric oxygen *ammonium polysulphide.*

Ammonium Hydrogen Sulphur + Oxygen ⟶ Ammonium Polysulphide + Water

$$(NH_4)_2SO_4 + 3O_2 \longrightarrow 2(NH_4)_2S_2O_3 + 3H_2O$$

The same yellow ammonium sulphate is obtained by digesting flowers of sulphur with ammonium hydrosulphide. The main product appears to be ammonium pentasulphide. On prolonged exposure to air the solution deposit sulphur and forms a colourless solution containing *ammonium thyosulphate.*

$$\text{Ammonium Petasulphide} + \text{Oxygen} \longrightarrow \text{Ammonium Thyosulphate} + \text{Sulphur}$$
$$2(NH_4)_2S_5 + 3O_2 \longrightarrow 2(NH_4)_2S_2O_3 + 6S$$

Ammonium Nitrate, (NH_4NO_3) : Ammonium Nitrate was first prepared by Glauber and was called *nitrium flammans.*

$$\text{Ammonium Hydroxide} + \text{Nitric Acid} \longrightarrow \text{Ammonium Nitrate} + \text{Water}$$
$$NH_4OH + HNO_3 \longrightarrow NH_4NO_3 + 3H_2O$$

In industry it can prepared by following methods process:

1. By passing Ammonia gas through 62% concentrated nitric acid, where ammonium nitrate produced which is separated after the evaporation of the liquid.
2. By the reaction of Calcium nitrate on ammonium sulphate.

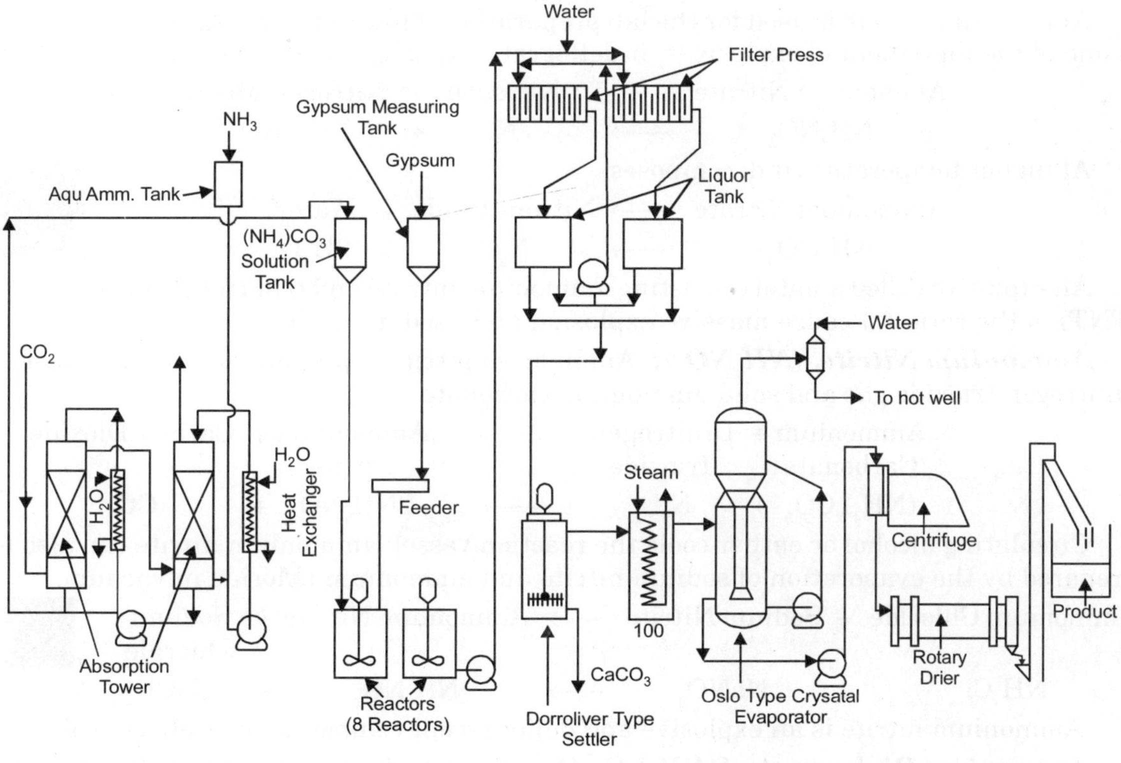

Fig. 10.31. The Industrial process for manufacture of Ammonium Sulphate

$$\text{Ammonium Sulphate} + \text{Calcium Nitrate} \longrightarrow \text{Ammonium Nitrate} + \text{Calcium Sulphate}$$
$$(NH_4)_2SO_4 + Ca(NO_3)_2 \longrightarrow NH_4NO_3 + 2CaSO_4$$

3. Also by using sodium nitrate instead of sodium chloride in ammonia-soda process

$$\underset{\text{Bicarbonate}}{\text{Ammonium}} + \underset{\text{Nitrate}}{\text{Sodium}} \longrightarrow \underset{\text{Nitrate}}{\text{Ammonium}} + \underset{\text{Bicarbonate}}{\text{Sodium}}$$

$$NH_4HCO_3 + NaNO_3 \longrightarrow NH_4NO_3 + NaHCO_3$$

Ammonium nitrate exists in five crystalline forms with definite temperature

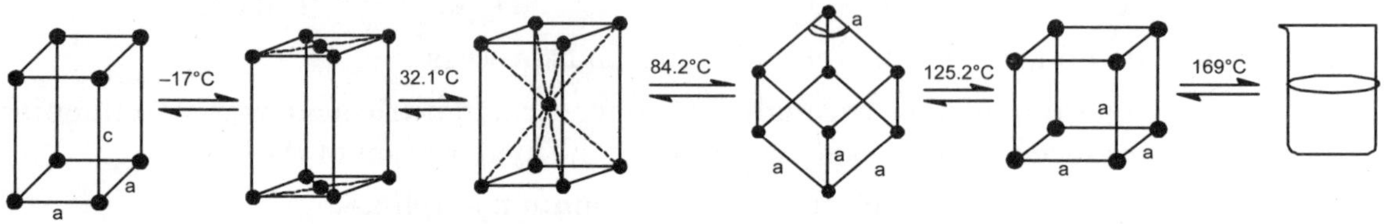

Ammonium nitrate in the anhydrous state melts at 169°C

Ammonium nitrate is used for the lab preparation of nitrous oxide. Ammonium nitrate is one of the ingredient of explosives, it deflograte[*] at 250°C with a yellow flame.

$$\text{Ammonium Nitrate} \longrightarrow \text{Nitrogen} + \text{Nitrogen Monoxide} + \text{Water}$$

$$NH_4NO_3 \longrightarrow N_2 + NO + 2H_2O$$

At higher temperature it decomposes

$$\text{Ammonium Nitrate} \longrightarrow \text{Nitrous Oxide} + \text{Water}$$

$$NH_4NO_3 \longrightarrow N_2O + 2H_2O$$

EXPLOSIVE

An explosive called amatol consisting ammonium nitrate and trinitrotoluene (TNT) in the ratio 4:1 cause massive explosion and used in bombs.

Ammonium Nitrite, (NH_4NO_2): Ammonium nitrite is prepared by the action of dinitrogen trioxide gas and solid ammonium carbonate.

$$\underset{\text{Carbonate}}{\text{Ammonium}} + \underset{\text{Trioxide}}{\text{Dinitrogen}} \longrightarrow \underset{\text{Nitrite}}{\text{Ammonium}} + \text{Carbon Dioxide}$$

$$(NH_4)_2CO_3 + N_2O_3 \longrightarrow 2NH_4NO_2 + CO_2$$

Circulating alcohol or eather cools the reaction vessel; ammonium nitrite can also be prepared by the evaporation of sodium nitrite and ammonium chloride in vacuum.

$$\text{Ammonium Chloride} + \text{Sodium Nitrite} \longrightarrow \text{Ammonium Nitrite} + \underset{\text{Chloride}}{\text{Sodium}}$$

$$NH_4Cl + NaNO_2 \longrightarrow NH_4NO_2 + NaCl$$

EXPLOSIVE

Ammonium nitrite is an explosive and deliquescent compound, as well as explosive.

Ammonium Dichromate, [$(NH_4)_2Cr_2O_7$] : It comes from ammonium chromate when it looses ammonia, also it is prepared by the reaction of ammonia and chromic trioxide.

$$\text{Chromic Trixide} + \text{Ammonia} \longrightarrow \text{Ammonium Dichromate}$$

$$2CrO_3 + NH_4OH \longrightarrow (NH_4)_2Cr_2O_7$$

It is formed as red crystals, when heated strongly it decomposes into chromic oxide and nitrogen, this can be proved by an experiment known as Chemical Volcano.

Exp : On an asbestos board some ammonium dichromate is piled and a filter paper moistened with spirit is tucked in the pile, when the pile is ignited, slowly it catches fire and just like natural volcano it irrupts.

Ammonium dichromate is used in the laboratory as a chemical reagent.

$$\text{Ammonium Dichromate} \longrightarrow \text{Chromic Oxide} + \text{Nitrogen} + \text{Water}$$
$$(NH_4)_2Cr_2O_7 \longrightarrow Cr_2O_3 + N_2 + 4H_2O$$

Microcosmic Salt, Ammonium Sodium Hydrogen Phosphate, [Na(NH)$_4$HPO$_4$]: It is also known as sodium ammonium hydrogen phosphate. It is called microcosmic salt because it is found in the urine, of the microcosm, man. It is prepared in the lab by the reaction of sodium phosphate and ammonium chloride in worm water in a ratio of 6 : 1 and then filtered off the residue.

$$\underset{\text{Phosphate}}{\text{Sodium}} + \underset{\text{Chloride}}{\text{Ammonium}} \longrightarrow \underset{\text{Chloride}}{\text{Sodium}} + \underset{\text{Hydrogen Phosphate}}{\text{Sodium Ammonium}}$$
$$NaHPO_4 + NH_4Cl \longrightarrow NaCl + Na(NH)_4HPO_4$$

When the salt is strongly heated it decomposes and looses ammonia and also sodium metaphosphate is produced with water.

$$\underset{\text{Hydrogen Phophate}}{\text{Sodium Ammonium}} \longrightarrow \text{Ammonia} + \text{Water} + \underset{\text{Metaphosphate}}{\text{Sodium}}$$
$$Na(NH)_4HPO_4 \longrightarrow NH_3 + H_2O + NaPO_3$$

Microcosmic salt is used in the lab for analysis.

FERTILIZER

*Ammonium Sulphate

Among the fertilizers used the most common and widely used is ammonium sulphate. This fertilizer is used for paddy cultivation. Ammonium Sulphate consists of 18 to 21% of nitrogen and is therefore a low concentration nitrogen fertilizer. The repeated use this fertilizer. The repeated use this fertilizer makes the soil acidic therefore, it is needed to neutralize the soil with application of slaked lime (calcium hydroxide) from time to time. There are different types of industrial processes for t manufacturing of the compound, among one is mentioned below (see Fig. 10.31).

From ammonia and Sulphuric Acid : This process produces high-grade ammonium sulphate from synthetic ammonium and concentrated sulphuric acid. These are allowed to react to produce ammonium sulphate, which is concentrated and granulated.

The process is worked in an evaporator called *Oslo Crystal Evaporator.* The evaporator operates at 75°C and 450 to 500 mm of Hg Vacuum the medium is kept acidic to get white crystals.

The reaction equation is stated below :

$$\text{Ammonia} + \text{Sulphuric Acid} \longrightarrow \text{Ammonium Sulphate}$$
$$2NH_3 + H_2SO_4 \longrightarrow (NH_4)_2SO_4$$

*Deflograte catches fire when heated.

CHAPTER

11

Sulphur and Sulphates

Introduction

Sulphur is one of the ancient substances known in the history of chemical science. From the time Egyptians sulphur was recognized. The use of burning sulphur was stated by Homer (900 B.C).

The alchemists recognized sulphur as the Principle of Combustibility and a constituent of metals. Lavoisier (1777) first stated that sulphur is a distinct element.

Occurrence

Sulphur occurs in free state as well as in the form of different compounds as sulphates and sulphides. Deposits of free sulphur are generally found in volcanic residue. The most important sources of elementary sulphur are the deposits of Sicily, Louisiana arid Texas in U.S.A and also in Japan. In India, sulphur deposits are found in the States of Bihar, Karnataka and Tamil Nadu, Himachal Pradesh and Rajasthan.

Sulphur mainly occurs in combined state in the form of ores, which are compounds of sulphur in combination with different metals, e.g. Galena, (PbS), Lead Sulphide, Blende (ZnS), Zinc Sulphide also known as Sphalerite, Cinnabar (HgS). Mercuric Sulphide, also found in gypsum and Anhydride of Gypsum $CaSO_4$, calcium sulphate.

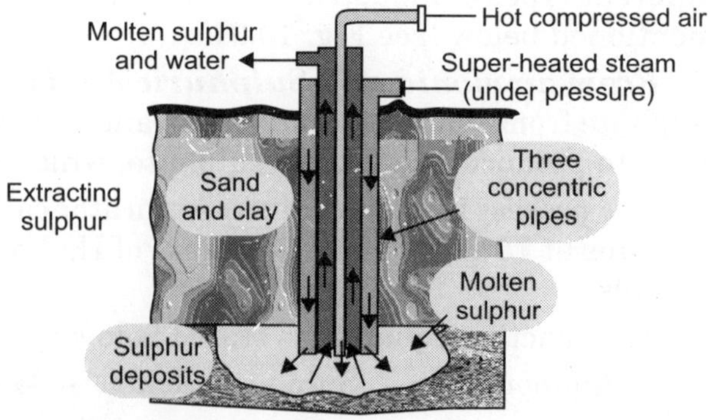

Fig. 11.1. The Frasch Process

Extraction of Sulphur

In American deposits sulphur occurs in depth of 212.8- 457.2m. Three concentric pipes are shrunk down into the sulphur bed (see Fig. 11.1).

Superheated water at 170°C tempera-ture and at a pressure of 12.168-15.21 pascal is forced down the outer pipes; the superheated water melts the sulphur. Now hot compressed air is passed through the central pipe. The simultaneous upward thrust of the compressed air and the super-heated water forces the molten sulphur (as foam of water sulphur and air, also known as flower of sulphur cools it solidifies and then it is isolated from water. Herman Frasch (1900) developed the process that is why this is known as Frasch Process.

Silician Process

Sulphur deposits are also found in Sicily is associated with 70% of clay and Gypsum. In order to recover pure sulphur, the sulphur bearing rocks are filled in the inclined floor of the circular Kiln also known as *Calcorni* (see Fig. 11.2) and ignited. A part of sulphur is bummed and acts as fuel. The rest of the sulphur melts and collects at the lowest bottom of the kiln, this method of recovering the sulphur is lost as fuel, and the value is 2 of the total value.

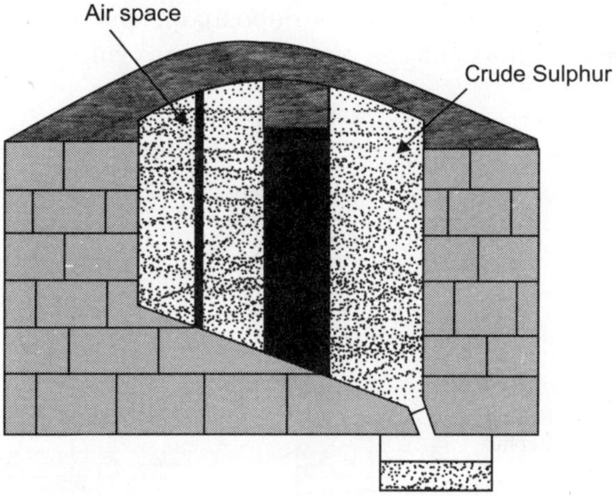

Fig. 11.2.

The Calcorni Kilns are replaced by Gill Kilns. These kilns consist of a series of chambers connected with one another by pipes and are arranged in a circular manner (see Fig. 11.3). A chimney is connected at the centre of the chambers. Each has an air inlet.

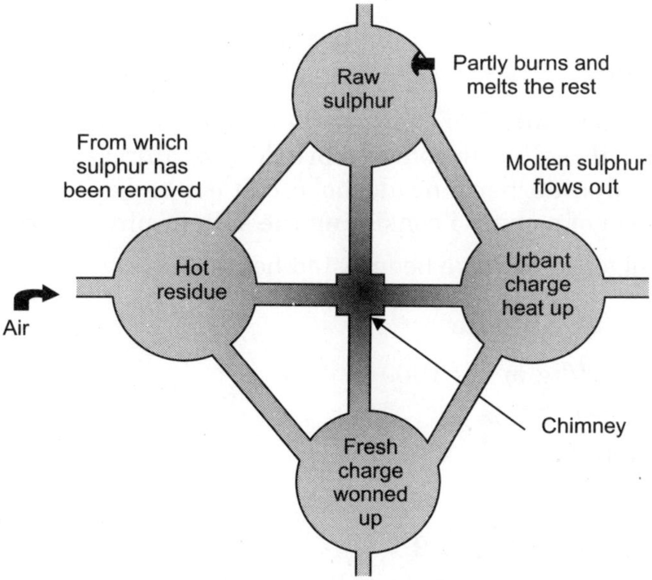

Fig. 11.3. The Diagram of Gill Kiln

Air enters the first chamber and gets heated up this air is then circulated inside the kiln thus saving the chambers, the recovery of the chambers is about 70%. The sulphur is cooled in the wooden moulds and is termed as *block sulphur*. This is refilled about 95%.

Refining Sulphur

Impure sulphur is heated to boiling point and vapours are collected in a brick chamber when sulphur vapours deposit on the walls of the chamber as flowers of sulphur. The molten mass at the base is cast into rolls of sulphur (see Fig. 11.4).

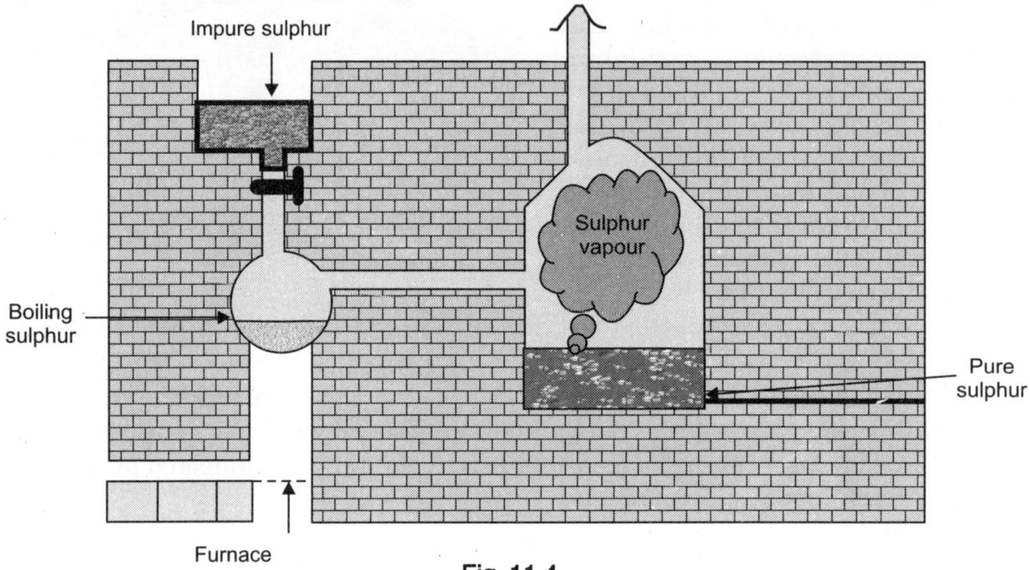

Fig. 11.4.

Allotrops of Sulphur

Sulphur occurs in various allotropic forms, which show the different physical properties. Ordinary sulphur is pale yellow in colour, a brittle crystalline solid, insoluble in water, but soluble in carbon disulphide, benzene alcohol etc. It is a bad conductor of electricity. Below it is shown in the form of a chart I consisting the salient physical properties of sulphur.

The properties of sulphur have been listed below.

Physical Properties of Sulphur

Density of Sulphur [/kg m^{-3}] : 1960

Molar volume [/cm^3] : 15.53

Bulk modulus [/GPa] : 7.7

Hardnesses

Mineral hardness [no units] : 2.0
Electrical resistivity [/10^{-8} W m; or mW cm]: > 10^{23}
Refractive index [no units] : 1.001111

Valence Shell Orbital Radii

The following are calculated values of valence shell orbital radii, R_{max}

Orbital	Radius [/pm]	Radius [/AU]
s	75.1	1.41911
p	85.5	1.61545

Thermodynamic Properties

State	ΔfH° /kJ mol^{-1}	ΔfG° /kJ mol^{-1}	S° /JK^{-1} mol^{-1}	C_pH /JK^{-1}mol^{-1}	$H°_{298.15}$-H°0 /kJ mol^{-1}
Solid (rhombic)	*0	0	*32.054 ± 0.050	22.6	*4.412 ± 0.006
Solid (monoclinic)	0.3				
Gas	*277.17 ± 0.15	238.28	*167.829 ± 0.006	23.67	*6.657 ± 0.001
Gas (S_2)	*128.60 ± 0.30	79.3	*228.167 ± 0.010	32.5	*9.132 ± 0.002
Gas (S_8)	102.3	49.7	430.9	156.4	31.51

Thermal Data

Melting point [/K] : 388.36 [or 115.21 °C (239.38 °F)]

Boiling point [/K] : 717.87 [or 444.72 °C (832.5 °F)] (liquid range: 329.51 K)

Critical temperature [/K] : 1314 [or 1041 °C (1906 °F)]

Atomic Dimensions

Bond length in SS [/pm] : 205

Atomic radius (empirical) [/pm] : 100

Atomic radius (calculated) [/pm] : 88

Covalent radius (empirical) [/pm] : 102

Van der Waals radius [/pm] : 180

Ionization Energies and Electron Affinity

Electron affinity [/kJ mol^{-1}] : 200

Ionization Energies

	Ionization energy number Enthalpy /kJ mol^{-1}
1st	999.6
2nd	2252
3rd	3357
4th	4556
5th	7004.3
6th	8495.8
7th	27107
8th	31719
9th	36621
10th	43177

Electronic Configuration

Ground state Electronic Configuration: [Ne].$3s^2.3p^4$

Shell Structure 2.8.6

Term symbol: $^3P_2$

Electronegativity

Pauling electro negativity [/Pauling units]: 2.58

Sanderson electro negativity [/Pauling units]: 2.96

Allred Rochow electro negativity [/Pauling units]: 2.44

Mulliken-Jaffe electronegativity [/Pauling units]: 2.69 (16.7% s orbital)

Allen Electronegativity [/Pauling units]: 2.589

Effective Nuclear Charges

1s	15.54				
2s	10.63	2p	11.98		
3s	6.37	3p	5.48	3d	
4s		4p		4d	4f
5s		5p		5d	
6s		6p			
7s					

Electron Binding Energies

Label	Orbital	eV
K	1s	2472
L_I	2s	230.9
L_{II}	$2p_{1/2}$	163.6
L_{III}	$2p_{3/2}$	162.5

We shall now discuss the allotropic form and their respective properties. Sulphur has different allotropic forms they are discussed below:

1. Rhombic Sulphur : This form of sulphur is also stable form of sulphur below 95.5°C (the transition point) and most of the other forms turn to this temperature if allowed to stand below this temperature. The melting point of rhombic sulphur depends on the method of heating the substance and on the nature of the liquid sulphur with which it is in equilibrium. If rhombic sulphur is heated very slowly it will convert to the monolithic sulphur and the melting point obtained will be that of the monolithic variety. If the rate of heating is increased some what, rhombic should ideally come into equilibrium with liquid sulphur only in the lambda form (λ-form) and melting point is 112.8°C. Rhombic sulphur is lemon yellow insoluble in water, slightly soluble in ethyl alcohol, diethyleather and benzene and very much soluble in carbon disulphide. Its density is 2.07g/cm^3 and its hardness is 2.5cm on Moh's Scale. Its molecular formula is S8 and its molecular configuration is a ring of eight covalently bonded sulphur atoms in the shape of wrinkled crown with bond distances of 2.12 angstrom and bond angles of 105°C (see Fig. 11.5).

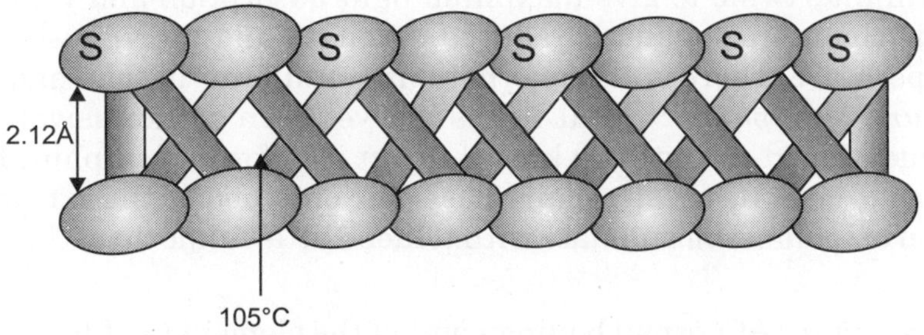

Fig.11.5. The Structure of Sulphur Molecule

2. Monoclinic Sulphur : Also called *prismatic sulphur* and *beta sulphur (P-sulphur)* is the stable modification of the element above the transition temperature and below the melting point. It crystallizes form molten sulphur in the needle like prisms, which are almost colourless. Its molecular configuration is also eight atom wrinkled-crowned structures and this form also soluble in carbon disulphide and insoluble in water.

3. Plastic Sulphur : Also called Gamma Sulphur (γ-Sulphur) is formed when molten sulphur at or near normal boiling is squeezed to the solid state (such as by pouring in cold water) (see Fig. 11.6).

This form of sulphur is partly soluble in carbon di sulphide. It was thought to be composed of two types of sulphur λ- sulphur and m sulphur ($S\lambda$ soluble in carbon disulphide) and ($S\mu$ insoluble in carbon disulphide)

Mu-sulphur is insoluble in carbon disulphide

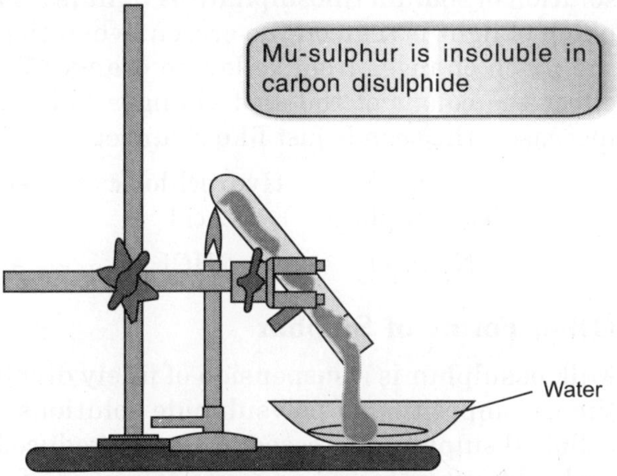

Fig. 11.6. The Plastic Sulphur Preparation

These forms probably exist in liquid sulphur also because plastic sulphur appears to be only the super-cooled liquid. After long standing at room temperatures this form of sulphur reverts to the rhombic form. Plastic sulphur exists as a long zigzag, chain of sulphur atoms and if it is strongly stretched it behaves like a rubber because the zigzag chains are obtained in straight lines and the material becomes fibrous and rigid.

4. Purple Sulphur : Is formed by the sudden cooling of sulphur vapour at an evaluated temperature (where sulphur exists as S_2) units. It is unstable and reverts to yellow sulphur upon being warmed to room temperature.

5. Liquid Sulphur : Liquid sulphur possesses the remarkable property when it is heated. Liquid sulphur is obtained by heating elementary sulphur at the melting point. Its colour is changed to dark reddish black as its viscosity increases and both colours has a maximum darkness when the viscosity is maximum at 200°C. Above this temperature the colour lightens and the viscosity diminishes. It is thought that at the melting point liquid sulphur chains which can intertwine to give maximum light adsorption and viscosity. Above this temperature the chain breaks and is reduced in length and the viscosity decreases with increasing temperature. There is also another form of sulphur in solid and molten sulphur called *pi-sulphur 7r- sulphur,* S_x. But it has not yet been so extensively studied as the forms already mentioned and little is known about it. A form of sulphur, Rho-sulphur (ρ-sulphur) believed to be S_6 has been repaired as being obtained by the extraction of acidified aqueous solution of sodium thiosulphate with toluene little is known of this form of sulphur, however.

6. Gaseous Sulphur : At normal boiling point of the element (444.6°C) gaseous sulphur is orange yellow colour. As the temperature increases the colour becomes deep red and then becomes lighter until 650°C it is straw yellow several molecular spices are in equilibrium in gaseous sulphur, S_8, S_6, S_4 and S_2. The proportions vary with the temperature. At the normal boiling point the vapour largely. Above 2000°C it is dissociated into sulphur atoms.

7. Colloidal Sulphur : It is obtained when hydrogen sulphide gas is passed through a cooled solution of sulphur dioxide in water. When dilute hydrochloric acid is added to the solution of sodium thiosulphate. A light is allowed to pass through the hydrochloric acid a patch of light is seen on the screen, when the thiosulphate solution is added the colour of the patch changes from yellow to orange. The shape also becomes round due to tyndall effect the colour of the spot changes like setting sun as the deposition of the sulphur increases, the seen is just like a sunset.

Sodium Thiosulphate + Hydrochloric Acid ⟶ Sodium Chloride + Sulphur Dioxide + Coloidal Sulphur + Water

$$Na_2S_2O_3 + 2HCl \longrightarrow 2NaCl + SO_2 + S + H_2O$$

Other Forms of Sulphur

Milk of sulphur is a suspension of finely divided amorphous sulphur in water, obtained by the decomposition of polysulphide solutions with acid. It is soluble in carbon disulphide colloidal sulphur depressed in water produced by the action of gaseous hydrogen sulphide and cold concentrated aqueous solutions of sulphur dioxide, or by decomposing sodium thiosulphate with sulphuric acid. It decomposes quite slowly in carbon disulphide.

Circulation of sulphur in all living matter takes place through a cycle called sulphur cycle, sulphur occurs in various forms in nature and also as a component of certain amino acid. It is abundant in soil and as main constituent of protein and through a series of microbial transformation ends up as sulphates usable by plants.

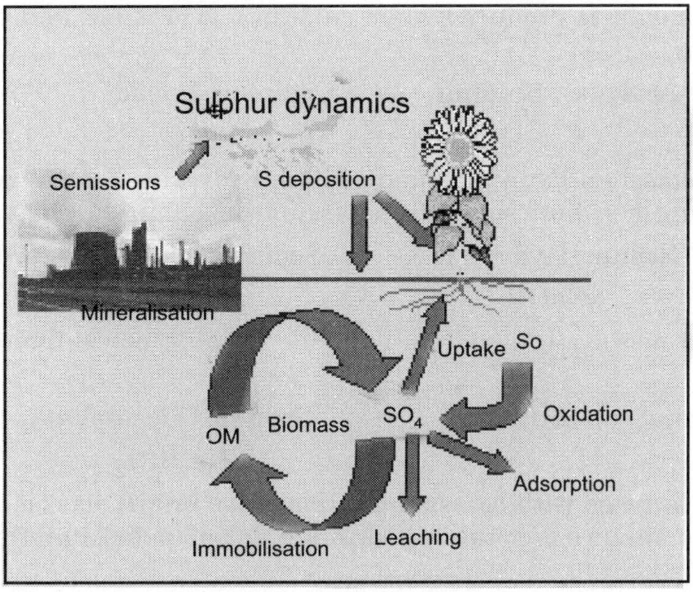

Fig. 11.7. The Sulphur Cycle

Sulphur containing proteins are degraded into their constituent amino acids by the action of variety of soil organisms, the sulphur of the amino acids is converted to hydrogen sulphide (H_2S) by another series of soil microbes. In the presence of oxygen, hydrogen sulphide is converted to sulphur and then to sulphate by sulphur bacteria. Eventually. the sulphate becomes hydrogen sulphide. Hydrogen sulphide rapidly gets converted into the water-soluble oxides of sulphur forming sulphurous acids and also sulphuric acids. These compounds contribute in large part to the acid rain that can kill sensitive aquatic organisms and damage marble monuments and stone buildings.

Chemical Properties of Sulphur

Sulphur is insoluble in water but crystalline sulphur is soluble in carbon di sulphide but amorphous sulphur is insoluble in carbon disulphide.

Sulphur is an inflammable substance and burns in air or oxygen with a blue flame to form sulphur dioxide and a little amount of sulphur trioxide. Almost all the elements except gold platinum and silver react with sulphur Boiling sulphur react with steam to form hydrogen sulphide and sulphur dioxide.

$$\text{Steam} + \text{Sulphur} \longrightarrow \text{Hydrogen Sulphide} + \text{Sulphur Dioxide}$$

$$2H_2O + 3S \longrightarrow 2H_2S + SO_2$$

Hot concentrated nitric acid reacts with sulphur to produce sulphuric acid

$$\text{Nitric Acid} + \text{Sulphur} \longrightarrow \text{Sulphuric Acid} + \text{Nitrogen Dioxide} + \text{Water}$$

$$6HNO_3 + S \longrightarrow H_2SO_4 + 6NO_2 + 2H_2O$$

Also sulphur dioxide is produced when sulphuric acid is hot and concentrated state reacts with sulphur.

Sulphuric Acid	+	Sulphur	$\longrightarrow$	Sulphur Dioxide	+	Water
$2H_2SO_4$	+	S	$\longrightarrow$	$3SO_2$	+	$2H_2O$

Solutions of potassium hydroxide and sodium hydroxide reacts with sulphur give a mixture of potassium or sodium sulphide, potassium or sodium sulphate also thiosulphate.

Sulphur	+	Sodium Hydroxide	$\longrightarrow$	Sodium Sulphide	+	Water
3S	+	6NaOH	$\longrightarrow$	$2Na_2S$	+	$3H_2O$

Sulphur gets a dissolved in sodium sulphide forming polysulphide, e.g. Na_2S_2 and Na_2S_4

Sodium Sulphite	+	Sulphur	$\longrightarrow$	Sodium Thiosulphate
Na_2SO_3	+	S	$\longrightarrow$	$Na_2S_2O_3$

When sulphur is fused with potassium carbonate, a brown mass is formed called liver of sulphur, this is a mixture of potassium di sulphide and potassium thiosulphate.

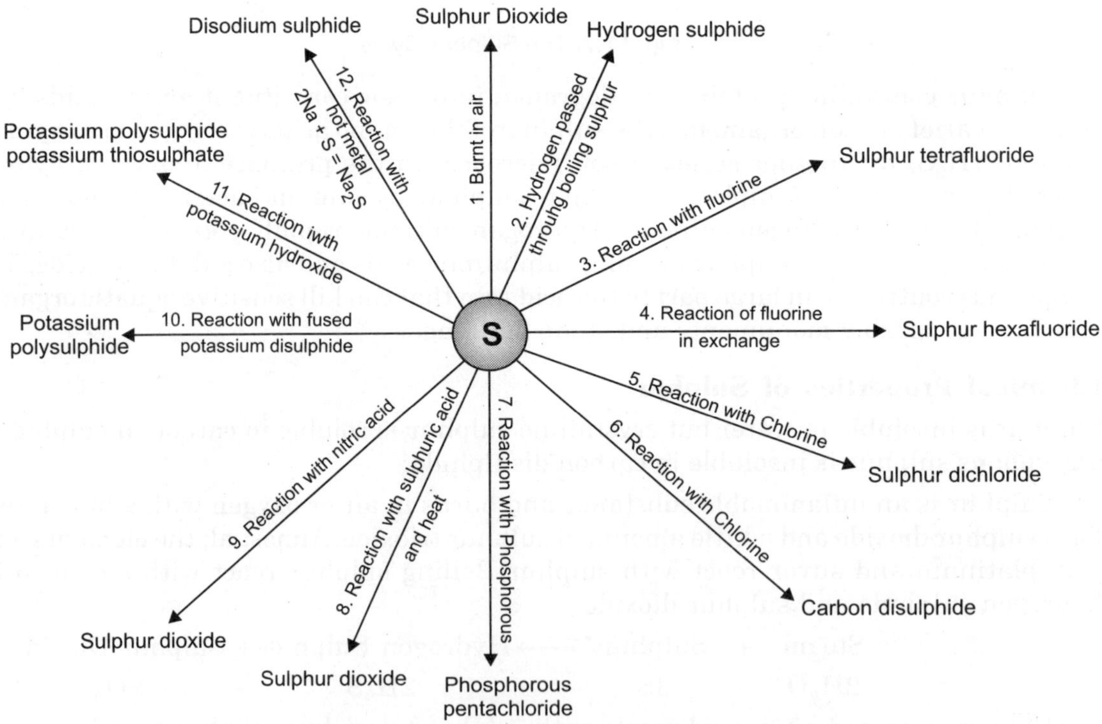

Fig. 11.8. The Reaction Cycle of Sulphur

Reactions (2) and (5) occur when hydrogen gas and chlorine gas is bubbled through hot sulphur. Reactions (6) and (7) and (12) take place on heating the components to high temperature. Reactions (8) and (9) take place on heating sulphur with sulphuric acids and t he reactions (10) and (11) take place a little above room temperature.

Sodium Sulphide + Sulphur $\longrightarrow$ Sodium Polysulphide

$$Na_2S + S \longrightarrow Na_2S_2$$

$$2Na_2S + 6S \longrightarrow Na_2S_4$$

Sulphur + Potassium Carbonate $\longrightarrow$ Potassium Disulphate + Potassium Thiosulphate + Carbon Dioxide

$$6S + 3K_2CO_3 \longrightarrow 2K_2S_2 + K_2S_2O_3 + CO_2$$

Compounds of Sulphur

We shall now study about the compounds of sulphur under this subhead, as we all know that sulphur reacts with many other elements and also compounds to *form* different compounds and many chemical complexes. We shall study them in detail.

I. Oxy-compounds of sulphur : These compounds of sulphur are produced when sulphur reacts with oxygen, they are sulphur dioxide, sulphur trioxide, etc.

SULPHUR DIOXIDE

Introduction

Sulphur dioxide is one of the main compounds of sulphur, which is formed, with the reaction of oxygen.

Occurence

Sulphur dioxide occurs in the volcanic ashes, also when sulphur pyrites are burned in air. Mainly sulphur dioxide is obtained by the reaction of burning sulphur in oxygen, when burning sulphur is introduced in the gas jar full of oxygen the sulphur bursts into flame and then starts burning in a blue flame forming fumes of sulphur dioxide.

Sulphur + Oxygen $\longrightarrow$ Sulphur Dioxide

$$S + O_2 \longrightarrow SO_2$$

It is also obtained by the reaction of sulphides and bisulphites of different compound with sulphuric and hydrochloric acid.

1. Calcium Bisulphite and dilute hydrochloric acid reacts to form sulphur dioxide and calcium chloride.

Calcium Bisulphite + Hydrochloric Acid $\longrightarrow$ Calcium Chloride + Sulphur Dioxide + Water

$$CaCO_3 + 2HCl \quad CaCl_2 + SO_2 + H_2O$$

2. Sodium thiosulphate also reacts with dilute hydrochloric acid to form sulphur and sulphur dioxide.

Sodium Thiosulphate + Sulphuric Acid ⟶ Sodium Sulphate + Water + Sulphur + Sulphur Dioxide

$$Na_2S_2O_3 + H_2SO_4 \longrightarrow Na_2SO_4 + H_2O + S + SO_2$$

3. Sodium Thiosulphate also reacts with hydrochloric acid to form sulphur and sulphur dioxide

Sodium Thiosulphate + Hydrochloric Acid ⟶ Sulphur + Sulphur Dioxide + Sodium Chloride + Water

$$Na_2S_2O_3 + 2HCl \longrightarrow S + SO_2 + 2NaCl + H_2O$$

4. By roasting sulphides such as zinc sulphide and iron pyrites this method is also the one used for preparing large volumes of sulphur dioxide needed in industry

(*i*) Zinc Sulphide + Oxygen ⟶ Sulphur Dioxide + Zinc Oxide

$$2ZnS + 3O_2 \longrightarrow 2SO_2 + 2ZnO$$

(*ii*) Iron Pyrites + Oxygen ⟶ Ferric Oxide + Sulphur Dioxide

$$4FeS_2 + 11O_2 \longrightarrow 2Fe_2O_3 + 8SO_2$$

5. Concentrated sulphuric acid when reacts with metals, like copper mercury and lead sulphur dioxide is produced.

(*i*) Copper + Sulphuric Acid ⟶ Sulphur Dioxide + Water

$$Cu + 2H_2SO_4 \longrightarrow SO_2 + 2H_2O$$

(*ii*) Mercury + Sulphuric Acid ⟶ Mercury Sulphate + Sulphur Dioxide + Water

$$Hg + 2H_2SO_4 \longrightarrow HgSO_4 + SO_2 + 2H_2O$$

6. Concentrated sulphuric acid reacts with non-metals like carbon and sulphur to form sulphur dioxide.

(*i*) Sulphuric Acid + Carbon ⟶ Sulphur Dioxidee+ Carbon Dioxide + Water

$$2H_2SO_4 + 2C \longrightarrow SO_2 + 2CO_2 + 2H_2O$$

(*ii*) Sulphur + Sulphuric Acid ⟶ Sulphur Dioxide + Water

$$6S + 4H_2SO_4 \longrightarrow 6SO_2 + 2H_2O$$

Laboratory Preparation of Sulphur Dioxide

Theory : Sulphur Dioxide in laboratory is prepared by the action of copper tunglings and sulphuric acid. The gas is collected by the upward displacement of air. (See Fig. 11.9).

Sulhpuric Acid + Copper ⟶ Sulphur Dioxide + Copper Sulphate + Water

$$2H_2SO_4 + Cu \longrightarrow SO_2 \Uparrow + CuSO_4 + 2H_2O$$

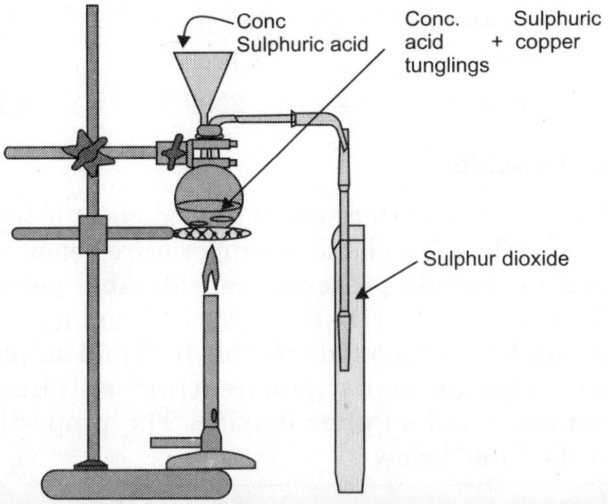

Apparatus	Reagents	Temperature	Pressure
Round bottom flask, stand, tripod burner, delivery, tube etc.	Concentrated	105°C	Normal

Fig. 11.9. The Lab Preparation of Sulphur Dioxide

Precaution : As sulphur dioxide is an unpleasant has it should not be inhaled directly or in directly, the gas is soluble in water and heavier than air, therefore, it is collected by the downward displacement of air.

Other Methods of Preparation of Sulphur Dioxide

Among the other methods the most important is the method in which sodium meta bi sulphate reacts with hydrochloric acid, the most important thing is this that no heat is required in this reaction. Fig. 11.10 shows the experimental set up for the reaction.

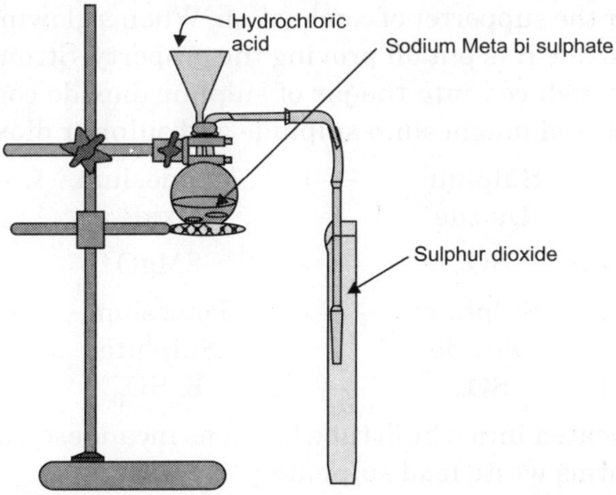

Fig. 11.10. The Preparation of Sulphur Dioxide from Sodium Meta Bisulphate

Sodium Metabi-sulphate	+	Hydrochloric Acid	⟶	Sodium Chloride	+	Sulphur Dioxide	+	Water
$Na_2S_2O_5$	+	2HCl	⟶	2NaCl	+	$2SO_2$	+	H_2O

Properties of Sulphur Dioxide

Sulphur dioxide is a colourless gas with a suffocating odour, it is heavier than air and can be easily condensed to a colourless liquid under temperature below –10°C the liquid freezes at – 72.7°C. Liquid sulphur dioxide possesses considerable polar character (dielectric constant 14.5 and dipole moment is 1.61D). It is a good solvent for non-metals like sulphur, phosphorous iodine etc, sulphite compounds soluble in liquid sulphur behaves like bases and the thyonil compounds as acids. On heating, sulphur dioxide undergoes disproportionation to sulphur and sulphur dioxide. The properties of sulphur dioxide (physical) are shown in the table below :

Molecular weight	64.06
Density	1.4542g/cc
Melting Point	–75.46°C
Boiling Point	–10.02°C
Dielectric constant	Viscosity
Dipole Moment	1.61D
Viscosity	0.039mNs/cm square
Surface Tension	227.3 Ncm^{-2}
Critical Temperature	157.2°C

Chemical Properties

1. Incombustible and non-supporter of combustion: In normal condition sulphur dioxide is neither combustible nor the supporter of combustion. When a glowing splinter is introduced in the jar of sulphur dioxide it is put off proving the property. Strongly burning potassium or magnesium when introduced into the jar of sulphur dioxide contains to burn forming potassium thiosulphate and magnesium sulphide and sulphur dioxide also dissociates.

(*a*)	Magnesium	+	Sulphur Dioxde	⟶	Magnesium Oxide	+	Magnesium Sulphide
	3Mg	+	SO_2	⟶	2MgO	+	MgS
(*b*)	Potassium	+	Sulphur Dioxide	⟶	Potassium Sulphite	+	Potassium Thiosulphate
	4K	+	SO_2	⟶	K_2SO_3		

Pure lead dioxide heated into a bulb tube becomes incandescent when sulphur dioxide is passed over it and forms white lead sulphate.

$$\text{Lead Dioxide} + \text{Sulphur Dioxide} \longrightarrow \text{Lead Sulphate}$$
$$PbO_2 + SO_2 \longrightarrow PbSO_4$$

Sulphur dioxide also reacts with sodium peroxide forming sodium sulphate and incandescence.

$$\text{Sodium Peroxide} + \text{Sulphur Dioxide} \longrightarrow \text{Sodium Sulphate}$$
$$Na_2O_2 + SO_2 \longrightarrow Na_2SO_4$$

With solutions of sodium hydroxide it forms sulphates and bi sulphates.

(*i*) Sodium Hydroxide + Sulphur Dioxide + Water ⟶ Sodium Sulphite + Water

$$2NaOH + SO_2 + H_2O \longrightarrow Na_2SO_3 + 2H_2O$$

(*ii*) Sodium Hydroxide + Sulphur Dioxide + Water ⟶ Sodium Bisulphite + Water

$$2NaOH + SO_2 + 3H_2O \longrightarrow NaHSO_3 + 3H_2O$$

2. *Thermal Dissociation:* It dissociates when heated strongly at temperature of 1200°C.

$$\text{Sulphur Dioxde} \xrightarrow{1200°C} \text{Sulphur} + \text{Oxygen}$$
$$SO_2 \longrightarrow S + O_2$$

3. *Acidic Nature:* Sulphur dioxide dissolves in water to give sulphurous acid, which ionizes as shown below:

Sulphur Dioxide + Water ⟶ Sulphurous Acid ⇌ Hydrogen Ion + Sulphur Dioxide

$$SO_2 + H_2O \longrightarrow H_2SO_3 + 2H^+ + SO_2^-$$

Sulphurous Acid is Stable in Dilute Acid if Heated it Looses Whole of its Sulphur Dioxide

4. *Unsaturated Nature:* Sulphur is hexavalent but as sulphur dioxide (SO_2) only four valence atoms of sulphur are satisfied by 2 atoms of oxygen. The remaining two valence atoms of sulphur remain unsaturated and render sulphur dioxide as unsaturated compound. It gives addition products with oxygen, chlorine and lead oxide or barium chloride.

Contains tetravalent Sulphur ⟶ Sulphur Trioxide (contains hexavalent sulphur)

$$2SO_2 + O_2 \longrightarrow 2SO_3$$

5. *Reducing Property:* In presence of moisture it can liberate nascent hydrogen or in presence of an oxidizing agent it can easily take away an atom of hydrogen.

(*i*) Sulphur Dioxide + Water ⟶ Sulphuric Acid + Nascent Hydrogen

$$SO_2 + 2H_2O \longrightarrow H_2SO_4 + 2H$$

(*ii*) Sulphur Dioxide + Water + Oxygen ⟶ Sulphuric Acid

$$SO_2 + H_2O + O_2 \longrightarrow H_2SO_4$$

(iii) Acidified Potassium permanganate is decolourised

(*iv*) Potassium Permanganate + Sulphur Dioxide + Water ⟶ Potassium Sulphate + Sulphuric Acid + Manganese Sulphate

$$2KMnO_4 + SO_2 + 2H_2O \longrightarrow K_2SO_4 + 2H_2SO_4 + 2MnSO_4$$

(v) Chlorine is changed to hydrochloric acid

Chlorine+Water+Sulphur Dioxide ⟶ Hydrochloric Acid +Sulphuric Acid

$$Cl_2 + 2H_2O + SO_2 \longrightarrow 2HCl + H_2SO_4$$

(*vi*) Potassium Dichromate + Sulphuric Dioxide + Sulphuric Acid ⟶ Potassium Sulphate + Chromic Sulphate + Water

$$K_2Cr_2O_7 + SO_2 + 2H_2SO_4 \longrightarrow K_2SO_4 + Cr(SO_4)_2 + H_2O$$

(vii) Iodine is converted to hydrogen iodide

Iodine +Water+ Sulphur Dioxde ⟶ Hydrogen Iodide +Sulphuric Acid

$$I_2 + 2H_2O + SO_2 \longrightarrow 2HI + H_2SO_4$$

(viii) Ferric sulphate is changed to ferrous sulphate

Ferric Sulphate + Water + Sulphur Dioxide ⟶ Ferrous Sulphate + Sulphuric Acid

$$Fe_2(SO_4)_3 + 2H_2O + SO_2 \longrightarrow 2FeSO_4 + 2H_2SO_4$$

6. *Bleaching and disinfecting properties in the presence of moisture sulphur dioxide reduces the* vegetable colourings matter to colourless reduction product. It is therefore often used as bleaching agent On standing in air, the colourless product is again converted to its original colour, due to reaction with atmospheric oxygen. for some cases sulphur dioxide forms a colourless addition product liberating sulphur dioxide, is plotted below:

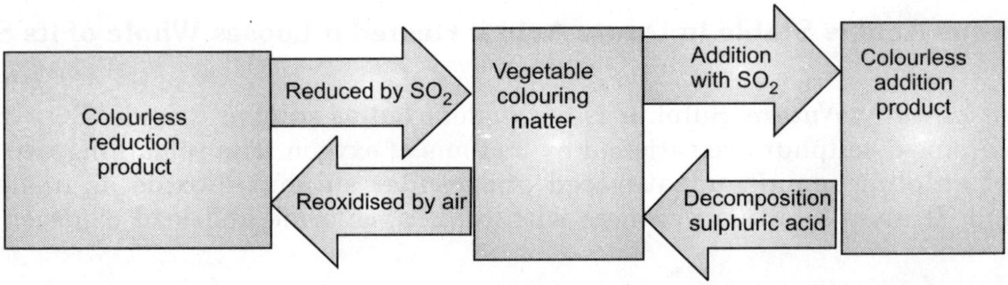

Fig. 11.11.

7. *Oxidising Property :* Its reaction with hydrogen sulphide, sodium, potassium and iron shows it is an oxidizing agent also

(*i*) Hydrogen Sulphate + Sulphur Dioxide ⟶ Sulphur + Water

$$2H_2S + SO_2 \longrightarrow 3S + H_2O$$

(*i*) Iron + Sulphur Dioxide ⟶ Ferric Oxide + Ferrous Sulphide

$$3Fe + SO_2 \longrightarrow 2FeO + FeS$$

Composition of Sulphur Dioxide

Volumetric Composition : In the method of determination the apparatus used is shown in the fig. 11.12. A piece of sulphur is placed on the platinum coil and some dry oxygen is enclosed in the bulb. The level of the mercury is equalized in the bulb and the volume of the oxygen is noted. The electric current is passed through the loop, which gets heated and burns the sulphur, which reacts with dry oxygen forming sulphur dioxide. On cooling no change in volume is notice showing thereby one volume of oxygen combines with sulphur dioxide.

$S + O_2$	$\longrightarrow$	SO_2
1 Volume		1 Volume
1 molecule		1 molecule
2 atoms		1 molecule

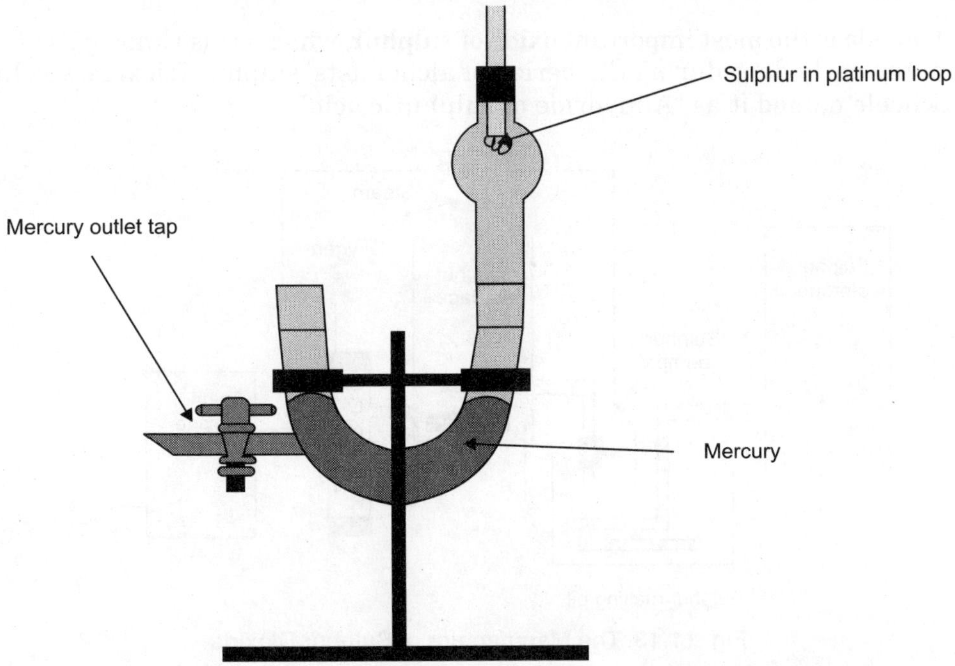

Fig. 11.12. The Apparatus for the Dietermination of the Composition of Sulphur Dioxide

Applying Avogadro's Law to the above we can say that one molecule of oxygen (=2 atoms) combines with sulphur (say x atoms) to produce one molecule of sulphur dioxide is S_xO_2 with molecular weight is $32x + 32$.

Molecular weight of sulphur dioxide as calculated from its vapour density (which is 32) data is found to be $2 \times 32 = 64$.

Forming an equation with the molecular weight as calculated from the equal S_xO_2 we get $32x + 32 = 64$ or $x = 1$

$\therefore$ Molecular formula of sulphur dioxide is SO_2

Gravimetric Composition : To determine the composition of sulphur dioxide by this process, a known weight of sulphur is burnt in a current of pure dry oxygen. Sulphur dioxide produced is absorbed in weighted caustic potash bulbs gives the weight of oxygen that combines with a fixed weight of sulphur. It is found that 32 parts by weight of oxygen combine with 32 parts by weight of sulphur.

Industrial Preparation of Sulphur Dioxide

In this process sulphur from the sulphur storage is send to the melting pit, where sulphur is heated up to the temperature of 126°C by steam for removing the impurities. Then it is sent to the furnace via sulphur pump, where it is heated at a temperature of 760°C, it reacts with oxygen forming sulphur dioxide. This sulphur dioxide is then sent to the boiler for further evaporation at 288°C after that sulphur dioxide is collected by filtration

Introduction

Sulphur trioxide is the most important oxide of sulphur, which is also known as *philosophic salt.* From the early days during the period of alchemists' sulphur trioxide was known. It wa P.B. Scheele named it as "Anhydride of sulphuric acid".

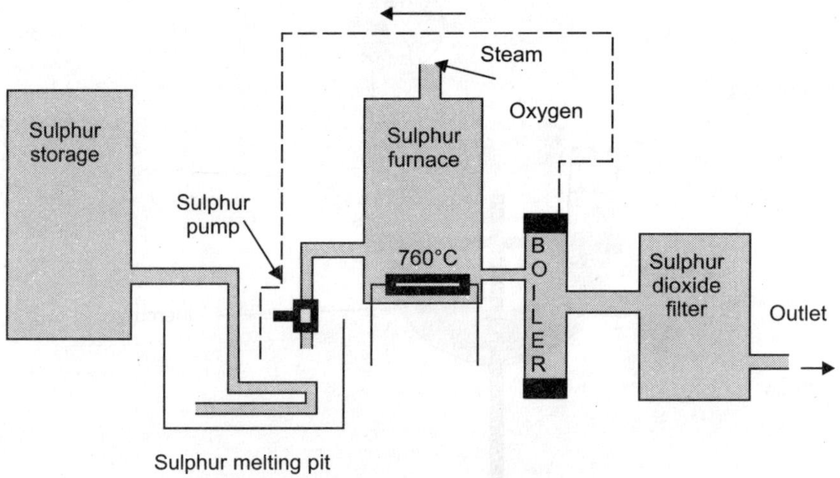

Fig. 11.13. The Manufacture of Sulphur Dioxide

Preparation

1. From direct reaction between sulphur dioxide and oxygen: Sulphur trioxide can be prepared by the direct combination of sulphur dioxide and oxygen.

$$\text{Sulphur Dioxide} + \text{Oxygen} \longrightarrow \text{Sulphur Trioxide}$$

$$2SO_2 + O_2 \longrightarrow 2SO_3$$

This reaction is reversible and much slow at ordinary temperature but in a presence of a catalyst called platinum asbestos or vanadium pentoxide and also at a temperature of 454°C, the reaction is brisk and an optimum yield of sulphur trioxide is obtained. This temperature must be kept constant because higher the temperature lesser than yield and at 1000°C the reaction stops.

2. Sulphur Trioxide by the Dehydration of Sulphuric Acid : This process prepares - the sulphur trioxide, where phosphorous pentoxide absorbs water.) The water contents from the sulphuric acid living its anhydrous part the sulphur trioxide.

Sulphuric Acid	+	Phosphorous Pentoxide	⟶	Metaphosphoric Acid	+ Sulphur Trioxide
H_2SO_4	+	P_2O_5	⟶	$2HPO_3$	+ SO_3

3. By heating Sodium pyrosulphate: when heated forms sodium sulphate and sulphur trioxide.

Sodium Pyrosulphate ⟶ Sodium Sulphate + Sulphur Trioxide

$$Na_2S_2O_7 \longrightarrow Na_2SO_4 + SO_3$$

Laboratory Preparation of Sulphur Trioxide

Theory: In laboratory sulphur trioxide is prepared by the reaction of sulphur dioxide and oxygen in the presence of a catalyst called platinum. The gas is collected by the in an "U" tube and cooled by freezing mixture and contact from the outside atmosphere is avoided. The apparatus set up is shown in Fig. 11.14.

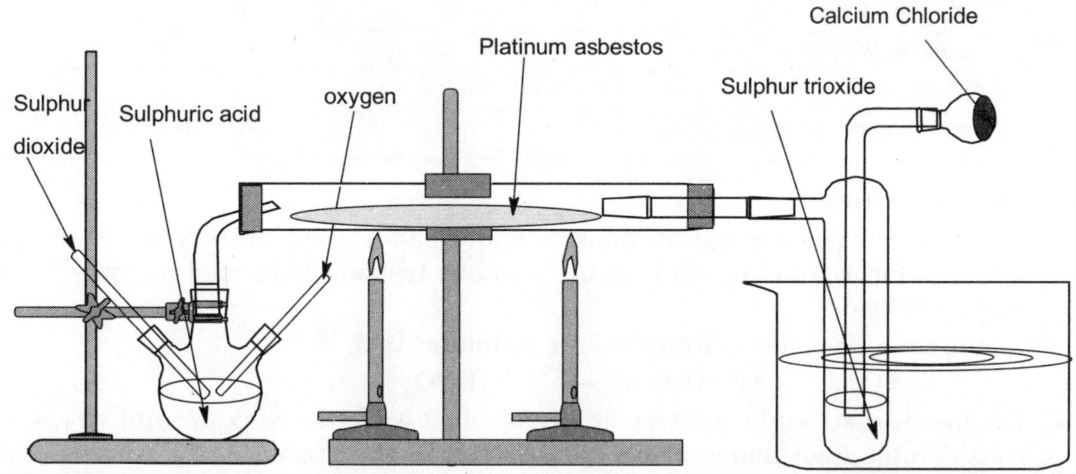

Apparatus	Reagent	Temperature	Catalyst
Woolf's bottle, Glass tube, burner, stand etc.	Sulphur, dioxide, oxygen oxygen, platinum sulphuric sulphuric acid, calcium chloride, freezing mixture,	454°C	Platinum

Fig. 11.14. The Experimental Set up for the Lab preparation of Sulphur Trioxide

PROPERTIES OF SULPHUR TRIOXIDE

Physical Properties : Sulphur Trioxide Mainly Exists in Three Forms

(1) Alpha-Sulphur Trioxide : In this form sulphur trioxide becomes a colourless liquid which has a freezing point at 15°C and melting point of 16.9°C, the shape of the compound is needle like crystals. The boiling point of this sulphur trioxide is 45°C, its specific gravity is 1.92289 (at 25°C). This form is obtained by repeated fractionation of sulphur trioxide obtained by any methods described above. The molecular weight is 80(approx.).

(2) Beta-Sulphur Trioxide: The alpha form if allowed to stand at 16°C for sometimes changes gradually into Beta-form. This form when solidifies at 32.5°C appears as asbestos like needle shape crystals. Its molecular weight is 160 which corresponds to the formula S206, when kept at a temperature between 50 to 100°C

(3) Gamma-Sulphur Trioxide : It is obtained form the Beta-sulphur trioxide on complete drying. It sublimes without melting under ordinary pressure.

The detail information on physical properties is listed in the table.

Melting point (α-sulphur trioxide)	16.9°C
Boiling Point (α-sulphur trioxide)	45°C
Critical Temperature (SO_3 gas)	68°C
Critical Pressure (SO_3 gas)	127cm³/mol
Heat of Fusion (at 16.8°C) (α-sulphur 8.60kJ/mol trioxide)	
Heat of Vapourization (at 45°C)	40. 69kJ/mol

Chemical Properties

(*i*) Sulphur trioxide strongly in moist air and has a great affinity for water, with which it forms sulphur acid. Hence sulphur trioxide is termed as anhydrous of sulphuric acid.

Sulphur Trioxide + Water ⟶ Sulphuic Acid

$$SO_3 + H_2O \longrightarrow H_2SO_4$$

(*ii*) On heating strongly heating it breaks into sulphur dioxide and oxygen, at considerable high temperature (100°C) and in the presence of a catalyst called platinum asbestos.

Sulphur Trioxide ⟶ Sulphur Dioxide + Oxygen

$$2SO_3 \longrightarrow 2SO_2 + O_2$$

(*iiii*) It is dissolved in concentrated sulphuric *acid, form oleum,* the ore of sulphuric acid, when diluted forms sulphuric acid.

(A) Sulphuric Acid + Sulphur Trioxide ⟶ Oleum

$$H_2SO_4 + SO_3 \longrightarrow H_2S_2O_7$$

(B) Oleum + Water ⟶ Sulphuric Acid

$$H_2S_2O_7 + H_2O \longrightarrow 2H_2SO_4$$

(*iv*) With Oxides of metals sulphur trioxide forms sulphates.

Sulphur Trioxide + Sodium Oxide ⟶ Sodium Sulphate

$$SO_3 + Na_2O \longrightarrow Na_2SO_4$$

(*v*) It reacts with carbon and forms carbon dioxide and liberates iodine from potassium iodide

Carbon + Sulphur Trioxide ⟶ Carbon Dioxide + Sulphur Dioxide

$$C + 2SO_3 \longrightarrow CO_2 + 2SO_2$$

(*vi*) With ammonia it forms Sulphonic Acid

Ammonia + Sulphur Trioxide ⟶ Sulphur Acid

$$NH_3 + SO_3 \longrightarrow 2SO_2$$

(*vii*) With Hydrogen Chloride it forms Chlorosulphonic Acid

Hydrogen Chloride (gas) + Sulphur Trioxide ⟶ Chlorosulphonic Acid

$$HCl + SO_3 \longrightarrow HSO_3Cl$$

INDUSTRIAL PREPARATION OF SULPHUR TRIOXIDE

Sulphur Trioxide is Prepared in Industry by the Following Processes

(*i*) *Badische Process* : In this process sulphur trioxide is prepared by the action of dry sulphur dioxide and oxygen, which is passed through an iron chamber known as converter. In which is passed through at a temperature of 412°C to 445°C in the preheater and then

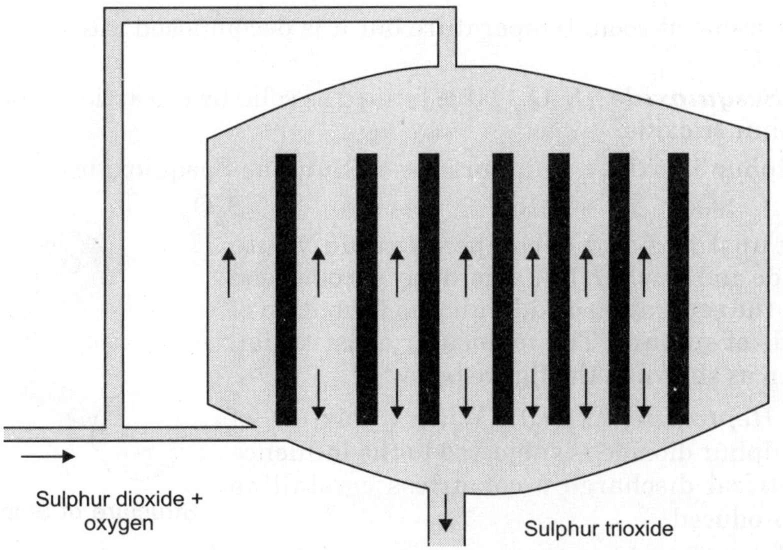

Fig. 11.15. The Industrial Preparation of Sulphur Trioxide by Badische Process

passed through the top in a converter. This consists of a number of vertical pipes packed with platinized asbestos. The gasses first pass around these pipes as shown in Fig. 11.16 and then drawn through the pipes filled with catalyst. This set up heats the incoming gas cooling at the same time, the gases react with the iron pipes and thus the temperature is properly maintained. If the temperature rises very high, cool gaseous mixture is introduced.

(ii) Schroder-Grillo Process : Here the catalyst is prepared by moistening Epsom salt, Magnesium Sulphate $MgSO_4 7H_2O$ with sodium or potassium chloride and heating in the presence of sulphur dioxide. The salt looses water and swells up to a voluminous contact mass on which very finely divided platinum is placed. This is put on shelves in iron converters, lagged outside (see Fig. 11.16) and when the process is Started it goes on without extra heating.

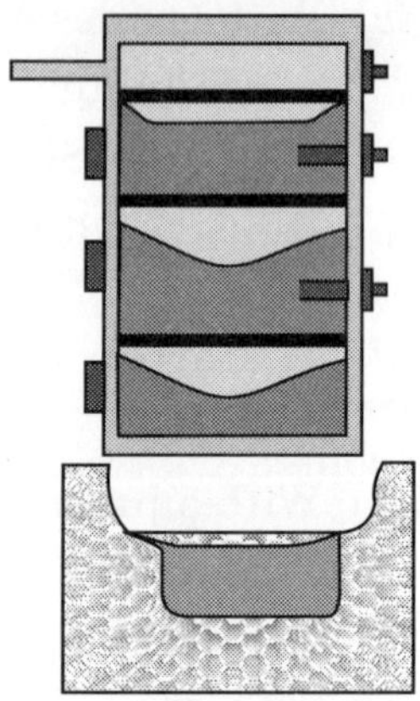

Fig. 11.16. The Schroder Grillo Process

Sulphur Monoxide, (SO) : Sulphur monoxide is a colourless gas. It is prepared by the action of electrical discharge on a mixture of sulphur dioxide and sulphur vapour under reduced pressure.

Sulphur Dioxide + Sulphur Vapour $\longrightarrow$ Sulphur Monoxide

$$SO_2 + S \longrightarrow 2SO$$

It is also formed when finely divided silver reacts with thionyl chloride.

Thionyl Chloride + Silver $\longrightarrow$ Sulphur Monoxide + Silver Chloride

$$SOCl_2 + 2Ag \longrightarrow SO + 2AgCl$$

The gas is stable at room temperature but it is decomposed into sulphur vapour and dioxide at 100°C.

Sulphur Sesquioxide, (S_2O_3) : It is funned as solid by the action of powdered sulphur on liquid sulphur trioxide.

Sulphur Trioxide + Sulphur $\longrightarrow$ Sulphur Sesquioxide

$$SO_2 + S \longrightarrow S_2O_3$$

It is very unstable and decomposes rapidly into sulphur trioxide and sulphur. It is vigorously decomposed by water with the separation of sulphur and formation of many oxyacids of sulphur. The molecular constitution may be written as shown in the figure below:

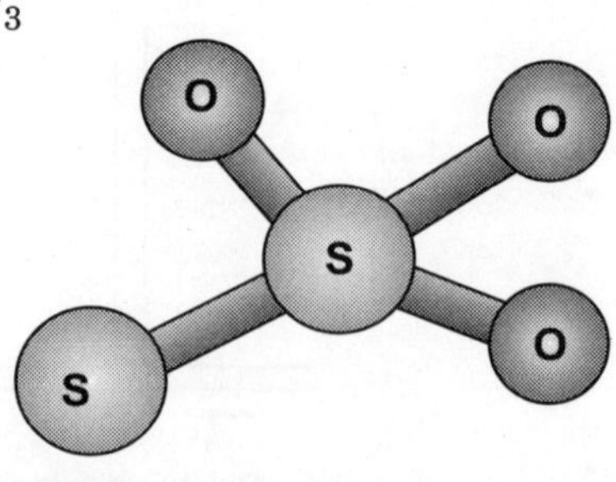

Structure of Sulphur Sesquioxide

Sulphur Heptoxide, (S_2O_7) : When a mixture of oxygen and sulphur dioxide is subjected to the influence of silent electrical discharge a colourless crystalline substance is produced

Sulphur Trioxide + Oxygen $\longrightarrow$ Sulphur Heptoxide

$$4SO_3 + O_2 \longrightarrow 2S_2O_7$$

With water it forms persulphuric acid and is therefore regarded as the anhydride of persulphuric acid.

$$\text{Sulphur Heptoxide+ Water} \longrightarrow \text{Persulphuric Acid}$$
$$2S_2O_7 + O_2 \longrightarrow H_2S_2O_8$$

With water it form persulphuric acid and is therefore regarded as the anhydride of persulphuric acid.

Sulphur Tetraoxide, (SO_4) : When sulphur dioxide mixed with an excess of oxygen in a constant electrical discharge at reduced pressure, at a temperature of liquid air. A white solid, sulphur tetraoxide is obtained. It is fairly stable in solution and does not form a per acid or hydrogen peroxide,

$$\text{Sulphur Dioxide + Oxygen} \longrightarrow \text{Sulphur Tetraoxide}$$
$$SO_2 + O_2 \longrightarrow SO_4$$

We shall now study the hydrides of sulphur, among them hydrogen sulphide is the most important.

COMPOUNDS OF SULPHUR AND HYDROGEN

Hydrogen Sulphide

Among the hydro-compounds of sulphur, hydrogen sulphide is the most important. It is also known as *sulphurated hydrogen gas.* Hydrogen sulphide gas was known to the alchemists from the early days by its smell (sulphurous vapour), but it was Sheele in the year 1777 who carefully examined the gas and stated that it was the compound of sulphur and hydrogen.

Occurrence

Hydrogen sulphide occurs in the volcanic gases also in several mineral waters. Natural gas contains hydrogen sulphide. Hydrogen sulphide is formed when animal or plant waste decomposes in the restricted supply of air. Sheele produced hydrogen sulphide by the action of acids on calcium pyrusulphide, manganese sulphide or ferrous sulphide. He also showed that the process of heating of sulphur m hydrogen could prepare it. Sheele observed the solubility of the gas in water and its oxidation to sulphur by air, nitric acid or chlorine. He also noted that the gas reacts with solutions of numerous metallic salts however, it was not until Borthollet did analyze it in the year 1796 that the gas was recognized as hydrogen sulphide.

Preparation

Hydrogen sulphide can be prepared by the following methods :

> Catalyst used in this preparation of pure hydrogen sulphide is pumice stone

1. *By passing Hydrogen gas over molten sulphur* in a bulb tube the resultant gas consists of hydrogen sulphide.

$$\text{Hydrogen + Sulphur} \rightleftharpoons \text{Hydrogen Sulphide}$$
$$H_2 + S \rightleftharpoons H_2S$$

The reaction is reversible because, when hydrogen sulphide is subjected to high temperature it decomposes into hydrogen and sulphur. The pure gas is prepared synthetically in presence of pumice stone as a catalyst. At 600°C, when the reaction is complete.

Water + Sulphur $\rightleftharpoons$ Hydrogen Sulphide + Sullphuric Acid

$4H_2O + S \rightleftharpoons H_2S + H_2SO_4$

2. *By boiling sulphur in water* when sulphur is boiled in water, the steam reacts with sulphur and traces of hydrogen sulphide are formed.

Antimony Sulphide + Hydrochloric Acid $\rightleftharpoons$ Antimony Chloride + Hydrogen Dulphide

$Sb_2S_3 + 6HCl \rightleftharpoons 2Sbl_3 + 3H_2S$

3. *By heating magnesium hydro sulphide* when magnesium hydrosulphide is heated at 60°C with water hydrogen sulphide is obtained

Magnesium Hydrosulphide + Water $\longrightarrow$ Hydrogen Sulphide + Magnesium Oxide

$Mg(HS)_2 + H_2O \longrightarrow 2H_2S + MgO$

Laboratory Preparation of Hydrogen Sulphide

Theory : In laboratory hydrogen sulphide is prepared by the action of dilute sulphuric acid on ferrous sulphide. The gas is collected by the upward displacement of air. (See Fig. 11.17).

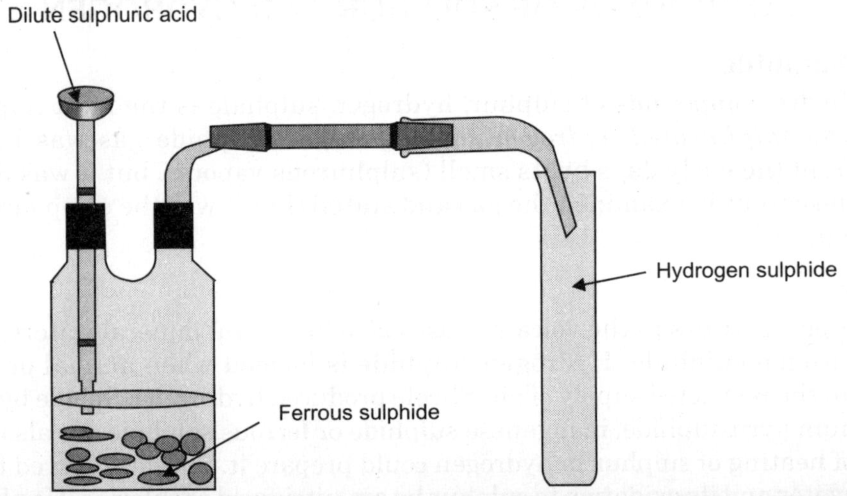

Apparatus	Reagents	Temperature	Pressure
Woulf's bottle, Thirstle funnel, gas jar corks etc.	Ferrous sulphide, dilute sulphuric acid	Normal	Normal

11. 17. The Lab Preparation of Hydrogen Sulphide

Reaction Equation

Ferrous Sulphide + Sulphuric Acid $\longrightarrow$ Hydrogen Sulphide + Ferrous Sulphate

$FeS + H_2SO_4 \longrightarrow H_2S + FeSO_4$

Precaution

1. The reaction is exothermic, so care must be taken.

2. The gas is poisonous therefore, direct inhaling must be avoided.
3. Care must be taken while putting ferrous sulphide, slight mishandle can break the Woulfs bottle.

POISON GAS

Physical, Hydrogen sulphide has an objectionable odour of rotten egg, and the gas is very toxic. The gas is colourless and fairly acidic in taste. It is very much soluble in water at different temperature, which is shown in the table below.

Temp°C	0	10	20	30	40	50	60	70	80	90	100
Solubility (100g of water)	4.9	6.72	8.7	10.9	13.2	15.5	17.8	20.3	22.8	24.1	24.7

Hydrogen sulphide has an ignition point of 250°C and an explosive range in air of 4.5 to 45.5%.

Although hydrogen sulphide is stable at room temperature, but at high temperature it breaks dow to its elements, hydrogen and sulphur.

Physical Properties

The physical properties are given in a tabular manner below:

Melting Point	–86.6°C
Boiling Point	–60°C
Critical Temperature	100°C
Heat of Vapourization	18.67 cal/mol
Heat of Fusion	23.80 cal/mol
Heat of Capacity	134
Van der Waals Constant	4.544 barL^2/mol(a)
	0.04339 L/mol

Checmial Properties

Hydrogen sulphide is chemically active and is very much useful reagent in modem chemistry in different ways. The chemical properties of hydrogen gas are being discussed below:

Hydrogen sulphide in complete dry state does not have any acidic property, but same hydrogen sulphide when forms an aqueous solution becomes weakly acidic.

Hydrogen sulphide reacts with many chemical compounds like light metal hydroxides, forming first normal sulphides and then hydrosulphides or acids.

Sodium Hydroxide + Hydrogen Sulphide ⟶ Sodium Sulphide + Water

$$2NaOH + H_2S \longrightarrow Na_2S + 2H_2O$$

Sodium sulphide reacts with hydrogen sulphide forming sodium hydrosulphide

Sodium Sulphide + Hydrogen Sulphide ⟶ Sodium Hydrosulphide

$$Na_2S + H_2S \longrightarrow 2NaHS$$

With sodium carbonate however hydrogen sulphide is able to produce sodium bi carbonate, as it is a weak acid. It cannot liberate carbon dioxide.

Sodium Carbonate	+	Hydrogen Silphide	⟶	Sodium Hydrosulphide	+	Sodium Bicarbonates
Na_2CO_3	+	H_2S		NaHS	+	$NaHCO_3$

Potassium and tin when heated with hydrogen sulphide react to form acidic sulphides. White tin and other metals forms normal sulphides.

Potassium	+	Hydrogen Sulphide	⟶	Potassium Bisulphide	+	Hydrogen
2K	+	$2H_2S$	⟶	2KHS	+	H_2

Dry hydrogen sulphide does not react with metals like silver and copper but in presence of little moisture the reaction is rapid.

Copper	+	Hydrogen Sulphide	+	Oxygen	⟶	Cuprous Sulphide	+	Water
4Cu	+	$2H_2S$	+	O_2	⟶	$2Cu_2S$	+	$2H_2O$

All metals except mercury, gold and platinum react with hydrogen sulphide, silver when introduced to hydrogen sulphide (moistened) silver sulphide is formed.

Silver	+	Hydrogen Sulphide	⟶	Silver Sulphide	+	Water
2Ag	+	$2H_2S$	⟶	Ag_2S	+	$2H_2O$

Hydrogen Sulphide in Laboratory

Hydrogen sulphide is the most essential chemical in any chemical laboratory. To use it conveniently, hydrogen sulphide generator it used. In other words it is known as Kipp's. We have already read about it we are now going to see the figure of the Kipp's apparatus Fig. 11.18.

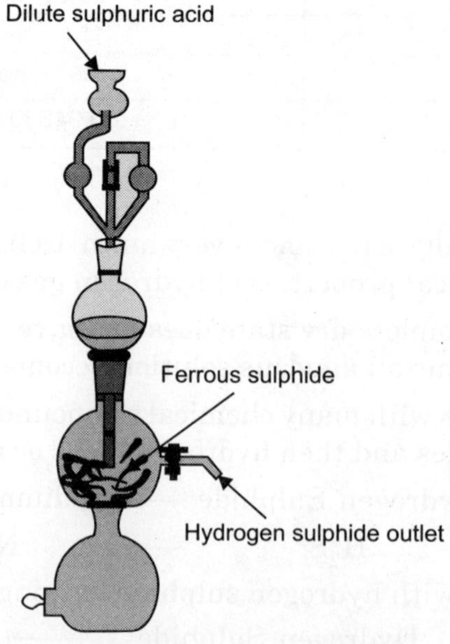

Fig. 11.18. The Kipp's Apparatus

In laboratory hydrogen sulphide is mainly used as reducing agent, while using this reagent some interesting and colourful chemical reactions are observed. Students are requested to perform these experiments because they will then only understand the most important properties of hydrogen sulphide.

Experiment	Observation	Inference
1. Hydrogen sulphide is passed through a solution of zinc oxide.	A black ppt is observed	This black ppt is due to zinc sulphide $ZnO + H_2S \rightarrow ZnS + H_2O$
2. Hydrogen Sulphide is bubbled through a solution of potassium permanganate (acidic)	The pink solution is turned colourless and a yellow ppt is produced	Potassium permanganate is converted to manganese sulphate. $2KMnO_4 + 3H_2SO_4 + 5H_2S \rightarrow K_2SO_4 + 2MnSO_4 + 8H_2O + 5S$
3. Hydrogen Sulphide is bubbled through an acidic solution of potassium dichromate	The solution turns green	It turns green due to the formation of chromic sulphate. $K_2Cr_2O_7 + 4H_2SO_4 + 3H_2S \rightarrow K_2SO_4 + Cr_2(SO_4)_3 + 7H_2 + 2S$
4. Hydrogen sulphide is bubbled into ferric chloride	The yellow solution turns colourness.	Ferric chloride is turned into ferrous chloride $FeCl_3 + H_2S \rightarrow FeCl_2 + 2HCl + S$
5. Hydrogen sulphide is bubbled in dilute nitric acid	Yellow ppt with brown fumes.	Yellow ppt is is sulphur and brown fumes are nitrogen dioxide. $2HNO_3 + H_2S \rightarrow 2H_2O + 2NO_2 + S$
6. Hydrogen sulphide is passed through sulphuric acid.	Yellow ppt is formed with the evolution of an unpleasant gas	Yellow ppt is sulphur and the acid is converted into sulphur dioxide. $H_2SO_4 + H_2S \rightarrow SO_2 + S + 2H_2O$
7. Hydrogen sulphide is bubbled through copper sulphate.	Black ppt is formed	The copper sulphate is converted to cuprous sulphide which is black $CuSO_4 + H_2S \rightarrow CuS + H_2SO_4$
8. Hydrogen sulphide is passed through a solution of antimony chloride.	Orange ppt is formed	There is a ppt of antimony sulphite $2SbCl_3 + 3H_2S \rightarrow 6HCl + Sb_2S_2$
9. Hydrogen sulphide is allowed to pass through a solution of mangnaese chloride and ammonium hydroxide.	A pink ppt is formed	The ppt is of manganous sulphide (the reaction takes place in alkaline medium.
10. Cadmium sulphide solution is exposed to hydrogen sulphide gas.	Yellow ppt is produced	There is a deposition of cadmium sulphide $CdSO_4 + H_2S \rightarrow H_2SO_4 + CdS$
11. Lead Acetate solution is added to a gas jar of hydrogen sulphide.	Black ppt is formed	Lead sulphide is produced, $Pb(CH_3COOH)_2 + H_2S \rightarrow PbS + 2CH_3COOH$
12. Hydrogen sulphide is passed through sodium nitroprusside solution	A voilet ppt is formed	$NaOH + H_2S \rightarrow Na_2S + H_2S$ $Na_2S + Na_2Fe(CN)_5NO \rightarrow Na_4Fe(CN)_5NO_5$

Hydrogen sulphide is readily oxidized by many oxidizing agent. The oxidation products depend upon the oxidizing agents, its concentration and its molal ratio to the hydrogen sulphide and also upon the conditions of reactions. Some of the oxidizing agents of hydrogen sulphide and their products are summarized in the table below:

Oxidizing agent	Condition	Chief products
Oxygen (air)	Flame, air in excess	Sulphur Dioxide
Oxygen (air)	Flame, hydrogen sulphide in excess	Sulphur
Sulphur Oxide	Concentrated acid	Sulphur and Sulphur Dioxide
Hydrogen Peroxide	Neutral solution	Sulphur
Sodium Peroxide	Dry system at high temperature	Sodium Sulphide, Sodium Disulphate
Ozone	Aqueous solution of hydrogen sulphide	Sulphur and sulphuric Acid
Nitric Acid	Concentrated aqueous solution of hydrogen sulphide	Sulphuric Acid
Iodine	Aqueous solution of excess chlorine	Sulphuric Acid

Composition of Hydrogen Sulphide

The following process determines the composition of hydrogen sulphide. In this process a known volume of hydrogen sulphide is heated in a closed vessel (see Fig. 11.19) in contact with metallic copper, which reacts with the gas to form cuprous sulphide and hydrogen is set free. After the reaction is over the volume of the gas remains unchanged on cooling.

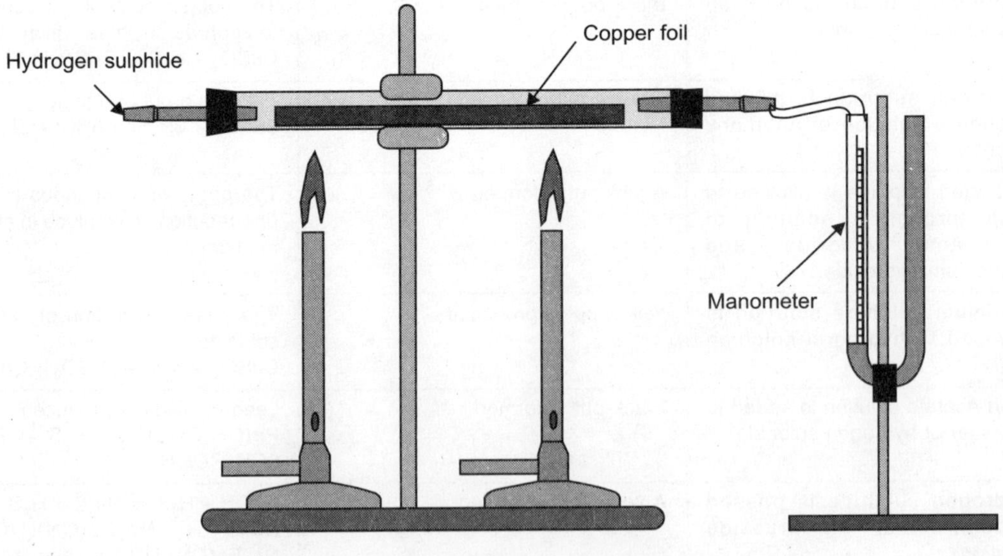

Fig. 11.19. The Volumetric Composition the Hydrogen Sulphide

This is demonstrated by opening one end of the cooled vessel under mercury, when no gas enters or escapes. Hydrogen sulphide therefore contains its own volume of hydrogen. Form this, with the use of Avogadro's Hypothesis and the knowledge of the vapour density of the gas, which is 17 the formula as H_2S.

To Prove that Hydrogen Sulphide Contains Hydrogen

Hydrogen sulphide is allowed to pass through a hot copper oxide which turns black due to formation of copper sulphide and the walls of the tube is found full of droplets which turns anhydrous copper sulphate blue, this proves that the liquid is water, the equation is shown below:

Copper Oxide	+	Hydrogen Sulphide	$\longrightarrow$	Copper Sulphide
CuO	+	H_2S	$\longrightarrow$	CuS

Industrial Preparation of Hydrogen Sulphide

In industry hydrogen sulphide is prepared by many processes but we shall discuss here only two of them.

1. *Hydrogen Sulphide from the reaction of hydrogen and Sulphur :* In this method hydrogen sulphide is mainly obtained by the reaction of hydrogen and sulphur vapour.

Hydrogen	+	Sulphur	$\longrightarrow$	Hydrogen Sulphide
$2H_2$	+	2S	$\longrightarrow$	H_2S

This reaction proceeds at a satisfactory rate about 500°C in the presence of a catalyst such as bauxite or aluminium hydro silicate minerals. This process yields about 96% manufacture of hydrogen sulphide. This method is only commercial if hydrogen is obtained locally as a waste.

2. *Girbotol Process :* This process is the most important process for the preparation of hydrogen sulphide. The gas is absorbed in a solution of a compound of hydrogen sulphide at low temperature, which readily dissociates on heating. During regeneration, the volatile hydrogen sulphide separates from the relatively non-volatile amine. Steam generated by boiling the amine solution is commonly employed to separate the hydrogen sulphide gas. The complete removal of the hydrogen sulphide from the amine solution during regeneration in turn permits complete purification of the gas in the counter current absorber. (See Fig. 11.20).

Triethanolohydroamine hydrogen sulphide	$\longrightarrow$	Hydrogen Sulphide	+	Triethanolhydromine
$(CH_2CH_2CH_2OH)_2NH.H_2S$	$\longrightarrow$	H_2S	+	$(CH_2CH_2CH_2OH)_2NH$

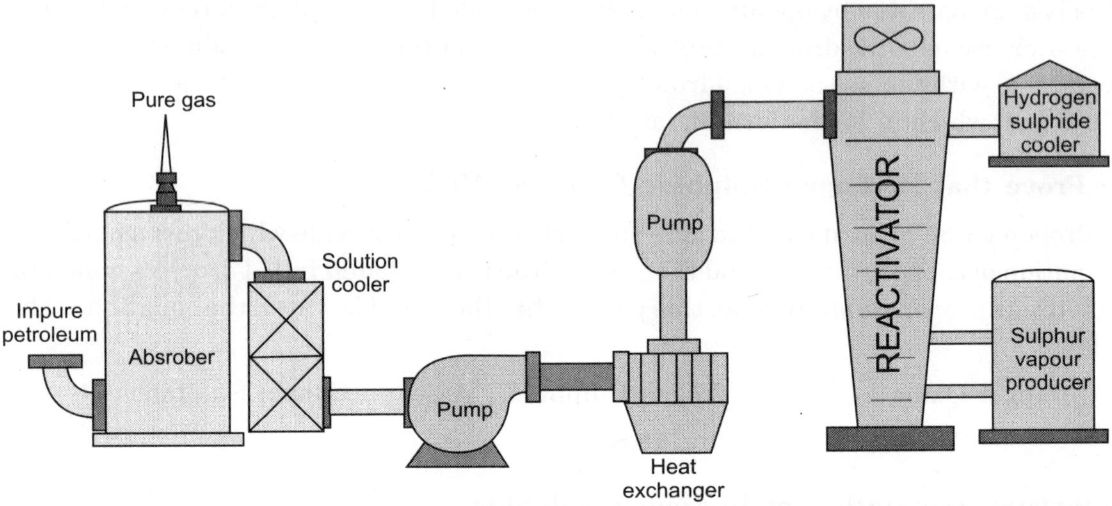

Fig. 11.20. The Industrial Process of Hydrogen Sulphide

Persulphides of Sulphur

If a solution of sulphur in aqueous sodium sulphide solution (the overall composition will Na_4S_5) run into cooled oil, constitution of this oil is unknown (it may be H_2S_2) or a mixture, say H_2S_2 and H_2S_3 containing dissolved sulphur). If it is distilled in quartz apparatus at a pressure of 2mm two volatile fraction of composition H_2S_n (where n = 2 to 5) are obtained. All the hydrogen polysulphides are so early decomposed by alkali that glass apparatus cannot be used for handling them unless it has been treated with dry hydrogen chloride to remove surface alkali; the drying may be phosphorous pentoxide or calcium chloride treated with hydrogen chloride.

Hydrogen Disulphide : H_2S_2 Hydrogen disulphide is a colourless liquid having a melting point of –81°C and boils at 71°C. It has a smell like that of hydrogen sulphide but also lachrymatory tree producing it is soluble in carbon di sulphide, benzene and eather. It dissolves sulphur but does not react with it. The molecular weight is normal. Electron diffraction experiments show that the molecules have chain structure.

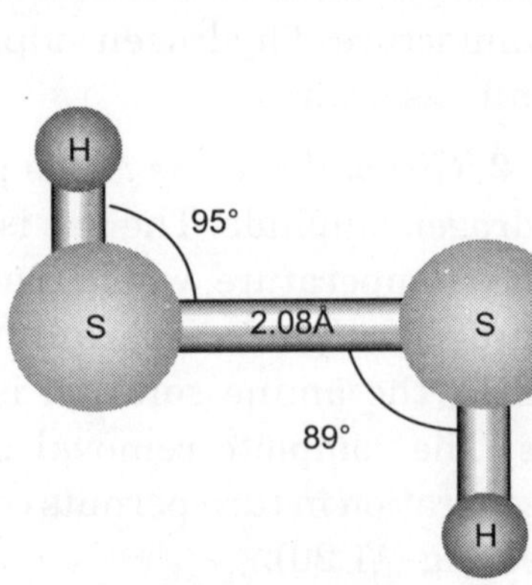

Structure of Hydrogen Disulphide Molecule

Compounds of Sulphur and Nitrogen

Sulphur and nitrogen have placed at the end of the diagonal in the periodic table also they are very close in electronegativity (S = 2.5 and N= 3.0). So they are bounded very easily among the sulphur nitrogen compounds.

Sulphur Dichloride + Ammonia ⟶ Tetrasulphur Tetranitride + Sulphur + Ammonium Chloride

$$6SCl_2 + 16NH_3 \longrightarrow S_4N_4 + 2S + 12NH_4Cl$$

Tetrasulphur Tetranitride, (S_4N_4): This compound is produced by the reaction of sulphur dichloride and ammonia, where sulphur and ammonium chloride is also produced.

Sulphur Dichloride + Ammonia ⟶ Tetrasulphur Tetranitride + Sulphur + Ammonium Chloride

$$6SCl_2 + 16NH_3 \longrightarrow S_4N_4 + 2S + 12NH_4Cl$$

It can be produced also from the reaction of disulphur dichloride (sulphur monochloride) and ammonia in the presence of carbon tetrachloride

Sulphur Monochloride + Ammonia —Carbon tetrachloride⟶ Tetrasulphur Tetranitride + Sulphur + Ammonium Chloride

$$6S_2Cl_2 + 16NH_3 \xrightarrow{CCl_4} S_4N_4 + 2S + 12NH_4Cl$$

Tetrasulphur Tetranitride also can be produced by the reaction of sulphur monochloride and ammonium chloride

Sulphur Monochloride + Ammonia ⟶ Tetrasulphur Tetranitride + Sulphur + Ammonium Chloride

$$6S_2Cl_2 + 16NH_3 \longrightarrow S_4N_4 + 2S + 12NH_4Cl$$

S_4N_4 is a solid in structure and has a melting point of 178°C; boiling point is 280°C the unique feature of this compound is that it is thermochromic which means that this compound changes its colour according to the temperature, e.g. at –186°C it is almost colourless and at 2°C it is yellow, when the temperature rises to 100°C the colour is red. It is more or less stable in air, but explodes when pressure is applied. The structure of the molecule is a hetrocyclic ring the shape of the molecule is like a cradle as shown in the Fig.

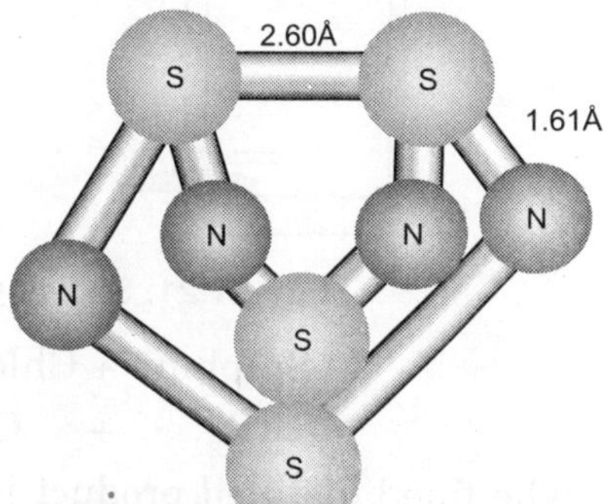

Tetrasulphur Tetranitride Molecule

Tetrasulphur Tetranitride is very slowly hydrolyzed but reacts with very rapidly with sodium hydroxide.

Tetrasulphur Tetranitride + Sodium Hydroxide + Water ⟶ Sodium Thyosulphate + Sodium Sulphide + Ammonia

$$S_4N_4 + 6NaOH + 3H_2O \longrightarrow Na_2S_2O_3 + 2NaSO_3 + 4NH_3$$

If Tetrasulphur tetrachloride is vapourized under reduced pressure and passed through silver wool, then di sulphur dinitrogen S_2N_2 is formed.

Tetrasulphur Tetranitride + Silver Wool ⟶ Disulphur Dinitrogen + Silver Sulphide + Nitrogen

$$S_4N_4 + 4Ag \longrightarrow S_2N_2 + 2Ag_2S + N_2$$

Disulphur dinitrogen is a crystalline solid, which is insoluble in water but soluble in many organic solvents like benzene, carbon di sulphide etc. it explodes when kept under pressure or heated.

Sulphur Halides

Under this subhead we are going to discuss the compounds of sulphur with chlorine, fluorine and bromine.

Sulphur Monochloride, (S_2Cl_2): Sulphur monochloride is prepared by the action of sulphur and chlorine, in the laboratory sulphur monochloride is prepared by passing a current of chloride over mohen sulphur and then collecting the distillate in a dry receiver. Which is cooled under distilled water, the experimental set up is shown in the Fig. 11.21.

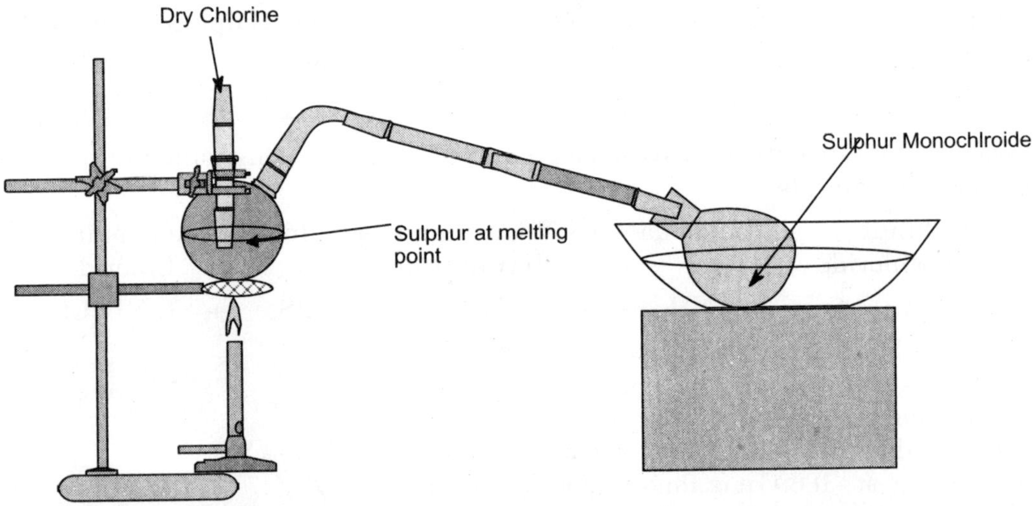

Fig. 11.21. Showing the Lab Preparation of Sulphur Monochloride

$$\text{Sulphur} + \text{Chlorine} \longrightarrow \text{Sulphur Monochloride}$$

$$S_2 + Cl_2 \longrightarrow S_2Cl_2$$

The fined distilled product is an amber colour liquid. The physical properties of the compound are stated in the table.

Sulphur monochloride fumes in moist air and has a very uncomfortable odour. In contact with water the liquid is slowly hydrolyzed, yielding hydrochloric acid, sulphur dioxide. The primary reactions are given below.

$$\text{Sulphur Monochloride} + \text{Water} \longrightarrow \text{Hydrochloric Acid} + \text{Hydrogen Sulphide} + \text{Sulphur Dioxide}$$

$$S2Cl_2 + 2H_2O \longrightarrow 2HCl + H_2S + SO_2$$

$$\text{Hydrogen Sulphide} + \text{Sulphur Dioxide} \longrightarrow \text{Water} + \text{Sulphur}$$

$$H_2S + SO_2 \longrightarrow 2H_2O + 3S$$

At the same time some thionic acids are also formed. The formula of the compound, sulphur monochloride is S_2Cl_2. The structure is shown in the Fig. below:

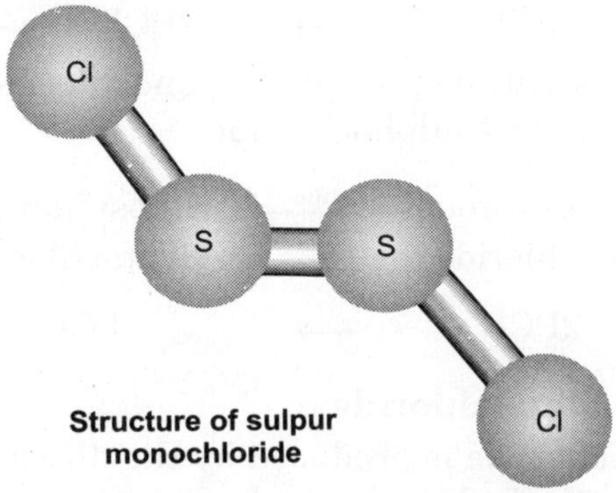

Structure of sulpur monochloride

Sulphur monochloride is used for dissolving sulphur, iodine and certain compounds. The compound has some chemical properties, which e being discussed below.

Sulphur monochloride generally reacts with ammonium chloride to form tetrasulphur tetrachloride.

$$\underset{\text{Monochloride}}{\text{Sulphur}} + \underset{\text{Chloride}}{\text{Ammonium}} \longrightarrow \underset{\text{Tetranitiride}}{\text{Tetrasulphur}} + \text{Sulphur} + \underset{\text{Acid}}{\text{Hydrochloric}}$$

$$6S_2Cl_2 + 4NH_4Cl \longrightarrow S_4N_4 + 8S + 16HCl$$

Sulphur monochloride is stable at room temperature but between 300-310°C the decomposition starts

$$\text{Sulphur Monochloride} \rightleftharpoons \text{Sulphur} + \text{Chlorine}$$

$$S_2Cl_2 \rightleftharpoons S_2 + Cl_2$$

Similarly the metallic sulphides also react with sulphur monochloride

$$\text{Ferric Oxide} + \underset{\text{Monochloride}}{\text{Sulphur}} \longrightarrow \underset{\text{Chloride}}{\text{Ferrous}} + \underset{\text{Dioxide}}{\text{Sulphur}} + \text{Sulphur}$$

$$2FeO + S_2Cl_2 \longrightarrow 2FeCl_2 + SO_2 + S_2$$

When Sulphur monochloride reacts with more chlorine in presence of water sulphuric and hydrochloric acid is formed.

$$\text{Sulphur Monochloride} + \text{Water} + \text{Chlorine} \longrightarrow \text{Sulphuric Acid} + \text{Hydrochloric Acid}$$

$$S_2Cl_2 + 8H_2O + 5Cl_2 \longrightarrow 2H_2SO_4 + 12HCl$$

Sulphur monochloride also reacts with potassium iodide to produce iodine and sulphur

$$\text{Sulphur Monochloride} + \underset{\text{Iodide}}{\text{Potassium}} \longrightarrow \text{Iodine} + \text{Potassium Chloride} + \text{Sulphur}$$

$$S_2Cl_2 + 2Kl \longrightarrow I_2 + 2KCl + S_2$$

When sulphur monochloride is allowed to react with phosphorous then both phosphorous trichloride and phosphorous sulphide are produced.

Sulphur Monochloride + Phosphorous Trichloride ⟶ Phosphorous + Phosphorous Sulphide

$$3S_2Cl_2 + 10P \longrightarrow 2PCl_3 + 2P_4S_3$$

In presence off iodine as catalyst sulphur monochloride reacts with phosphorous trichloride to produce phosphorous sulphochloride

Sulphur Monochloride + Phopshorous Trichloride $\xrightarrow{\text{Iodine}}$ Phosphorous Pentachloride + Phosphorous Sulphochloride

$$3S_2Cl_2 + 2PCl_3 \xrightarrow{I_2} PCl_5 + 2PCl_3$$

Manufacture of Sulphur Monochloride

In industries sulphur monochloride is produced by the direct combination of elements in *Batch Process,* carried out in vertical steel vessels. The process consists of charging refined sulphur to a reactor that contains sufficient sulphur chloride form a previous batch to cover the Spurger through which the chlorine is to be introduced. The batch is heated between 55°C to 60°C, to encourage the reaction, chlorine is then introduced to the batch until the desired degree of chlorination has been obtained. Since the reaction is excessive loss of this reactant on the other hand sulphur dichloride may also formed as an intermediate product (see Fig. 11.22).

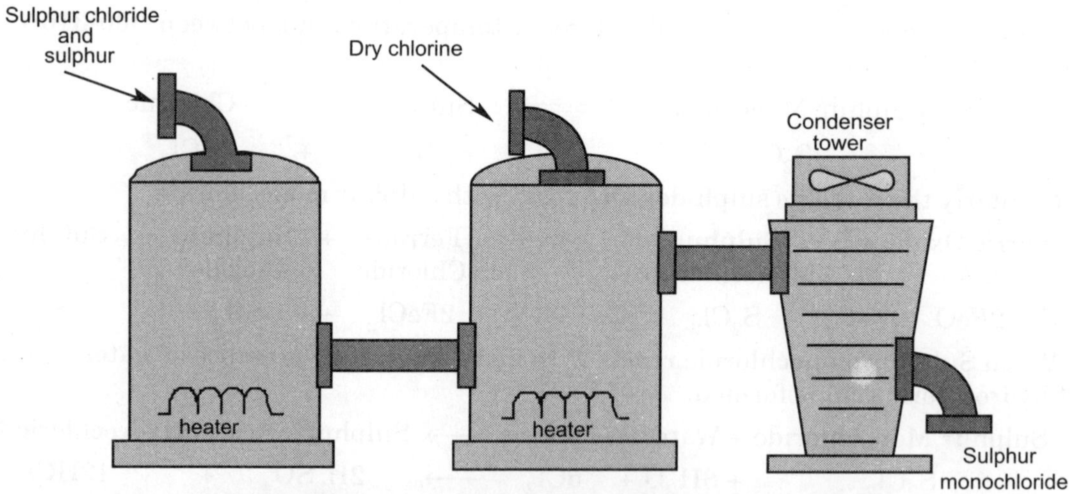

Fig. 11.22. The Industrial preparation of Sulphur Monochloride

Sulphur Dichloride (SCl_2) : This compound can be prepared by the reaction of hydrogen sulphide and chlorine.

Hydrogen Sulphide + Chlorine ⟶ Sulphur Dichloride + Hydrogen Chloride

$$H_2S + 2Cl_2 \longrightarrow SCl_2 + 2HCl$$

Sulphur dichloride chemically behaves like the mixture of sulphur monochloride and chlorine.

Sulphur Monochloride + Chlorine ⟶ Sulphur Dichloride

$$S2Cl_2 + 2Cl_2 \longrightarrow SCl_2$$

Sulphur Dichloride also reacts with sulphur trioxide and chlorosulphuric acid to produce thionyl chloride and sulphur dioxide.

Sulphur Dichloride + Sulphur Trioxide ⟶ Sulphur Fluoride + Thionyl Chloride

$$SCl_2 + SO_3 \longrightarrow SO_2 + SOCl_2$$

Sulphur Dichloride + Chlorosulphuric Acid ⟶ Sulphur Dioxide + Thionyl Chloride + Hydrogen Chloride

$$SCl_2 + ClSO_3H \longrightarrow SO_2 + SOCl_2 + 2HCl$$

Manufacture of Sulphur Dichloride

Sulphur dichloride is prepared by the reaction of sulphur monochloride and chlorine. The reaction is very slow, and particularly its later states is limited to temperatures below 40°C. It is carried out in the same equipment used to produce the monochloride. The commercial product often a small percentage of dissolved chlorine.

Sulphur Tetrachloride, Sulphur Trichlorosulphonium Chloride (SCl_4) : This compound of sulphur is usually prepared by the reaction of sulphur dichloride and chlorine at low temperature (about –9°C) and high pressure.

Sulphur Dichloride + Chlorine ⟶ Sulphur Tetrachloride

$$SCl_2 + Cl_2 \longrightarrow SCl_4$$

Sulphur tetrachloride melts at -30°C into a red liquid. It is a very unstable compound, dissociates into chlorine and sulphur dichloride. Sulphur tetrachloride reacts with water and rapidly forms sulphur dioxide and hydrochloric acid.

Sulphur Tetrachloride + Water ⟶ Sulphur Dioxide + Hydrochloric Acid

$$SCl_4 + 2H_2O \longrightarrow SO_2 + 4HCl$$

Sulphur Tetrachloride also reacts with chlorosulphuric acid to yield a salt named sulphur oxytetrachloride.

Sulphur Tetrachloride + Chlorosulphuric Acid ⟶ Sulphur Oxythetrachloride + Hydrochloric Acid

$$SCl_4 + ClSO_2H \longrightarrow S_2Cl_3Cl_4 + 4HCl$$

It reacts with ammonia to form tetrasulphur tetranitride.

Sulphur Tetrachloride + Ammonia ⟶ Sulphur Tetranitride + Hydrogen Chloride + Nitrogen

$$12SCl_4 + 16NH_3 \longrightarrow 3S_2N_4 + 48HCl + 2N_2$$

Disulphur Difluoride (S_2E_2) : It is known to be toxic gas that fumes in air and also has a disagreeable smell. It was first prepared by the reaction of silver di fluoride and sulphur.

POISON

Sulphur + Silver Difluoride ⟶ DiSulphur Difluoride + Silver

$$2S + AgF_2 \longrightarrow S_2F_2 + Ag$$

Di sulphur Difluoride readily gets hydrolyzed in humid air and in water to yield sulphur, sulphur dioxide and hydrogen fluoride.

Disulphur Difluoride	+ Water	$\longrightarrow$	Sulphur +	Hydrogen Fluoride +	Sulphur Dioxide
$4S_2F_2$	$+ 4H_2O$	$\longrightarrow$	$6S$ +	$8HF$ +	$2SO_2$

Di sulphur difluoride does not react with heavy metals like iron, steel, platinum but it reacts with glass and silica even at low temperatures, producing sulphur dioxide silicon tetrafluoride and thionyl fluoride.

Sulphur Difluoride (SF_2) : It is most unstable compound of sulphur and fluorine, this compound is prepared by the thermal decomposition of sulphur monofluoride it has not yet been prepared in pure state also.

Sulphur Tetrafluoride (SF_4) : It is a colourless gas with an irritating smell it can be condensed to a liquid (freezing point: —124°C). It does not attack glass and gets hydrolyzed by water and other hydroxides yielding sulphites and fluorides.

Sulphur Tetrafluoride + Water → Sulphur + Hydrogen Fluride

$$SF_4 + 2H_2O \rightarrow SO_2 + 4HF \quad \text{(Slow reaction)}$$

Sulphur Tetrafluoride + Sodium Hydroxide → Sodium Sulphite + Sodium Fluoride + Water

$$SF_4 + 6NaOH \rightarrow Na_2SO_3 + 4NaF + 3H_2O \quad \text{Fast reaction}$$

Sulphur tetrafluoride can also be prepared by the following reactions:

Sulphur	+ Fluorine (+Nitrogen)	$\longrightarrow$	Sulphur Tetrafluoride		
S	+ $2F_2$ (+N_2)	$\longrightarrow$	SF_4		
Sulphur	+ Cobaltic Fluoride (+Calcium Fluoride)	$\longrightarrow$	Sulphur Tetrafluride	+	Cobaltus Fluoride
S	+ $4COF_3$ (+CaF_2)	$\longrightarrow$	SF_4	+	$4CoF_2$
Sulphur Tetrachloride	+ Silver Fluordie	$\longrightarrow$	Sulphur tetrafluoride	+	Silver Chloride
$3SCl_4$	+ 12AgF	$\longrightarrow$	$3SF_4$	+	12 AgCl
Cobaltic Fluoride	+ Sulphur	$\longrightarrow$	Sulphur Tetrafluoride	+	Cobaltus Fluoride
$4COF_3$	+ S	$\longrightarrow$	SF_4	+	$4COF_2$

Sulphur hexafluoride is inert chemically

Sulphur Hexafluoride (SF_6) : The only hexa-halide of sulphur is sulphur hexafluoride (SF_6), which is the ultimate product of the fluorination of sulphur, it is stable and also inert. Sulphur hexafluoride does not get reacted with hydroxyl ions, even at red-hot condition, and also does not combines with hydroxides or oxygen in fused state also. It only reacts with hydrogen at 400°C. This inertness is due to two is inert causes (1) Steric Crowding (2) Absence of suitable orbital for accommodating electron pairs from nucleophiles to form semi coordinated SF_6. Sulphur hexafluoride is prepared by the direct interaction of burning sulphur in fluorine gas

$$\text{Sulphur} + \text{Florine} \longrightarrow \text{Sulphur Hexafluoride}$$
$$S_2 + 6F_2 \longrightarrow 2SF_6$$

Though sulphur hexafluoride is inert it gets decomposed in the presence of sodium

$$\text{Sulphur Hexafluoride} + \text{Sodium} \longrightarrow \text{Sodium Sulphide} + \text{Sodium fluoride}$$
$$2SF_6 + 8Na \longrightarrow Na_2S + 6NaF$$

Also it gets decomposed in the presence of hydrogen sulphide:

$$\text{Sulphur Hexafluoride} + \text{Hydrogen Sulphide} \longrightarrow \text{Hydrogen Fluoride} + \text{Sulphur}$$
$$SF_6 + 3H_2S \longrightarrow 6HF + 4S$$

Sulphur Pentafluoride (SF_5) : This compound of sulphur is highly toxic, a colourless liquid. Sulphur pentafluoride is prepared by the photo-effective reaction of sulphur pentachlorifuloride and hydrogen.

$$\text{Sulphur Pentachlorofluoride} + \text{Hydrogen} \longrightarrow \text{Sulphur Pentafluoride} + \text{Hydrogen Chloride}$$
$$2SClF_5 + H_2 \longrightarrow 2SF_5 + 2HCl$$

Sulphur pentafluoride decomposes at 150°C to sulphur hexafluoride and sulphur tetrafluoride

$$\text{Sulphur Pentafluoride} + \text{Sulphur Hexafluoride} \longrightarrow \text{Sulphur Tetrafluoride}$$
$$2SF_5 + SF_6 \longrightarrow SF_4$$

Similarly, it reacts with chlorine and bromine forming sulphur chlorofluoride and sulphur pentabromofluoride.

$$\text{Sulphur Pentafluoride} + \text{Bromine} \longrightarrow \text{Sulphur Penta bromofluoride}$$
$$2SF_5 + 2Br \longrightarrow 2SBrF_5$$
$$\text{Sulphur Pentafluoride} + \text{Chlorine} \longrightarrow \text{Sulphur Pentafluorochloride}$$
$$2SF_5 + 2Cl \longrightarrow 2SClF_5$$

Sulphur Monobromide (S_2Br_2) : It is the only definite compound of bromine and sulphur, it is also known as disulphur dibromide. Sulphur monobromide is prepared by the direct reaction of equatomic mixture of sulphur and bromine; at a temperature of 102°C in a sealed vessel

$$\text{Sulphur} + \text{Bromine} \longrightarrow \text{Sulphur Monobromide}$$
$$S_2 + Br_2 \longrightarrow 2Cl$$

Sulphur monobromide gets dissolved in carbon di sulphide and itself is the solvent for sulphur. It is similar to sulphur monochloride but unstable.

Oxyhalides of Sulphur

Sulphur has some oxyhalides also, some of them are discussed below:

Thyonil Chloride ($SOCl_2$) : Peroz and Bloch first obtained this compound in 1849 by the interation of sulphur dioxide and phosphorous pentachloride

Sulphur Dioxide	+	Phosphorous Pentachloride	⟶	Thionyl Chloride	+	Phophorous Oxychloride
SO_2	+	PCl_5	⟶	$SOCl_2$	+	$POCl_3$

Thionyl chloride is also prepared by the reaction of sulphur and chlorine monoxide at a temperature of –12°C.

Thionyl chloride manufactured by the reaction of sulphur trioxide and sulphur monochloride at 75°C to 80°C and also passing a steam of chloride into the mixture to reconvert the separated sulphur into sulphur monochloride.

Sulphur Trioxide	+	Sulphur Monochloride	⟶	Thionyl Chloride	+	Sulphur Dioxide	+	Sulphur
SO_3	+	S_2Cl_2	⟶	$SOCl_2$	+	SO_2	+	S

Thionyl Chloride fumes in moist air and is decomposed in water forming hydrochloric acid and sulphurous acid, for this reaction thionyl chloride is also known as *Acid chloride.*

Thionyl chloride when heated up to 150°C it starts decomposing into chlorine sulphur dioxide and sulphur monochloride.

Thionyl Chloride	⟶	Chlorine	+	Sulphur Monochloride	+	Sulphur Dioxide
$4SOCl_2$	⟶	$3Cl_2$	+	S_2Cl_2	+	$2SO_2$

There are numerous reactions shown below :

1. Thionyl chloride reacts with sulphur at 180°C to form sulphur monochloride and sulphur

Thiony Chloride	+	Sulphur	⟶	Sulphur Monochloride	+	Sulphur Dioxide
$SOCl_2$	+	S	⟶	S_2Cl_2	+	SO_2

2. Thionyl chloride reacts with phosphorous at high temperature forming phosphorous trichloride, sulphur monochloride.

Thionyl Chloride	+	Phosphorous	⟶	Phosphorous Tetrachloride	+	Sulphur	+	Sulphur Dioxide
$2SOCl_2$	+	P	⟶	PCl_4	+	S	+	SO_2

3. Thionyl Chloride reacts with tellurium in excess to form tellurium tetrachloride sulphur dioxide and sulphur.

Thionyl Chloride	+	Tellurium	⟶	Tellurium Tetrachloride	+	Sulphur	+	Sulphur Dioxide
$SOCl_2$	+	Te	⟶	$TeCl_4$	+	S	+	SO_2

4. Thionyl chloride also reacts with metallic sulphides these reactions often take at 150°C – 200°C

Thionyl Chloride + Ferrous Sulphide ⟶ Sulphur Monochloride + Iron + Sulphur Dioxide

$$SOCl_2 + FeS \longrightarrow S_2Cl_2 + Fe + SO_2$$

5. Metallic oxide e.g. ferric oxide reacts with thionyl chloride producing ferrous chlorides and sulphur dioxide

Ferric Oxide + Thionyl Chloride ⟶ Mercurous Chloride + Sulphur Dioxide

$$FeO + 4SOCl_2 \longrightarrow 3HgCl_2 + 2SO_2$$

6. Mercury reacts with sulphuryl chloride in excess producing mercurous chloride, sulphur dioxide and sulphur monochloride, at 150°C

Mercury + Thionyl Chloride ⟶ Mercurous Chloride + Sulphur Dioxide + Sulphur Dioxide

$$3Hg + 4SOCl_2 \longrightarrow 3HgCl_2 + 2SO_2 + S_2Cl_2$$

7. In the presence of a catalyst aluminium chloride thionyl chloride reacts with hydrogen sulphide forming hydrogen chloride and sulphur dioxide and sulphur

Thionyl Chloride + Hydrogen Sulphite ⟶ Sulphur + Hydrogen Chloride + Sulphur Dioxide

$$SOCl_2 + FeS \longrightarrow S_2Cl_2 + HCl + SO_2$$

8. Sulphur trioxide reacts with thionyl chloride forming pyrosulphuryl chloride and sulphur dioxide

Thionyl Cloride + Sulphur Trioxide ⟶ Phyrosulphuryl Chloride + Sulphur Dioxide

$$SOCl_2 + 2SO_3 \longrightarrow S_2O_5Cl_2 + SO_2$$

9. Silver nitrate reacts with thionyl chloride at a considerable low temperature of about –78°C, where sulphuryl bromide and hydrogen chloride are formed

Silver Naitrate + Thionyl Chloride ⟶ Sulphuryl Nitrogen Dioxychloride + Silver Chloride

$$AgNO_3 + SOCl_2 \longrightarrow SO_2NO_2Cl + AgCl$$

10. Hydrogen Bromide reacts with thionyl chloride at a considerable low temperature of about -78°C, where sulphuryl bromide and hydrogen chloride are formed

Hydrogen Iodine + Thionyl Chloride ⟶ Iodine+ Sulphur + Hydrogen Chloride + Sulphur Trioxide

$$6HI + 3SOCl_2 \longrightarrow 3I_2 + 2S + 6HCl + SO_3$$

11. Hydrogen iodide reacts with thionyl chloride, liberating iodine, sulphur and also producing hydrogen chloride and sulphur dioxide

Hydrogen Iodine + Thionyl Chloride ⟶ Iodine + Sulphur + Hydrogen Chloride + Sulphur Trioxide

$$6HI + 3SOCl_2 \longrightarrow 3I_2 + 2S + 6HCl + SO_3$$

12. Dinitrogen Tetraoxide reacts with thionyl chloride forming nitrogen oxychloride

Dinitrogen Tetraoxide	+	Thionyl Chloride	$\longrightarrow$	Nitrogen Oxychloride	+	Sulphur Trioxide
N_2O_4	+	$SOCl_2$	$\longrightarrow$	2NOCl	+	SO_3

13. Barium peroxide reacts with thionyl chloride producing barium chloride, sulphur dioxide and sulphuryl chloride.

Barium Peroxide	+	Thionyl Chloride	$\longrightarrow$	Sulphur Dioxide	+	Sulphuryl Chloride	+	Barium Chloride
BaO_2	+	$2SOCl_2$	$\longrightarrow$	SO_2	+	SO_2Cl_2	+	$BaCl_2$

Thionyl chloride with organic compounds consisting hydroxy compounds, with organic acids it forms acid chlorides, it also reacts with phenols and alcohols to forms chlorides of respective derivatives. Equimolal quantities of thionyl chloride and acid produce the acid chloride depending upon the ratio of acid to thionyl chloride.

The reaction of thionyl chloride with Grinard reagent yields the corresponding sulphoxide. As these reactions are totally organic they are not being discussed here.

Sulphuryl Chloride, Sulphonic Oxychloride (SO_2Cl_2): Sulphuryl chloride also known as sulphonic oxychloride is a colourless liquid with an extremely pungent odour. Sulphuryl Chloride is prepared by the action of sulphur dioxide and chlorine in the presence of carbon as catalyst.

Sulphur Dioxide + Chlorine $\xrightarrow{\text{Carbon}}$ Sulphuryl Chloride

$$SO_2 + Cl_2 \xrightarrow{C} SO_2Cl_2$$

> Sulphuryl chloride is highly toxic and the inhalation of its vapour could be fatal, also avoid contact with skin

This reaction is carried out in a water-cooled vessel where the catalyst is moistened with sulphonyl chloride being formed. The product is then condensed in a condenser. Sulphuryl chloride in complete dry state does not corrode metals like aluminium magnesium or nickel a room temperature. But it gets slowly hydrolyzed with water. Sulphuryl chloride gets decomposed to sulphuric acid and hydrogen chloride.

Sulphuryl Chloride + Water $\longrightarrow$ Sulphuric Acid + Hydrogen Chloride

$$SO_2Cl_2 + 2H_2O \longrightarrow H_2SO_4 + 2HCl$$

Sulphuryl chloride is stable at room temperature but it decomposes at high temperature

Sulphur Dioxide + Chlorine $\longrightarrow$ Sulphuryl Chloride

$$SO_2 + Cl_2 \longrightarrow SO_2Cl_2$$

At a temperature of 200°C it reacts with sulphur forming sulphur monochloride and sulphur dioxide

Sulphur Dioxide + Chlorine $\xrightarrow{200°C}$ Sulphuryl Chloride

$$SO_2 + Cl_2 \xrightarrow{200°C} SO_2Cl_2$$

With hydrogen sulphide sulphuryl chloride forms hydrogen chloride sulphur monochloride, sulphur, sulphur dioxide.

Sulphuryl Chloride + Hydrogen Sulphide ⟶ Hydrogen Chloride + Sulphur Dioxide + Sulphur

$$SO_2Cl_2 + H_2S \longrightarrow HCl + CO_2 + S$$

Sulphuryl Chloride + Hydrogen Sulphide ⟶ Hydrogen Chloride + Sulphur Monochloride + Oxygen

$$SO_2Cl_2 + H_2S \longrightarrow HCl + SO_2 + O_2$$

Sulphuryl chloride liberates bromine and iodine from corresponding bromides and iodides and bromides.

Sulphuryl Chlrodie + Potassium Broxide ⟶ Potassium Chloride + Bromine + Sulphur Dioxide

$$SO_2Cl_2 + 2KBr \longrightarrow HCl + Br_2 + SO_2$$

Iodine reacts with sulphuryl chloride in the presence of aluminium chloride as a catalyst to form iodine chlorides.

Sulphuryl Chloride + Iodine $\xrightarrow{\text{Aluminium Chloride}}$ Iodine Monochloride + Sulphur Dioxide

$$SO_2Cl_2 + I_2 \xrightarrow{AlCl_3} ICl + SO_2$$

Sulphuryl Chloride + Iodine Chloride ⟶ Iodine Trichloride + Sulphur Dioxide

$$SO_2Cl_2 + ICl \longrightarrow ICl_3 + SO_2$$

Phosphorous chlorides react with sulphuryl chloride producing phosphorous oxychloride and thionyl chloride.

Phophorous Pernatchloirde + Sulphuryl Chloride ⟶ Thionyl Chloride + Chloirine + Phosphorous Oxychloride

$$PCl_5 + SO_2Cl_2 \longrightarrow SOCl_2 + Cl_2 + POCl_3$$

Sulphuryl chloride reacts with zinc in the presence of eather producing zinc chloride and zinc sulphoxylate.

Zinc + Sulphuryl Chloride ⟶ Zinc Chloride + Zinc Sulphoxylate

$$2Zn + SO_2Cl_2 \longrightarrow ZnCl_2 + ZnSO_2$$

Sodium acetate reacts with sulphuryl chloride producing acetyl chloride so it should kept in cool dark place away from heat.

The commercial material is a pale yellow liquid containing at least 99% sulphuryl chloride. Sulphuryl chloride is used to convert organic acids to the corresponding acid chlorides and acids anhydrides. It is used as a selective chlorinating agent particularly in the production of chlorophenol, which is used as disinfectants. It is also used in the synthesis of various pharmaceuticals dyes and surfactants.

Pyrosulphuryl Chloride, Disulphuryl Chloride ($S_2O_5Cl_2$): It is a dense colourless liquid. At room temperature pyrosulphuryl chloride is somewhat stable, at a temperature above 185°C it decomposes into sulphur dioxide, sulphur trioxide and chlorine.

Pyrosulphuryl Choride ⟶ Sulphur Dioxide + Sulphur Trioxide + Chlorine

$$S_2O_5Cl_2 \longrightarrow SO_2 + SO_3 + Cl_2$$

Pyrosulphuryl Chloride reacts with potassium chromate forming chromyl chloride.

Pyrosulphuryl Chloride + Potassium Chromate ⟶ Potassium Pyrosulphate + Chromyl Chloride

$$S_2O_5Cl_2 + K_2Cr_2O_4 \longrightarrow K_2S_2O_7 + CrO_2Cl_2$$

Pyrosulphuryl chloride may be prepared by the reaction of wide range of chlorides, with sulphur trioxide, e.g. sulphur monochloride, thionyl chloride phosphorous pentachloride, phosphorous oxychloride, silicon tetrachloride, sodium chloride and carbon tetrachloride. But, however, these processes are not carried out because these reactions are pretty complicated and they are also costly for commercial purpose, so the usual method is to add oleum to hot carbon tetrachloride and obtaining pyrosulphuryl chloride with carbon oxychloride.

Oleum + Carbon Tetrachloride ⟶ Chlorosulphonic Acid + Carbon Oxychloride

$$H_2S_2O_7 + CCl_4 \longrightarrow 2HClSO_3 + COCl_2$$

Immediately the chlorosulphonic acid, which was, produced starts reaching with hydrolyzed to original hydrate.

Chlorosulphuric Acid + Carbon Tetrachloride ⟶ Carbon Oxychloride + Hydrogen Chloride + Pyrosulphuryl Chloride

$$2HClSO_3 + CCl_3 \longrightarrow COCl_2 + 2HCl + S_2O_5Cl_2$$

The reacted mass contains chlorosulphuric acid, which is again treated with sodium chloride to remove the acid and the resulted is then fractionally distilled to obtain a clear and refined pyrosulphuryl chloride. This process has a disadvantage also, *a poisonous gas called phosgene is obtained as a by-product.*

Another more convenient and also in use to prepare pyrosulphuryl chloride involves the reaction between sulphur monochloride and sulphur trioxide. The process is done in the same process as the sulphur trioxide, was prepared by batch process. The sulphur trioxide and sulphur monochloride is then slowly added to liquid sulphur trioxide at 35°C -40°C temperature the addition is continued, until the 890/oto 95% sulphur trioxide has been used. The final equation is

Sulphur Monochloride + Sulphur Trioxide ⟶ Pyrosulphuryl Chloride + Sulphur Dioxide

$$S_2Cl_2 + 5CO_2 \longrightarrow S_2O5Cl_2 + 5SO_2$$

A third process is also available which involves the reaction of sulphur trioxide vapour with sulphur dichloride to form pyrosulphuryl chloride at a temperature of 110°C. the industrial reactor is fitted with a reflux condenser to return any remaining reactants which have vapourize from the reaction mass. The pyrosulphuryl chloride produced is collected in the reactor. The crude product is fractionally distilled to remove the sulphur dioxide, sulphur trioxide, sulphur dichloride and thionyl chloride. A product partly of 99% is obtained. The reaction equation is

Sulphur Monochloride	+	Sulphur Trioxide	$\longrightarrow$	Pyrosulphuryl Chloride	+	Sulphur	+	Thionyl Chloride
$4S_2Cl_2$	+	$4SO_3$	$\longrightarrow$	$2S_2O_5Cl_2$	+	4S	+	$2SOCl_2$

Sulphur Oxytetrachloride ($S_2O_5Cl_4$) : Sulphur oxytetrachloride also known as Trichlorosulphonium Chlorosulphate, it is a pungent, colourless crystalline solid, sulphur oxytetrachloride is prepared at a very low temperature of –12°C or even lower. When sulphur tetrachloride and chlorosulphonic acid reacts with each other sulphur oxytetrachloride is produced.

Sulphur Tetrachloride	+	Chlorosulphonic Acid	+	Oxygen	$\longrightarrow$	Sulphur Oxytetrachloride	+	Hydrogen Chloride
SCl_4	+	$HClSO_3$	+	O_2	$\longrightarrow$	$S_2O_5Cl_4$	+	HCl

It slowly decomposes at room temperature to thionyl to chloride. Sulphur oxytetrachloride hydrolyzes to sulphurous acid. It is believed that, sulphur oxytetrachloride is the trichlorosulphonium salt of chlorosulphonic acid.

Thionyl Bromide ($SOBr_2$) : It is an orange coloured liquid prepared by the action of hydrogen bromide on thionyl chloride.

Hydrogen Bromide	+	Thionyl Chlride	$\longrightarrow$	Thonyl Bromide	+	Hydrogen Chloride
2HBr	+	$SOCl_2$	$\longrightarrow$	$SOBr_2$	+	2HCl

Thionyl bromide is quite unstable at room temperature, as it decomposes very slowly into sulphur monobromide, bromine and sulphur dioxide.

We shall now look a chart given below consisting of physical and chemical data of the sulphur halides.

The Physical Properties of Sulphur Halides

Name

	Thionyl Chloride	Sulphuryl Chloride	Sulphuryl Fluoride	Thiony Bromide	Sulphuryl Oxytetra-chloride	Pyrosul-phuryl Chloride
Formula	$SOCl_2$	SO_2Cl_2	SOF_2	$SOBr_2$	$S2O_5Cl_4$	$S_2O_5Cl_5$
State	Liquid	Liquid		Liquid	Soild	Liquid
Colour		Colourless		Orange yellow	Yellow	Colour-less
Melting Point (°C)	–104.5	–46			57	–3
Boiling Point (°C)	79	69.1		138		152
Critical Temperature (°C)	18.67					
Heat of Vaporization (KCl/mol^{-1})		31.4				35.1
Specific gravity	2.697					

Acids of Sulphur

Sulphur when reacts with oxygen forming compounds known as oxides which we have already read and these oxides when reacts with water on reaction produces many important acids. Some of these acids are important and some of them are unstable and thus are not suitable for study. A list is provided for a quick look at these acids and their structural formula.

The Acids of Sulphur and their Structural Formula

Acids	Formula	Structure
HYdrosulphuric Acid/Dithionous Acid	$H_2S_2O_4$	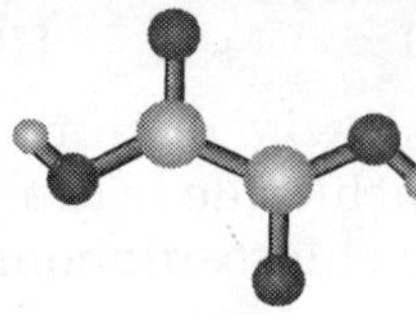
Sulphurous Acid	H_2SO_3	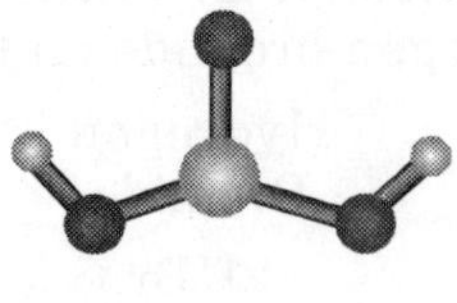
Thiosulphurous Acid	$H_2S_2O_2$	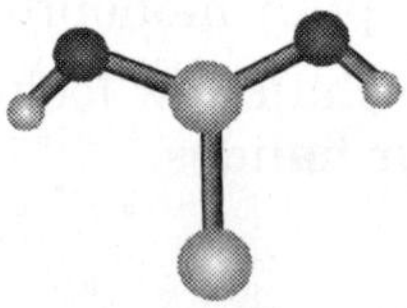
Pyrosulphurous Acid *(Hydroxysulfanesulfonic acid oxide)*	$H_2S_2O_5$	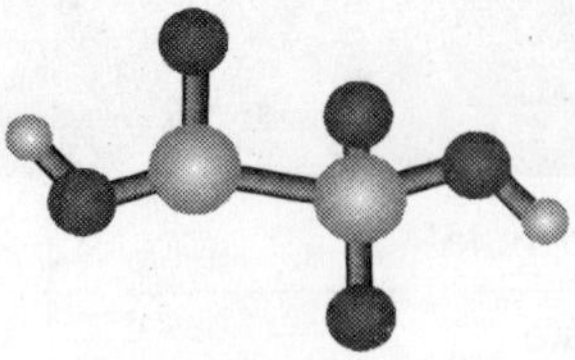
Sulphuric Acid	H_2SO_4	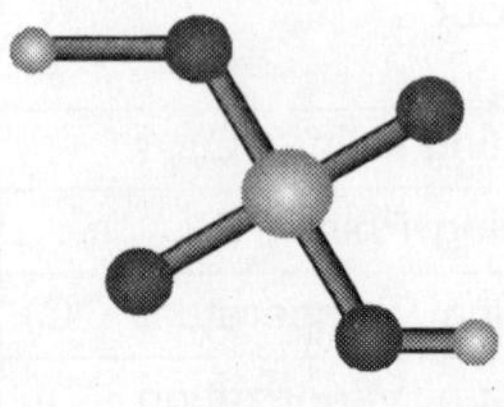

Acids	Formula	Structure
Pyrosulphuric Acid (Oleum)	$H_2S_2O_7$	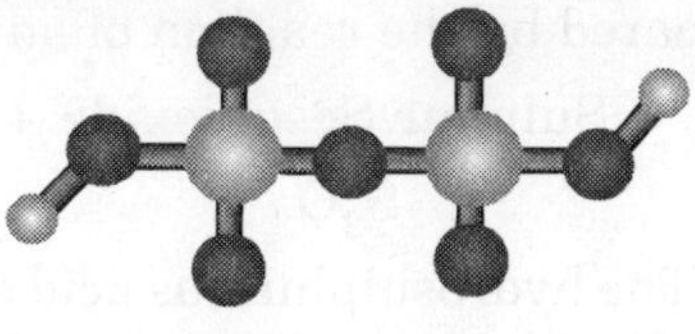
Dithionic Acid (Hydrosulphurous Acid)	$H_2S_2O_6$	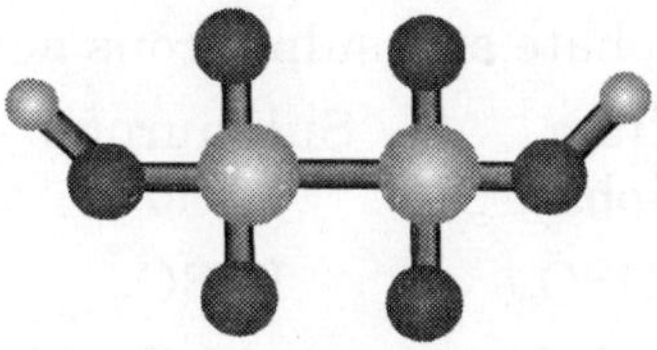
Peroxymonosulphuric Acid Trioxidance-1-sulfonic acid	H_2SO_5	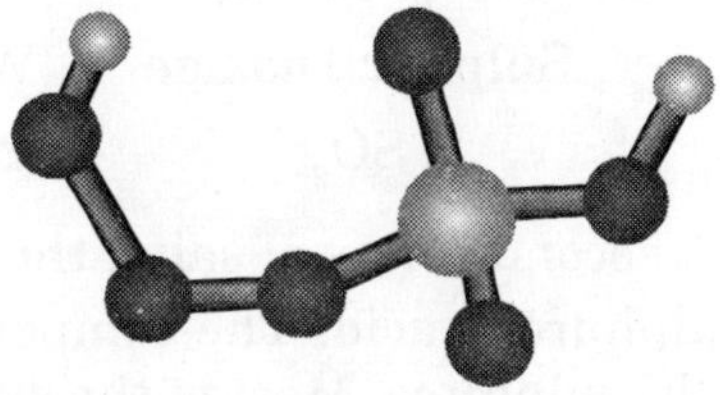
Thiosulphuric Acid	$H_2S_2O_3$	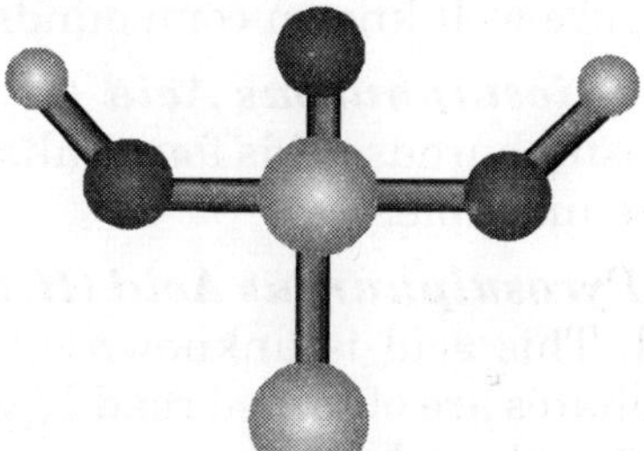
Sulphymic Acid	H_3SO_3N	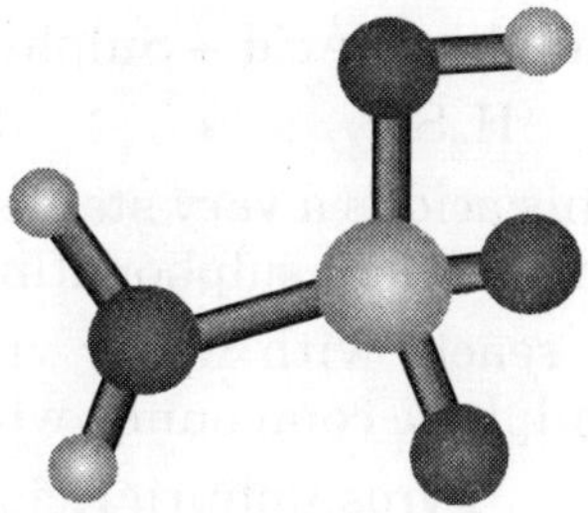

Hydrosulphurous Acid ($H_2S_2O_4$): In is also known as dithionous acid, it is quite stable in the anhydrous state but starts dissociating when reacts with water, this acid is prepared by the reaction of sulphur sesquioxide and water.

Sulphur Sesquioxide + Water ⟶ Hydrosulphurous Acid

$$S_2O_3 + H_2O \longrightarrow H_2S_2O_4$$

The hydrosulphurous acid does not exist in the aqueous state, the only stable compound is sodium hydrosulphite or sodium dithionate, which is prepared by the reaction of sodium bisulphate and sulphurous acid in the presence of zinc.

Sodium Bisulphate + Sulphurous Acid + Zinc ⟶ Sodium Dithionate + Zinc Bisulphate + Water

$$2NaHSO_3 + H_2SO_3 + Zn \longrightarrow Na_2S_2O_4 + ZnSO_3 + 2H_2O$$

Sulphurous Acid (H_2SO_3): When sulphur dioxide is dissolved in water, sulphurous acid is produced. It is a very weak acid yet its compounds are strong and stable and many are found in crystalline form

Sulphur Dioxide + Water ⟶ Sulphurous Acid

$$SO_2 + 2H_2O \longrightarrow H_2SO_3$$

A second series of salts, the hydrogen sulphites ($KHSO_3^-$) are also known compounds of sulphurous acid. The compounds of sulphurous acids are called 'sulphides' 'sulphites' and 'bi-sulphites. Most of the sulphites except that of rubidium cesium etc. are not soluble in water. Some of the important compounds are potassium hydrogen sulphate ($KHSO_3$) etc., are well-known compounds.

Thiosulphurous Acid ($H_2S_2O_7$): Thiosulphurous acid is also known from its salts. Thiosulphurous acids form salts in the presence of sodium and potassium salts, which are quite unstable.

Pyrosulphurous Acid ($H_2S_2O_2$): Pyrosulphurous acid is also known as disulphurous acid. This acid is unknown either in free state or in aqueous state, but its compound sulphates are obtained readily when hydrogen sulphide and other sulphites decompose or react with each other.

Pyrosulphuric Acid ($H_2S_2O_7$): Pyrosulphuric acid is the product when sulphur trioxide reacts with sulphuric acid. This acid is known as oleum, the ore of sulphuric acid.

Sulphuric Acid + Sulphur Trioxide ⟶ Pyrosulphuric Acid

$$H_2SO_4 + SO_3 \longrightarrow H_2S_2O_7$$

This acid is a very stable acid and extremely fuming having a melting point of 35.1°C. It is an excellent sulphonating agent when heated and looses sulphur trioxide continuously.

It reacts with water vigorously liberating high quantity of heat. It forms stable pyrosulphate compounds when reacts with metallic hydroxide.

Pyrosulphuric Acid + Water ⟶ Sulphuric Acid

$$H_2S_2O_7 + H_2O \longrightarrow 2H_2SO_4$$

It also reacts with potassium hydroxide forming potassium pyrosulphate

Sodium Hydroxide	+	Pyrosulphuric Acid	$\longrightarrow$	Sodium Pyrosulphate	+	Water
2NaOH	+	$H_2S_2O_7$	$\longrightarrow$	$Na_2S_2O_7$	+	H_2O
Potassium Hydroxide	+	Pyrosulphuric Acid	$\longrightarrow$	Sodium Pyrosulphate	+	Water
2KOH	+	$H_2S_2O_7$	$\longrightarrow$	$K_2S_2O_7$	+	H_2O

Thiosulphuric Acid ($H_2S_2O_3$): Thiosulphuric Acid is prepared by the reaction of hydrogen sulphide and sulphur dioxide

Hydrogen Sulphide	+	Sulphur Dioxide	$\longrightarrow$	Thiosulphuric Acid
H_2S	+	SO_2	$\longrightarrow$	$H_2S_2O_3$

It is also prepared by the reaction of sodium thiosulphate and hydrogen chloride.

Sodium Thiosulphate	+	Hydrogen Chlroide	$\longrightarrow$	Sodium Chloride	+	Thiosulphuric Acid
Na_2S_2	+	HCl	$\xrightarrow{78°C}$	2NaCl	+	$H_2S_2O_3$

Thiosulphuric acid when reacts with elements forms thiosulphate yet it is a weak acid.

Dithionic Acid ($H_2S_2O_6$): Dithionic acid till now has not yet been obtained in pure state, the only process where barium dithionate reacts with sulphuric acid forming dithionic acid.

Barium Ditionate	+	Sulphuric Acid	$\longrightarrow$	Barium Sulphate	+	Dithionic Acid
BaS_2O_6	+	H_2SO_4	$\longrightarrow$	$BaSO_4$	+	$H_2S_2O_6$

Dithionic acid forms stable compounds called thionates, when manganese dioxide reacts with dithionic acid manganese dioxide is produced.

Dithionic Acid	+	Maganese Dioxide	$\longrightarrow$	Manganese Dithionate	+	Hydrogen Peroxide
$H_2S_2O_6$	+	2KOH	$\longrightarrow$	$K_2S_2O_3$	+	H_2O_2

Potassium thionats is obtained by the reaction of potassium hydroxide and thionic acid

Ditionic Acid	+	Potassium Hydroxide	$\longrightarrow$	Potassium Thionate	+	Water
$H_2S_2O_6$	+	2KOH	$\longrightarrow$	$K_2S_2O_3$	+	$2H_2O$

Sulphuric Acid, Vitri Oleum (H_2SO_4): This is the most important acid of the sulphur also known as the *'yardstick of the industrial development'*. Sulphuric Acid, is not at all a new acid it was known to the alchemists, also they used to call it as Vitri Oleum (or oil of mirror, because of its syruppy appearence).

Sulphuric Acid is such a compound, which has helped to produce many chemicals and many important acid of the chemical laboratory.

We shall now study the chemistry of the sulphuric acid.

Hence the sulphur trioxide can prepared in the laboratory by many processes, two of which has been discussed earlier. Here we will again discuss the process with the industrial preparation of sulphuric acid.

Laboratory Preparation of Sulphuric Acid

Theory: Sulphuric acid is prepared in the laboratory by the reaction of sulphur trioxide and water, this sulphur trioxide is prepared by the reaction of sulphur dioxide and oxygen in the presence of a catalyst (in this case vanadium pentoxide).

The Physical Properties of Sulphur Halides

Name	Sulphur Monochloride	Sulphur Diochloride	Sulphur Tetrachloride	Sulphur Difluoride	Sulphur Tetrafluoride	Sulphur Hexafluoride	Sulphur Monobromidse
Formula	S_2Cl_2	SCl_2	SCl_4	S_2F_4	SF_4	SF_6	S2Br_2
State	Liquid	Liquid	Liquid	Gas	Gas	Gas	Liquid
Colour	Yellow	Red	Yellowish Brown	Colourless	Colouress	Colouress	Red
Meling Point (°C)	–80	–122	–30	–105.5	124	–124	–46
Boiling Point (°C)	138	59		–99	–40	–40	54
Critical Temperature (°C)					45	91	
Critical Pressure (Pa)					3.82278		
Critical Molar Volume (cm^3)					199		
Heat of vapourisation ($Kcal/mol^{-1}$)	64.6	67.6		35.1	8.99	26.44	
Heat of Fusion (K cal/mol^{-1})					5.02		23.08
Di-electric Constant	4.79	9.25	2.945		1.81		9.25
Surface Tension ($dyne/cm^2$)	40.79				11.63		1.676
Viscosity ($dymne/cm^2$)	2.015		.801				1.801
Specific gravity	1.6733	1.621				2.6355	1.7
Specific conductance ($ohm^{-1}cm^{-1}$)		2×10^{-6}					
Heat of Capacity ($deg^{-1}g^{-1}$)					77.6		
Vander Waal's Constant ($barL^2$) a L/mol b	7.857(a) 0.087(b)						4.544(a) 0.4339(b)

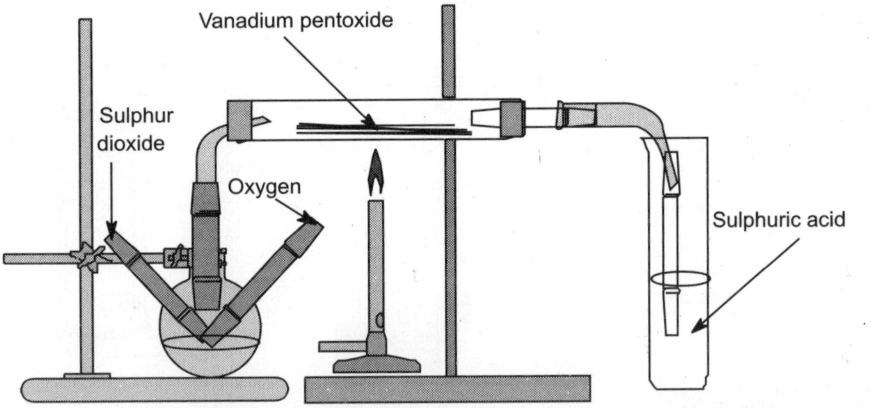

Apparatus	Reagents	Temperature	Pressure	Catalyst
Woulf's Bottle Reaction tube, delivery tube etc.	Sulphur dioxide, oxygen calcium chloride	410°C	Normal	Vanadium Pentoxide

Fig. 11.23. The Lab Preparation of Sulphuric Acid

Reaction Equation

$$\text{Sulphur Dioxide} + \text{Oxygen} \xrightarrow[\text{Pentoxide}]{\text{Vanadium}} \text{Sulphur Trioxide}$$

$$2SO_2 + O_2 \xrightarrow{V_2O_5} 2SO_3$$

$$\text{Sulphur Trioxide} + \text{Water} \longrightarrow \text{Sulphuric Acid}$$

$$SO_3 + H_2O \longrightarrow H_2SO_4$$

Precaution: *The sulphur trioxide must be carefully dissolved otherwise there is a chance of explosion because the produced acid is extremely fuming and the reaction is exothermic.*

Experiments 2

Theory : Sulphur Trioxide can also be prepared by the action of nitrogen dioxide and sulphur dioxide. Yielding part of its oxygen to dioxide again and this cycle go on.

Procedure : A large Flask fitted with a cork through which four inlet tubes are passed and also a small outlet tube is passed, three of the inlet tube comes from the wash bottles containing sulphuric acids and fourth from the flask where water is boiled. Through the wash bottle sulphur dioxide enters in the reaction flask (F) by boiling water in the flask (E) more steam is passed in the reaction flask. When white crystals react with steam liberating red nitrogen dioxide and forming fairly concentrated sulphuric acid, which collect at the bottom of the flask. The complex reactions that take place can be represented as.

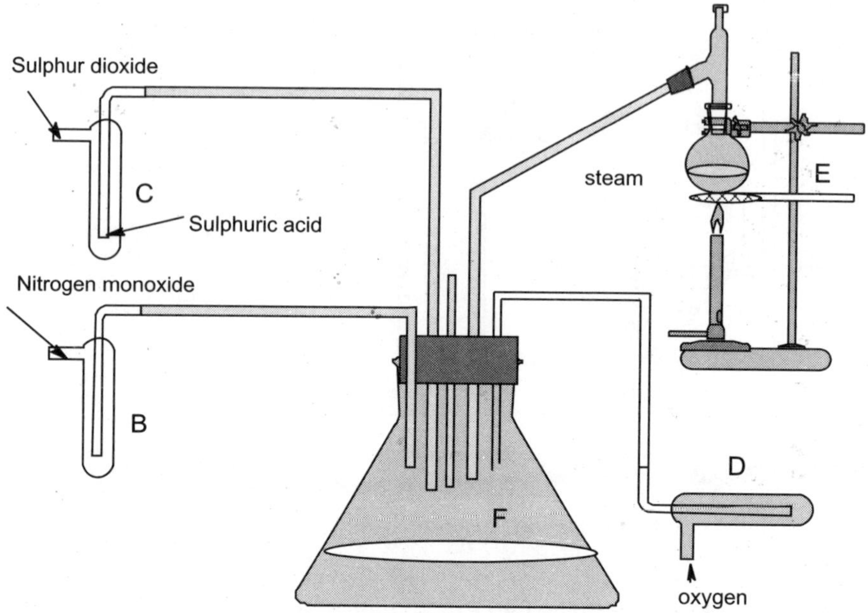

Fig. 11.24. The Preparation of Sulphuric Acid

Properties of Sulphuric Acid

Physical Properties : Sulphuric acid is an oily viscous liquid, colourless, tastes sour, we shall now study the physical properties of sulphuric acid.

Molar Mass	98.080
Freezing Point	10.31°C
Boiling Point	297°C
Density	1.8345 g cm^{-3}
Di-Electronic Constant (E_o)	110 (At 20°C)
Specific Conductance	1.04 $ohm^{-1}m^{-1}$
Viscosity	0.2455 Pois (at 20°C)
Standard Enthalpy of Formation	–814.0 kJ/mol
Gibb's Standard Enthalpy	–690 kJ/mol
Heat of Capacity	138.9
Heat of Fusion	10.71 kJ/mol
Heat of Vapourization	–193.3 $Kcal/mol^{-1}$
Ionization Constant	2.7×10^{-4} mol^2 dm^{-6}

Chemical Properties: Sulphuric Acid reacts mostly with all the metals liberating hydrogen and producing sulphates the compounds of its own, when reacts with metals it forms sulphates, e.g. with zinc it forms zinc sulphate and hydrogen.

Zinc + Sulphuric Acid $\longrightarrow$ Zinc Sulphate + Hydrogen

$$Zn + H_2SO_4 \longrightarrow ZnSO_4 + H_2$$

Also with tin it produces stannous sulphate and hydrogen

Tin + Sulphuric Acid $\longrightarrow$ Stannous Sulphate + Hydrogen

$$Sn + H_2SO_4 \longrightarrow SnSO_4 + H_2$$

De-hydrating property of Sulphuric Acid : Sulphuric Acid has a great affinity towards water even it absorbs water from the compounds.

Exp: When sulphuric acid is added to sugar it is converted to black carbon

Sugar + Sulphuric Acid $\longrightarrow$ Carbon (Sulphuric Acid + Water)

$$C_{12}H_{32}O_{11} + H_2SO_4 \longrightarrow 12C + (H_2SO_4) + H_2O$$

Exp: When sulphuric acid is added to formic acid and slightly heated.

Formic Acid + Sulphuric Acid $\longrightarrow$ Carbon Monoxide + (Water + Acid)

$$H_2CO_2 + H_2SO_4 \longrightarrow CO + (H_2O + H_2SO_4)$$

It is extremely necessary to keep the concentrated or even dilute sulphuric acid away from cloths etc.

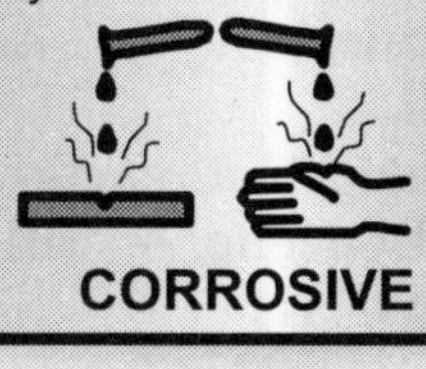

This are the Most Important Reactions of Sulphuric Acid

Sulphuric Acid also reacts with the hydroxides of sodium and potassium also calcium producing their sulphates.

Sodium Hydroxide + Sulphuric Acid $\longrightarrow$ Sodium Hydrogen Sulphate + Water

$$NaOH + H_2SO_4 \longrightarrow NaHSO_4 + H_2O$$

Potassium Hydroxide + Sulphuric Acid $\longrightarrow$ Potassium Hydrogen Sulphate + Water

$$KOH + H_2SO_4 \longrightarrow NaHSO_4 + H_2O$$

Gases like hydrogen when passes through the acid sulphur dioxide is liberated

Sulphuric Acid + Hydrogen $\longrightarrow$ Sulphur Dioxide + Water

$$H_2SO_4 + H_2 \longrightarrow SO_2 + 2H_2O$$

When nitrogen is allowed to pass through sulphuric acid, nitrous acid with the evolution of sulphur dioxide is produced.

Sulphuric Acid + Nitrogen $\longrightarrow$ Nitrous Acid + Water

$$H_2SO_4 + N_2 \longrightarrow 2HNO_2 + 2H_2O$$

Nitrous acid is produced with the liberation of sulphur dioxide

Sulphuric acid reacts with phosphorous producing phosphoric acid with deposition of sulphur.

Phosphorous + Sulphuric Acid ⟶ Sulphur Dioxide + Phosphoric Acid + Sulphur

$$2P + 3H_2SO_4 \longrightarrow SO_2 + 2H_3PO_4 \quad S$$

Sulphuric acid liberates bromine from potassium bromide and also, iodine from potassium iodide.

Potassium Bromide + Sulphuric Acid ⟶ Potassium Hydrogensulphate + Bromine + Water + Sulphur Dioxide

$$2KBr + 3H_2SO_4 \longrightarrow 2KHSO_4 + Br_2 + 2H_2O + SO_2$$

Potassium Iodide + Sulphuric Acid ⟶ Bromine + Water + Sulphur Dioxide

$$2KI + 3H_2SO_4 \longrightarrow Br_2 + 2H_2O + SO_2$$

Concentrated Sulphuric Acid when react with phosphorous pentachloride chlorosulphuric acid is produced.

Phosphorous Pentachloride + Sulphuric Acid ⟶ Chlorosulphuric Acid + Phosphorous Oxychloride + Water

$$PCl_5 + H_2SO_4 \longrightarrow HSO_3Cl + POCl_3 + H_2O$$

Sulphuric Acid reacts with benzene forming benzene sulphonic acid

Benzene + Sulphuric Acid ⟶ Benzene Sulphonic Acid + Water

$$C_6H_6 + H_2SO_4 \longrightarrow HSO_3Cl + H_2O$$

Sulphuric acid in fuming state reacts with hydrochloric acid forming chlorosulphuric acid

Sulphuric Acid (anhydride) + Hydrochloric Acid ⟶ Chlorosulphuric Acid

$$SO_3 + HCl \longrightarrow HSO_3Cl$$

Chlorosulphuric Acid forms colourless liquid boiling at 157.5°C. It is readily hydrolyzed to sulphuric acid and hydrochloric acid and hydrochloric acid.

Chlorosulphuric Acid + Water ⟶ Hydrochloric Acid + Sulphuric Acid

$$HSO_3Cl + H_2O \longrightarrow HCl + H_2SO_4$$

MOLECULAR STRUCTURE OF SULPHURIC ACID

Sulphuric Acid has a tetrahderal structure consisting of four oxygen atoms are arranged around the central sulphur atom. The oxygen atoms are single bonded to the hydrogen atom.

Industrial Preparation of Sulphuric Acid

Sulphuric Acid in industry is manufactures mainly by two processes, *(i) Chamber Process and (ii) Contact Process*

Chamber Process or Lead Chamber Process : The process is named so; because it uses three to twelve chambers where majority of this process is carried out (see fig.11.24), only three of the has been shown here to avoid complication. At first iron pyrites (FeS_2) also known as ferrous disulphide is burnt in excess of air. So that oxygen concentration is slightly higher that of sulphur dioxide.

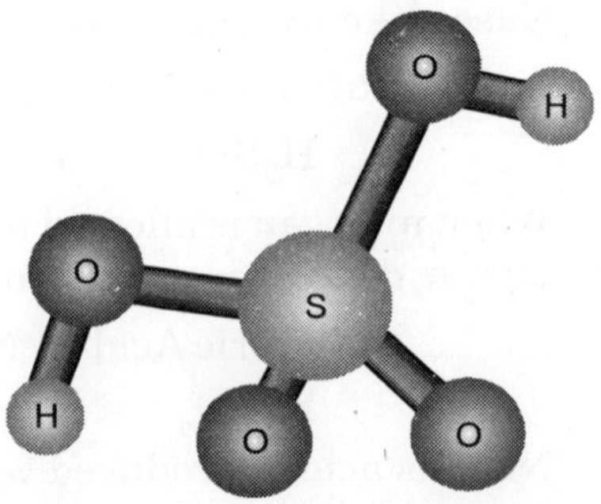

The Structure of Sulphuric Acid

The gases are sent to combustion chamber where they are made dust free, sufficient amounts of nitrogen that is nitrogen oxide and dioxide are introduced into the gas mixture. The gases are then entered to the base of the Glover tower, where it meets the downward floe of the Gay-Lussac acid or nitrous vitriol. The gas is cooled here and picks up the part of the moisture (steam) for the acid making process. The nitrogen content from the nitrous vitriol gets heated up which is supplied from the gas. The gas from the top of the glover tower is send to the Lead chambers, these vary in number as said earlier form three to twelve, depending upon the capacity of the plant. Water is sprayed from the top of the chambers, by means of atomizer. Here sulphur dioxide, nitrogen oxides, water and oxygen react to form sulphuric acid, which collects at the bottom of the chambers due to the condensation of the hot gaseous mixture. The gas still has some nitrogen content in it, which is removed by washing the gas with cold gloval acid. The residual gas mainly consists of nitrogen and oxygen, which is discharged through the stack in the air.

Lead Chamber Process

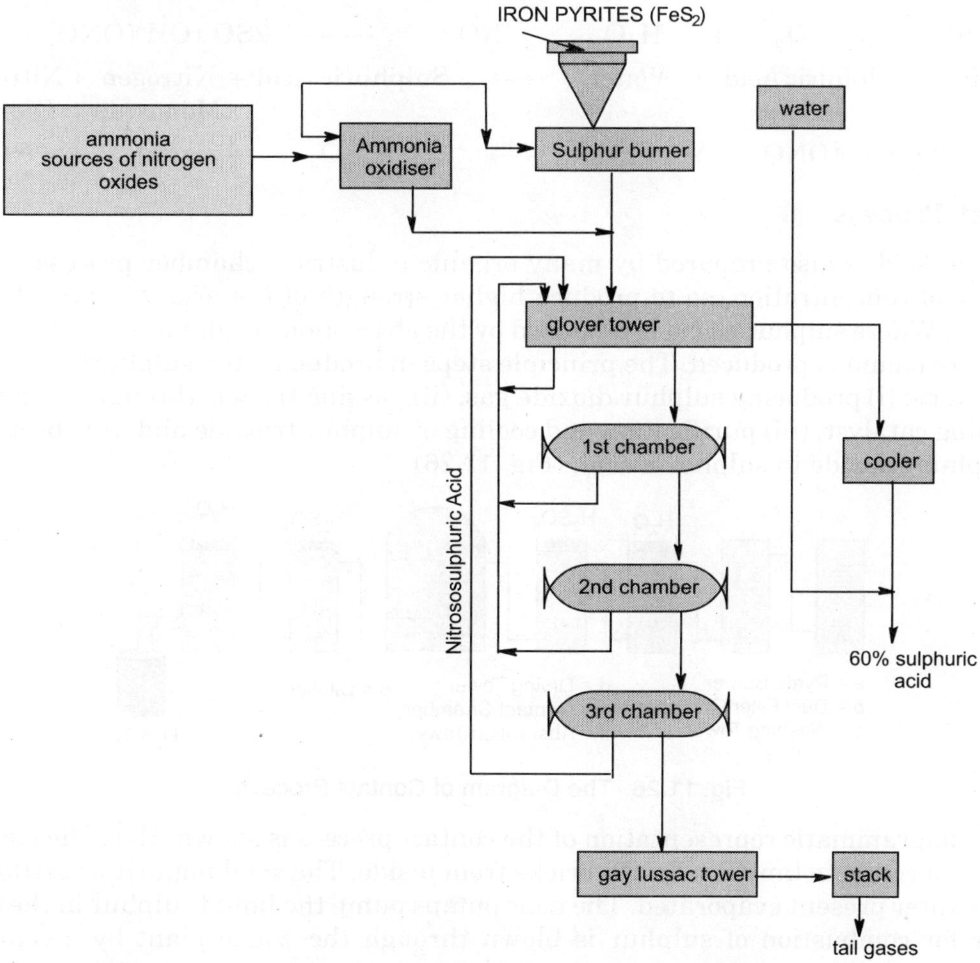

Fig. 11.25. The Lead Chamber Process for the manufacturing of Sulphuric Acid

Reactions of Lead Chamber Process

Nitrogen Monoxide + Oxygen ⟶ Nitrogen Dioxide

$$NO + 1/_2O_2 \longrightarrow NO_2$$

Sulphur Dioxide + Nitrogen Dioxide ⟶ Sulphur Trioxide + Nitrogen Monoxide + 22500 cal energy

$$SO_2 + NO_2 \longrightarrow SO_3 + NO$$

THESE ARE VERY IMPORTANT REACTIONS, STUDENTS MUST REMEMBER

Later it is known that in between these reactions nitrosulphuric acid is also formed the equation is follows :

Sulphur Dioxide + Oxygen + Nitrogen Dioxide + Nitrogen Monoxide ⟶ Nitrososulphuric Acid

$$SO_2 + O_2 + H_2O \quad NO \longrightarrow 2SO_2(OH)(ONO)$$

Nitrososulphuric Acid + Water ⟶ Sulphuric Acid + Nitrogen Monoxide + Nitrogen Dioxide

$$2SO_2(OH)(ONO) + H_2O \longrightarrow H_2SO_4 + NO + NO_2$$

Contact Process

Sulphuric Acid is also prepared by many organic industries; chamber process produces only 66% of concentration, so to produce higher strength of the acid. *Contact Process* is employed. Where sulphuric acid is produced by the absorption of sulphur trioxide sulphuric acid where oleum is produced. The principle steps in producing the sulphuric acid follows in four steps: (i) producing sulphur dioxide gas, (ii) passing the gas through the converter containing catalyst, (iii) purification and cooling of sulphur trioxide and (iv) absorption of the sulphur trioxide in sulphuric acid. (Fig. 11.26)

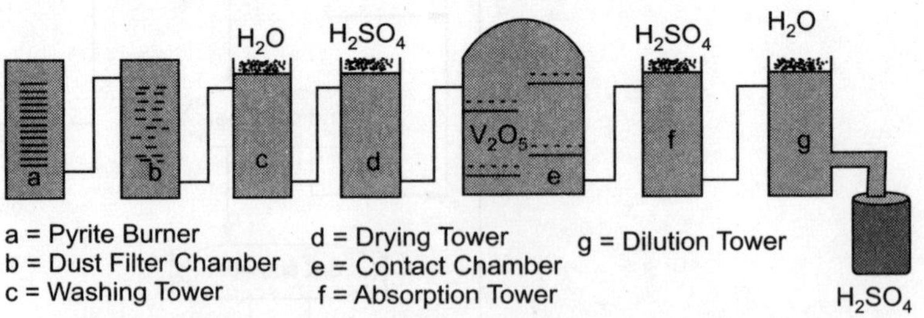

Fig. 11.26. The Diagram of Contact Process

The diagrammatic representation of the contact process is shown. Here the sulphur is melted in a concrete tank lined with bricks from inside. The solid impurities settles down and the water present evaporated. The gear pumps pump the liquid sulphur in the burner. The air for combustion of sulphur is blown through the whole plant by a centrifugal electrically driven lower. It passes, before its entrance into the sulphur burner, through drying tower, where it flows in the reverse direction of the flow of the sulphuric acid. The blower is usually situated between the drying tower and the burner.

The combustion gases coming from the burner pass through the waste heat boiler where the necessary steam for melting the sulphur and feeding the sulphur turbo pump is generated. They then pass through a filter. The gases enter the primary converter at about 400°C temperature, which is the ignition temperature of the catalyst (vanadium pentoxide). This converter contains a relatively thin layer of the catalyst. The reaction being exothermic, the gases leave at about 570°C. The gases are then cooled in the first heat exchanger at about same temperature as the previous converter. This converter has deeper catalyst bed and conducts the final conversion. The gases coming at about 460°C pass through the second heat exchanger and the gases gets condensed where the temperature is reduced to about 100°C. For the starting up of the plat, an oil fuel preheater is placed between filter and the heat exchanger to start the reaction. This is cut off as soon as the catalyst reaches the ignition temperature.

The gases containing more than 95% of the sulphur as sulphur trioxide are absorbed in oleum. The rate of absorbent of acid is so regulated that the strength of the acid does not increase more than 1%. All the sulphur trioxide, however, cannot be absorbed in such strong sulphuric acid; hence, it is passed through other absorber where 97% acid is sprayed to form 98% acid. Considerable amount of heat is produced during the absorption hence the acid is sent to the coolers. The absorption and the drying tower have a common circulation of acid in closed circuit with an extraction corresponding to the product.

Persulphuric Acid (H_2SO_5 and $H_2S_2O_8$): Anhydrous persulphuric acid can be prepared by the reaction of chlorosulphuric acid and anhydrous hydrogen peroxide.

Hydrogen Peroxide	+	Chlorosulphuric Acid	⟶	Peroxysulphonic Acid	+	Hydrogen Chloride
H_2O_2	+	$HClO_3$	⟶	H_2SO_5	+	HCl

Peroxysulphuric Acid, can also be prepared by the reaction of ionic oxidation of sulphuric acid is put under electrolysis. A platinum wire surrounded by a glass tube is used as the anode (+) while a spiral wire wound round the glass tube acts as cathode (%) (Fig. 11.27). The whole apparatus is immersed in freezing mixture. The addition of a small amount of hydrochloric acid facilitates the reaction. A current of I ampere is allowed to pass during the electrolysis. Hydrogen appears at the cathode and a solution of perdisulphuric acid is obtained.

Cathode: – $H^+ + e^- \rightarrow H\ H + H = H_2$

Anode:– $HSO_4^- + HSO_4^- \rightarrow H_2S_2O_8$

Sulphamic Acid (H_3SO_3N): It is also known as amidosulphuric acid and aminosulphuric acid. It was a laboratory curiosity for about 100 years until; new and practical methods of manufactures were discovered in 1935. Rose is credited for first preparation, after this E. Bergland prepared the pure acid. Sulphamic acid is commercially prepared by the action of oleum and urea.

Applications of Sulphuric Acid

Contamination of Lubricants & Coolants is the #1 Cause of Machinery Failure and Inferior Product Quality

Hydraulic Lubricants & Coolants

FILTREX provides the necessary state-of-the art technology capable of delivering the ue in filtering efficiencies. This high standard of excellence is achieved with the absolute minimum of maintenance andnegligible operator interface plus virtually no costly down time

Petroleum

Meeting the ever growing demands of the petroleum industry. FILTREX applications include: amine feedstocks, reduced crudes, fuel & motor oils, injection fluids, synthetic lubricants and completion fluids.

The Paper Industry

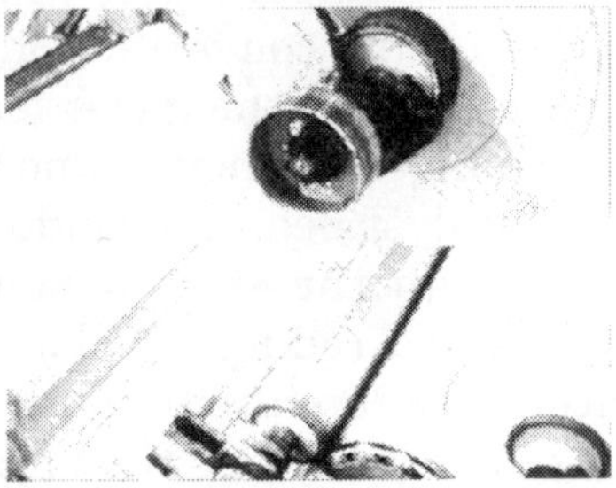

FILTREX Corporation meets the exacting demands of the Paper industry. Applications in this field include:Intake water, coolant water, waste water, white water, green & black liquor, mill water, dyes, pump seal water, decker shower water, clay slurry and all wet end additives.

Waste & Waste Water

Typical FILTREX applications include bottled water, R.O. membrane protection, pond water, cooling tower water, gas & oil drilling make-up water, D.I. water, waste water, poultry wash water, bottling wash water and dairy wash water.

Food & Beverages

FILTREX leads the way in thefiltration processes of food & beverage production. Including: corn syrup, dextrose, caro, jelly, juices, edible oils, milk sugar, city & well water, extracts, chocolates, soybeans, and beer & wine production.

Process & Chemical Fluids

The FILTREX Systems have produced outstanding results with Processed & Chemical Fluids, such as resins, paints, photographic solutions, ink manufacture, adhesives, herbicides & pesticides, sodium sulphates, cellulose nitrates, sodium chlorides, sulphuric acid and many, many more.

The Different uses of Sulphuric Acid

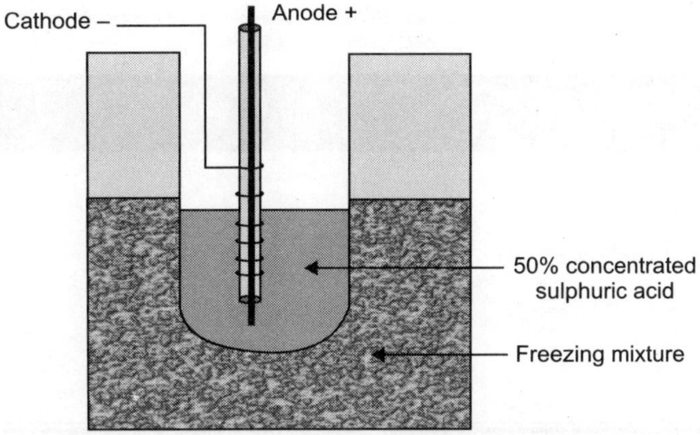

Fig. 11.27. The Preparation of Perdisulphuric Acid

$$\text{Oleum} + \text{Urea} \longrightarrow \text{Sulphamic Acid} + \text{Carbon Dioxide}$$
$$H_2S_2O_7 + N_2H_4CO \longrightarrow 2NH_3SO_3 + CO_2$$

Properties : Although easy to handle dry, the acid is highly ionized in solution and is comparable to other strong acid. The pH (acidic strength) of 15 of the acid is 1.18. Sulphamic acid is moderately stable in water 14.66 g in 100g of water at 0°C and 47.88 gm in 100g at 80°C. it is also stable in formaldehyde and practically insoluble in 705 to 100% pure sulphuric acid. It is a melting point 205°C and boiling point at 210°C.

Chemical Properties : Sulphamic acid starts decomposing at 215°C and at 260°C it is dissociated to sulphur trioxide and nitrogen, water and sulphur dioxide. At room temperature a dilute aqueous solution of the acid is stable for months. At higher temperature it is converted to ammonium sulphate.

$$\text{Sulphamic Acid} + \text{Water} \longrightarrow \text{Ammonium Sulphate}$$
$$NH_2SO_3H + H_2O \longrightarrow NH_4HSO_4$$

In liquid ammonia, sulpahmic acid behaves as dibasic acid, hydrogen of the amino group being replicable by Na, K etc. giving rise to salts like sodium ammonium sulphate Na_2NHSO_3.

$$\text{Sodium} + \text{Sulphamic Acid} \longrightarrow \text{Sodium Amonosulphate} + \text{Water}$$
$$Na + 2NH_3SO_3 \longrightarrow NaNHSO_3 + H_2O$$

Chlorine bromine and chlorates oxides sulpamic acid to sulphuric acid and nitrogen

$$\text{Chlorine} + \text{Sulphamic Acid} \longrightarrow \text{Sulphuric Acid} + \text{Water}$$
$$Cl_2 + 2NH_3SO_3 \longrightarrow H_2SO_4 + H_2O$$

Pure nitrous oxide is obtained by the reaction of concentrated nitric acid with sulphamic acid.

$$\text{Sulphamic Acid} + \text{Nitric Acid} \longrightarrow \text{Sulphuric Acid} + \text{Nitrogen} + \text{Water}$$
$$2NH_3SO_3 + NHO_3 \longrightarrow H_2SO_4 + N_2 + H_2O$$

This the most outstanding reaction in sulphamic acid and it is used industrially to destroy nitrates in the manufacture of dyes, and purification of sulphuric acid. With this we end the chemistry of sulphur.

CHAPTER 12

Halogens

Introduction

These elements occupy the VIIth group of the periodic table, the word, 'Halogen' comes from the Greek word *Halos* = Sea Salt, *Geneous* = Produce. These elements are derived from seawater. They are named as *fluorine, chlorine, bromine,* and *iodine*. This group is known as halogen family in chemistry we are now going to study and discuss about these elements and their chemical and as well as physical properties. This group has some unique features they are:

1. They are strongly similar in their physical and chemical properties.
2. The physical and the chemical properties change with the change of their atomic weight.
3. These elements have a high electronegative nature
4. Mono-valent in most of their nature.

FLUORINE

Introduction and Occurrence

Fluorine means, "to flow" the element is a gas and not available in the free state, only available in the formula of compounds or ores, they are Fluorspar (CaF_2), Cryolite (Na_3AIF_6), Fluorapatite ($CaF_2.3Ca(PO4)_2$).

Fluorine in nature is not available in free state but in laboratory it can be prepared in free state, by some methods one of them is *Moissan's Method.*

Davy first detected the occurrence of the Fluorine in the year 1813 and Moissan first isolated it in the year 1886 from potassium hydrogen fluoride. In this method potassium hydrogen fluoride is dissolved in hydrofluoric acid. The electrolysis is carried out in a "U" tube made up of platinum and iridium alloy. The tube is provided with two side tubes, the "U" tube is supported and two electrodes of same alloy are introduced into the tube (see Fig. 12.1). The whole apparatus is immersed in the bath of boiling methyl chloride (boiling point –23°C). The electric current is allowed to pass through the liquid mixture of potassium hydrogen fluoride and hydrofluoric acid vapour. In order to purify it the gas is at first passed through platinum or copper spiral cooled in methyl chloride. When most of the hydrofluoric acid vapour is condensed in the spiral and then the gas is again passed through the same tube consisting of sodium fluoride is fused condition here the last traces of hydrofluoric acid vapours are absorbed by the sodium fluoride forming sodium hydrogen fluoride.

Sodium Fluoride + Hydrogen Fluoride ⟶ Sodium Hydrogen Fluoride

$$NaF + HF \longrightarrow NaHF_2$$

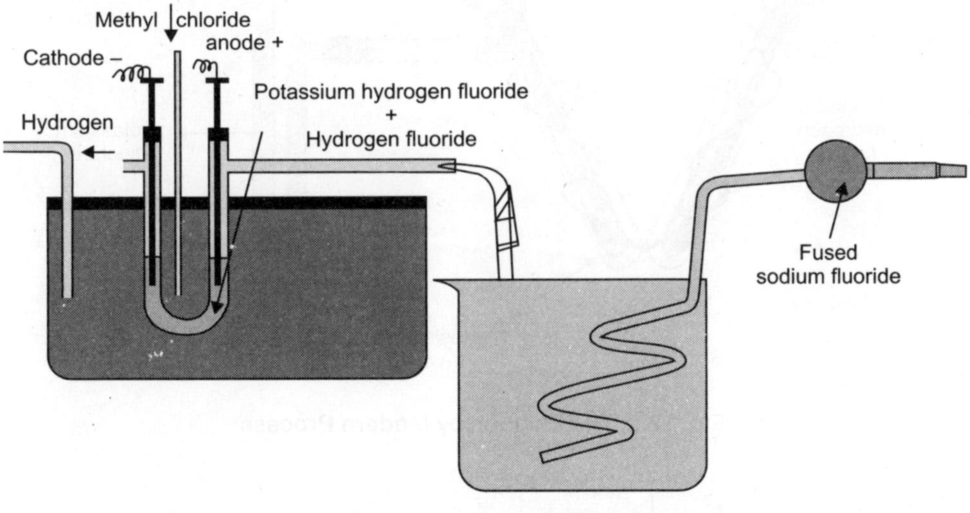

Fig. 12.1. The Preparation of Fluorine Moissans Method

Fluorine is also prepared by the method of the electrolysis of potassium hydrogen fluoride in a "V" tube as shown in Fig. 12.2. This vessel is made up of copper. The vessel is sealed with copper caps through which the graphite electrodes are introduced which are filled with bakelite cement. The vessel is filled with extremely pure and dry potassium hydrogen fluoride, which is melted by passing an electric current through the resistance wire, wound round the "V" tube. When the electric is passed through the molten potassium hydrogen fluoride, hydrogen and fluorine are separated at cathode and anode respectively. The evolved fluorine is freed from hydrogen fluoride by passing through dry potassium fluoride.

Fluorine can also be prepared by another method where the electrolyte is same but the environment is different. The whole apparatus consists of a rectangular steel tank double walled with stainless steel coating through which the steam is allowed to pass (see Fig. 12.3). The tank is fitted with airtight removeable cover, which carries carbon rod used as the positive electrode (+) two slitted cathode (–) negative which, are made up of copper grills are immersed in the tank. The cathode is slitted because to allow a free flow of the electrolyte inside the tank. The two electrodes are coated with a special substance called Polyethra fluroethylene (P.T.F.E) to protect the electrodes from fluorine. The steam for the electrolysis keeps the electrolyte heated. But once the electrolysis is started no more continuous heating is necessary to keep the temperature steady at 95°C. During the reaction some of the hydrogen fluoride is lost which is compensated by introducing more hydrogen fluoride into the mixture.

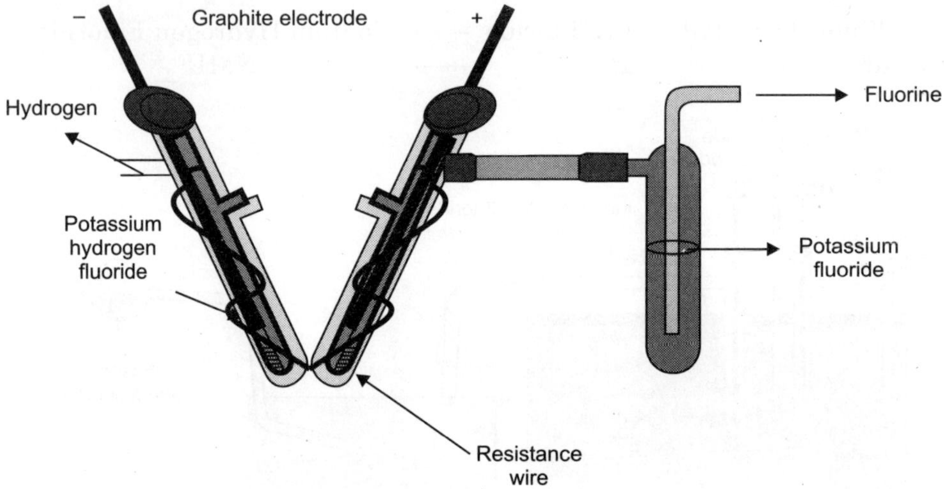

Fig. 12.2. The Fluorine by Modern Process

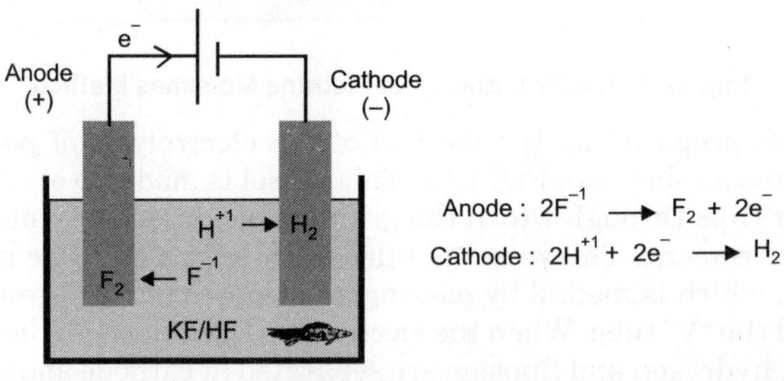

Fig. 12.3. The Industrial Preparation of Fluorine by Electrolysis Method

Properties of Fluorine

Physical Properties: Fluorine is a yellow coloured gas, slightly paler than chloride with a characteristic pungent odour. It can be condensed into a liquid yellow in colour. The physical properties are shown below:

Colour of the normal physical state	Yellow
Isotope	19
Atomic Radius	.72 Å
Ionic Radius	1.36 Å
Electronegativity (Pauling's Scale)	4.0
Ionization energy	17.4eV
Boiling Point	188°C
Freezing Point	–223°C
Density (of Liquid fluorine)	1.1 g/cc^{-1}
Standard Electrode Reduction potential	2.85V
Heat of Vapourization	1.64 Kcal/mol^{-1}
Head of Fusion	.51 kJ/mol
Heat of Dissociation	38K cal/mol^{-1}

Chemical Properties: Chemically fluorine is the most reactive of elements. It can directly react with almost all elements forming fluorides. It also reacts with many organic compounds, breaking the down to fluorides.

Most of the chemical properties with relevant chemical reactions are shown below:

1. Combination with Metals: Fluorine reacts with lighter metals named sodium, potassium at normal temperature.

$$\text{Sodium} + \text{Fluorine} \longrightarrow \text{Sodium Fluride}$$
$$2Na + F_2 \longrightarrow 2NaF$$
$$\text{Potassium} + \text{Fluorine} \longrightarrow \text{Potassium Fluoride}$$
$$2K + F_2 \longrightarrow 2KF$$

These metals spontaneously burn in the gas. With metals like magnesium, aluminium, nickel etc fluorine reacts at a slightly elevated temperature at about 100°C.

$$\text{Aluminium} + \text{Fluorine} \longrightarrow \text{Sodium Fluoride}$$
$$2Al + 3F_2 \longrightarrow 2AIF_3$$

It reacts with metals like gold, platinum at much high temperature at about 300°C to form fluorides.

$$\text{Gold} + \text{Fluorine} \longrightarrow \text{Aurum Fluoride}$$
$$Au + 3F_2 \longrightarrow 2AuF_3$$

2. Action on Non-metals: Non-metals such as carbon, sulphur, and phosphorus boron burn in fluorine forming fluorides.

$$\text{Phosphorus} + \text{Fluorine} \longrightarrow \text{Phosphorus Pentafluoride}$$
$$2Al + 10F_2 \longrightarrow 2PF_5$$
$$\text{Carbon} + \text{Fluorine} \longrightarrow \text{Aurum Fluoride}$$
$$C + 3F_2 \longrightarrow 2AuF_3$$

Boron + Fluorine ⟶ Boron Trifluoride

$$2B + 3F_2 \longrightarrow 2BF_3$$

3. Reaction with other Compounds

(*a*) Fluorine reacts with sodium chloride and liberates chlorine from it,

Sodium Chloride + Fluoride ⟶ Sodium Fluoride + Chlorine

$$2NaCl + F_2 \longrightarrow 2NaF + Cl_2$$

(*b*) It also reacts with potassium iodide to liberate iodine.

Potassium Iodide + Fluorine ⟶ Potassium Fluoride + Iodine

$$2KI + F_2 \longrightarrow 2KF + CI_2$$

(*c*) Fluorine reacts with hydroxides of sodium and potassium forming fluorides and also fluorine oxides.

Sodium Hydroxide + Fluorine ⟶ Sodium Fluoride + Fluorine Monoxide + Water

$$2NaOH + 3F_2 \longrightarrow 2BF_3 + F_2O + H_2O$$

(*d*) Fluorine reacts with ammonium forming nitrogen trifluoride (NF_3)

Ammonia + Fluorine ⟶ Hydrogen Fluoride + Nigrogen Trifluoride

$$NH_3 + 3F_2 \longrightarrow 3HF + NF_3$$

(*e*) Sulphur dioxide and hydrogen sulphide are both decomposed by fluorine

Sulphur Dioxide + Fluorine ⟶ Sulphur Difluoride + Oxygen

$$SO_2 + F_2 \longrightarrow SF_2 + O_2$$

At the same time some oxygen trioxide is also produced.

4. Action on Glass and Quartz: Fluorine readily attacks glass producing silicon tetrafluoride,

(glass)

Silicon Dioxide + Hydrogen Fluoride ⟶ Silicon Tetrafluoride + Water

$$SiO_2 + 3HF \longrightarrow SiF_4 + 2H_2O$$

Exp: Pour some molten wax on a clean glass plate and set to dry then make some design or curve some name on it. Now put some calcium fluoride and sulphuric acid and heat the mixture (Keeping the mixture in a lead dish). Hydrogen fluoride is evolved, immediately cover dish with the waxed glass. After sometimes the glass should be removed and the clean the wax then the clear etched mark of the fluorine will be visible.

Etching of glass

5. Action of Hydrocarbon: Fluorine reacts with hydrocarbons violently separating carbon and producing hydrogen fluoride. Methane is decomposed to carbon and hydrogen fluoride.

Methene + Fluorine ⟶ Carbon + Hydrogen Fluoride

$$CH_4 + 2F_2 \longrightarrow C + 4HF$$

To avoid explosion nitrogen fluorine mixture is allowed to react with hydrocarbon. For this purpose a mixture of hydrogen fluoride and nitrogen is passed through a metal tube packed with copper gauge, at a temperature between 160-350°C. Fluorine reacts first with copper gauge, forming copper fluoride, which acts as a catalyst afterwards. The resulting product is a complex organic substance. Mainly from the compounds refrigerants are used.

With excess of chlorine fluorine reacts to form chlorine trifluoride.

$$\text{Chlorine} + \text{Fluorine} \longrightarrow \text{Chlorine Monofluride}$$

$$Cl_2 + F_2 \longrightarrow 2ClF$$

With bromine fluorine reacts to form bromine trifluoride.

$$\text{Bromine} + \text{Fluorine} \longrightarrow \text{Bromine Pentafluoride}$$

$$2Br + 3F_2 \longrightarrow 2BrF_3$$

Fluorine also attacks iodine for form Iodine pentafluoride and with excess fluorine iodine heptafluoride is produced at a temperature of 300°C.

$$\text{Iodine} + \text{Fluorine} \longrightarrow \text{Iodine Pentafluoride}$$

$$2I + 5F \longrightarrow 2IF_5$$

8. Oxides of Fluorine: Fluorine has mainly two oxides which are mentioned below Fluorine monoxide, Oxygen Difluoride (F_2O) When fluorine is passed through sodium hydroxide fluorine monoxide is produced.

$$\text{Fluorine} + \text{Sodium Hydroxide} \longrightarrow \text{Sodium Fluoride} + \text{Florine Monoxide} + \text{Water}$$

$$3F_2 + 4NaOH \longrightarrow 4NaF + F_2O + H_2O$$

$$\text{Fluorine} + \text{Oxygen} \longrightarrow \text{Fluorine Dioxide}$$

$$F_2 + O_2 \longrightarrow F_2O_2$$

It is unstable and decomposes to fluorine and oxygen at –100°C

Uses of Fluorine

1. Fluorine is now extremely used for the preparation of fluorine compounds. Difluorochloroethene CF_2C_{12}, is used as the non-toxic and non-flammable refrigerant gas.
2. A new plastic material called Teflon, which is resistant to the action of acids, and other strong chemical reagents thus used as leak proof. It is a polymerized tetrafluoroethylene ($F_2C = CF_2$).
3. Fluorine compounds such as sodium fluoride and cryolite are used as insecticides. A small amount of fluoride in drinking water is believed to prevent tooth decay.
4. Uranium Hexafluoride is used to separate the isotopes of the uranium.
5. Sulphur hexafluoride having a high dielectric constant is used as the X-ray protection for the X-ray camera and other high voltage devices.

Short Notes on Freon 12

Freon 12 is as said earlier used as the refrigerant also known as the difluorochlorethane CF_2C_{12} the physical property of the Freon 12 is stated below:

Molecular Weight	102.923
Freezing Point	–135°C
Boiling Point	8.95°C
Critical Temperature	178.43°C
*T.L.V	10ppm

Hydrogen Fluoride, Fluorinated Hydrogen (HF)

Preparation: Hydrogen is prepared by many processes they are mention blows, besides this we shall also study the physical and chemical properties of hydrofluoric acid.

1. Direct Synthesis: Hydrogen fluoride can be prepared by the direct reaction of hydrogen and fluorine

$$\text{Hydrogen} + \text{Fluorine} \longrightarrow \text{Hydrogen Fluoride}$$
$$H_2 + F_2 \longrightarrow 2HF$$

2. Reaction of Fluorine and Water

$$\text{Water} + \text{Fluorine} \longrightarrow \text{Hydrogen Fluoride} + \text{Oxygen}$$
$$H_2 + F_2 \longrightarrow 4HF + O_2$$

3. Laboratory Preparation of Hydrogen Fluoride

Theory: Hydrogen fluoride can be prepared by the reaction of calcium fluoride and sulphuric acid in a dish (See Fig. 12.4).

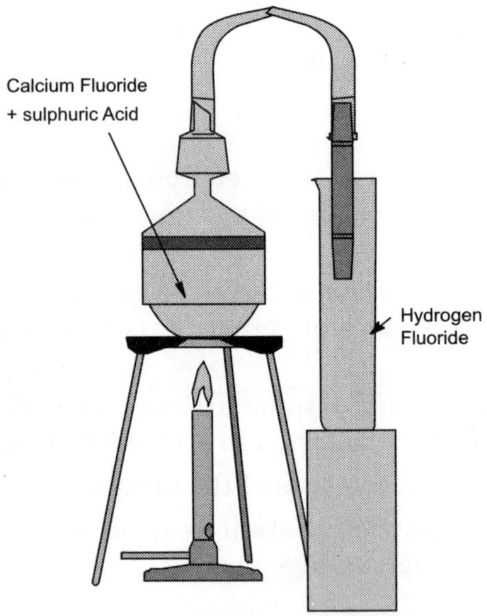

Precaution: *The whole reaction should be carried out in the lead or copper vessel and not in the glass because hydrogen fluoride attacks glass. The gas should be collected by the upward displacement of air in a lead jar.*

Fig. 12.4. The Lab Preparation of Hydrogen Fluoride

*T.I.V: This is the limit value which is known as the maximum safe concentration in air in the work place, expressed in the weight average in parts per million (ppm) in a 8 hour per day.

Reaction Equation

Calcium Fluoride + Sulphuric Acid ⟶ Hydrogen Fluoride + Calcium Sulphate

$$CaF_2 + H_2SO_4 \longrightarrow 2HF + CaSO_4$$

Preparation of Anhydrous Hydrogen Fluoride : This preparation is mainly done in industries, which firstly includes the preparation of potassium hydrogen fluoride, (KHF_2) by the reaction of potassium carbonate and fluorine.

Water + Potassium Carbonate + Fluorine ⟶ Potassium Hydrogen Fluoride + Carbon Dioxide + Oxygen

$$2H_2O + 2K_2CO_3 + 4F_2 \longrightarrow 4KHF_2 + 2CO_2 + 2O_2$$

Potassium Hydrogen Fluoride ⟶ Hydrogen Fluoride + Potassium Fluoride

$$KHF_2 \longrightarrow HF + HF$$

Anhydrous hydrogen fluoride is prepared by heating completely dry potassium hydrogen fluoride in a platinum retort, which is connected to a platinum "U" tube, this platinum tube kept in a chamber full of ice (see Fig. 12.5).

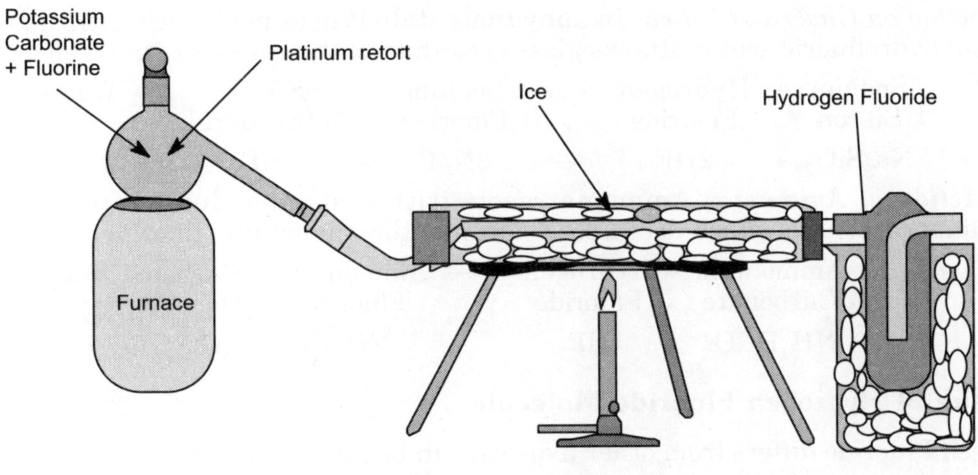

Fig. 12.5. The Industrial Preparation of Anhydrous Hydrogen Fluoride

The evolved hydrogen fluoride gas is connected in the form of a colourless liquid inside the platinum tube.

Anhydrous hydrofluoric acid is a colourless, strongly fuming liquid. The acid is highly poisonous and dangerous to handle. It attacks skin and forms ulcerated sores. In anhydrous state it is non-electrolyte. In the presence of a little water the gas reacts violently. The properties of hydrogen fluoride are shown below:

Physical Properties of Hydrogen Fluoride

Freezing Point	–90°C
Boiling Point	20°C
Critical Pressure	190.632Pa
Critical Molar Volume	$69 cm^3$
Heat of Vapourization	1.21 $kcal/mol^{-1}$
Heat of Fusion	–4.58 at –89.56°C

Chemical Properties

1. Acidic Nature: Hydrogen fluoride when reacts with water it forms a weak acid. At this state it is also conducts electricity. It neutralizes hydroxides of compounds of sodium, potassium forming salt and water.

Potassium Hydrooxide + Hydrogen Fluoride ⟶ Potassium Fluoride + Water

$$KOH + HF \longrightarrow KF + H_2O$$

2. Reaction of Fluoride Ion: Like all fluorides it forms no precipitate with barium chloride forming insoluble barium fluoride.

Barium Chloride + Hydrogen Fluoride ⟶ Barium fluoride + Hydrogen Chloride

$$BaCl_3 + 2HF \longrightarrow BaF_2 + 2HCl$$

3. Stability: Hydrogen fluoride is a stable compound and is not easily decomposed by heat and any other substances.

4. Action on Glass and Silica: In anhydrous state it does not attack with glass but in the sta of hydrofluoric acid it attacks glass according to the equation stated below:

Sodium Silicon + Hydrogen Flouride ⟶ Sodium Fluoride + Silicon Tetrafluoride + Water

$$Na_2SiO_3 + 2HF \longrightarrow 2NaF + SiF_4 + 3H_2O$$

5. Action of Ammonia: Ammonia reacts with hydrogen fluoride in the form of ammonium carbonate, which decomposes forming the ammonium fluoride.

Ammonium Carbonate + Hydrogen Fluoride ⟶ Ammonium Fluoride + Carbon Dioxide + Water

$$NH_4HCO_3 + HF \qquad NH_4F + CO_2 + H_2O$$

Structure of Hydrogen Fluoride Molecule

Hydrogen Fluoride differs from other hydracids in being polymeric.

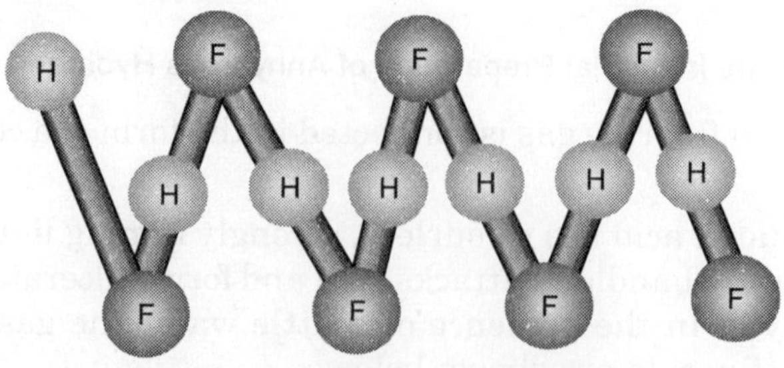

Structure of Hydrogen Fluoride Molecule

Uses of Hydrogen Fluoride

Hydrogen fluoride has acquired great important in chemical industry in recent years. Its important uses are listed below:

(*a*) The aluminium industry is undoubtedly the biggest consumer of hydrogen fluoride for the preparation of aluminium.

(*b*) Hydrogen fluoride is also used for the preparation of refrigerants like Freon (Fluorinated Hydrocarbon)

(*c*) Petroleum industries also consume hydrogen fluoride for the alkalization process.

(*d*) For the preparation of uranium hexafluoride, hydrogen fluoride is also used.

(*e*) Hydrogen fluoride is also in etching of glasses.

(*f*) Hydrogen fluoride is also used in alcohol and dyeing industries.

(*g*) It is also used for the preparation of fluorine, preservation of steels and also for the manufacture of Teflon.

Analysis of Fluorine

1. Fluorine is detected in the form of compounds when they react with barium chloride; barium fluoride is formed which is insoluble in nitric acid.
2. When a fluoride is heated with sulphuric acid, hydrogen fluoride is produced which etches the glass.
3. **Lassaingne's test:** This is the confirmatory test, here at first the compound containing fluoride is heated with sodium metal in an ignition tube to a red hot condition. Then this mixture is dipped in distil water inside a mortar, there it is grinded and the glass pieces are filtered off. This solution is treated with zirconoxy chloride, with alzerin red, which turns the red solution to yellow.

CHLORINE

Introduction

Scheele discovered chlorine in 1774 from muriatic acid (in Latin muria means brine, sodium chloride solution). Chlorine was obtained as greenish yellow gas (the gas got its name for the colour, Chlorous means Greenish-yellow) and was also named as 'dephilogistical acid air' it meant muriatic acid air. The muriatic acid deprived of the hydrogen, which was considered to by phlogiston. After the over throw of the phlogiston theory, chlorine was considered as the compound of oxygen and muriatic acid, and hence it was named as "oxymuriatic acid'. Until Davy in the year 1810 defined it as an element and he also proposed the name chlorine.

Occurrence

Chlorine in the compound state is highly abundant in the nature. The most important source of chlorine is sodium chloride which is also known as 'common salt or rock salt. Seawater contains about 2 to 35 of chlorine. Chlorine is also available in potassium chloride called Sylvine and Carnallite $KCl.MgCl_2\ 6H_2O$.

Preparation

Chlorine can be prepared by many ways they are as follows:

1. Chlorine can be prepared by electrolysis of sodium chloride in a fused state.

$NaCl \rightleftharpoons Na^+ + Cl^-$

$H^+ - e \rightarrow H^+$

$Cl + Cl = Cl_2$

2. Chlorine is also prepared by the reaction of potassium dichromate and hydrochloric acid.

Potassium Dichromate + Hydrogen Acid ⟶ Potassium Chloride + Chlorine + Water + Chromium Chloride

$K_2Cr_2O_7 + 14HCl \longrightarrow 2KCl + 3Cl_2 + 7H_2O + 2CrCl_3$

3. Chlorine is also prepared by the action of hydrochloric acid on potassium permanganate

Potassium Permanganate + Hydrochloric Acid ⟶ Potassium Chloride + Manganous Chloride + Chloride + Water

$2KMnO_4 + 16HCl \longrightarrow 2KCl + 2MnCl_2 + 5Cl_2 + 8H_2O$

4. Chlorine is also prepared by the electrolysis of silver chloride and Sodium Chloride

Silver Chloride ⇌ Silver + Chlorine

$2AgCl \rightleftharpoons 2Ag + Cl_2$

Sodium Chloride ⇌ Sodium + Chlorine

$2NaCl \longrightarrow 2Na \quad Cl_2$

5. Chlorine is also produced by the action of hydrochloric acid on red lead.

Red Lead + Hydrochloric Acid ⟶ Lead Chloride + Water + Chlorine

$Pb_3O_4 + 8HCl \longrightarrow 3PbCl_2 + 4H_2O + Cl_2$

6. When cupric chloride is heated chlorine is produced.

Cupric Chloride ⟶ Copper + Chlorine

$CuCl_2 \longrightarrow Cu + Cl_2$

7. When hydrochloric acid reacts with bleaching powder, chlorine is produced.

Bleacing Power + Hydrochlric Acid ⟶ Calcum Chloride + Water + Chlorine

$CaOCl_2 + 2HCl \longrightarrow CaCl_2 + H_2O + Cl_2$

Laboratory Preparation of Chlorine

Theory: Chlorine is prepared by the reaction of hydrochloric acid and manganese dioxide and the mixture is heated in a flask. The evolved gas is passed through water and sulphuric acid to get free from hydrochloric acid and moisture (see Fig. 12.6).

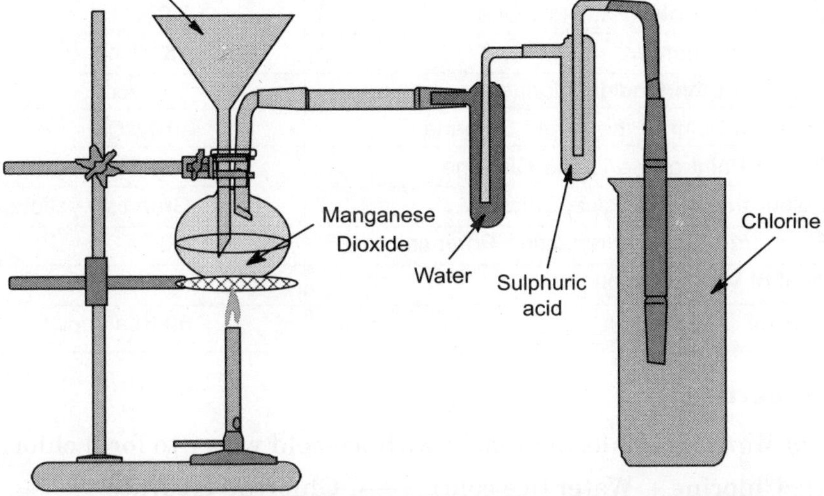

APPARATUS	REAGENTS	TEMPERATURE
Flask, burner, traps, delivery tube, gas jar, corks etc.	Conc. hydrochloric acid, Manganese dioxide, sulphuric acid, distil water	100-110°

Fig. 12.6. The Lab Preparation of Chlorine

Reaction Equation

Hydrochloric Acid + Manganese Dioxide ⟶ Chlorine + Manganese Chloride + Water

$$4HCl + MnO_2 \longrightarrow Cl_2 \Uparrow + MnCl_2 \Downarrow + 2H_2O$$

Precaution: *Chlorine is a suffocating gas be careful do not inhale directly.*

PROPERTIES OF CHLORINE

Physical Properties

Chlorine is a greenish-yellow gas, which is soluble in water and heavier, than air, therefore, the gas is collected by the upward displacement of air the physical properties are given below:

Atomic Mass	35.4527
Atomic number	17
Isotopes	35, 47
Electronic Structure	$2p^2 3s^2 3p^5$

Atoms Radius	.99Å
Ionic Radius	1.81 Å
Electronegativity (Pauling's Scale)	3.0 Å
Ionization Energy	13.01 cal
Density of the Liquid Chlorine	3.2 g/cc
Freezing Point of the Liquid Chlorine	–102°C
Boiling Point of the Liquid Chlorine	–34.3°C
Colour and the Physical State	Greenish yellow gas
Standard Electrode reduction Potential	1.36V
Heat of Vapourization	4.42 Kcal mol^{-1}
Heat of Dissociation	50 Kcal mol^{-1}

Chemical Properties

1. Action of Water: (*i*) Chlorine reacts with ice-cold water to form chlorine hydrate

$$\text{Chlorine} + \text{Water (ice cold)} \longrightarrow \text{Chlorine Hydrate}$$

$$Cl_2 + 6H_2O \longrightarrow Cl_2H_2O$$

(*ii*) At higher temperature it forms hydrochloric acid and also hydrochlorous acid.

$$\text{Chlorine} + \text{Water} \xrightarrow{\text{Boiling water}} \text{Hydrochloric Acid} + \text{Hydrochlorous Acid}$$

$$2Cl_2 + H_2O \longrightarrow HCl + HClO$$

2. Photochemical Effect: Chlorine in the sunlight forms oxygen and hydrogen chloride when it reacts with water.

$$\text{Chlorine} + \text{Water} \xrightarrow{\text{sunlight}} \text{Hydrogen Chloride (gas)} + \text{Oxygen}$$

$$2Cl_2 + 2H_2O \longrightarrow 4HCl + O_2$$

3. Action with Elements: When chlorine reacts with phosphorous, phosphorous trichloride is produced.

$$\text{Phosphorous} \quad \text{Chlorine} \longrightarrow \text{Phosphorous Trichloride}$$

$$2P + 3Cl_2 \longrightarrow 2PCl_3$$

$$\text{Phosphorous} + \text{Chlorine} \longrightarrow \text{Phosphorous Pentachloride}$$

$$4P + 5Cl_2 \longrightarrow 2P_2Cl_5$$

When Antimony reacts with chlorine antimony trichloride. Antimony reacts with chlorine antimony trichloride is produced.

$$\text{Antimony} + \text{Chlorine} \longrightarrow \text{Antimony Trichloride}$$

$$2Sb + 3Cl_2 \longrightarrow 2SbCl_3$$

When copper is heated with chloride cupric chloride is produced.

$$\text{Copper} + \text{Chlorine} \longrightarrow \text{Cupric Chloride}$$

$$Cu + Cl_2 \longrightarrow CuCl_2$$

Sodium violently reacts with chloride and sodium chloride is produced, zinc, aluminium and iron also react with chloride producing respective chlorides.

$$\text{Sodium} + \text{Chlorine} \longrightarrow \text{Sodium Chloride}$$
$$2Na + Cl_2 \longrightarrow 2NaCl$$
$$\text{Iron} + \text{Chlorine} \longrightarrow \text{Ferrous Chloride}$$
$$Fe + Cl_2 \longrightarrow FeCl_2$$
$$\text{Aluminium} + \text{Chlorine} \longrightarrow \text{Aluminium Trichloride}$$
$$Fe + Cl_2 \longrightarrow 2AlCl_3$$

4. *Action with other compounds:* With excess of ammonia it produces ammonium trichloride which is an explosive material.

$$\text{Ammonia} + \text{Chlorine} \longrightarrow \text{Nitrogen Trichloride} + \text{Hydrogen}$$
$$2NH_3 + 3Cl_2 \longrightarrow 2NCl_3 + 3H_2$$

With Carbon dioxide chlorine produces carbonyl chloride.

$$\text{Carbon Dioxide} + \text{Chlorine} \longrightarrow \text{Carbonyl Chloride}$$
$$CO_2 + 3Cl_2 \longrightarrow COCl_2$$

Chlorine when reacts with sulphur dioxide sulphuryl chloride is produced.

$$\text{Sulphur Dioxide} + \text{Chlorine} \longrightarrow \text{Sulphuryl Chloride}$$
$$SO_2 + Cl_2 \longrightarrow SO_2Cl_2$$

Chlorine reacts with turpentine oil forming black particles of carbon and hydrogen chloride

$$\text{Turpentine Oil} + \text{Chlorine} \longrightarrow \text{Carbon} + \text{Hydrogen Chloride}$$
$$C_{10}H_6 + Cl_2 \longrightarrow 10C + 6HCl$$

Chlorine reacts with sodium hydroxide producing sodium oxychloride and sodium chloride in cold and dilute state.

$$\text{Sodium Hydroxide} + \text{Chlorine} \longrightarrow \text{Sodium Chloride} + \text{Sodium Oxychloride} + \text{Water}$$
$$2NaOH + Cl_2 \longrightarrow NaCl + NaOCl + H_2O$$

Sodium Carbonate reacts with chlorine to form carbon dioxide and sodium oxychloride.

$$\text{Sodium Carbonate} + \text{Chlorine} \longrightarrow \text{Sodium Chloride} + \text{Sodium Oxychloride} + \text{Carbon Dioxide}$$
$$3Na_2CO_3 + 3Cl_2 \longrightarrow 3NaCl + 3NaOCl + 3CO_2$$

With hydrogen sulphide chlorine reacts to liberate sulphur.

$$\text{Hydrogen Sulphide} + \text{Chlorine} \longrightarrow \text{Sulphur} + \text{Hydrogen Chloride}$$
$$H_2S + Cl_2 \longrightarrow S + 2HCl$$

Potassium Bromide reacts with chlorine producing to form bromine and potassium chloride.

Potassium Bromide + Chlorine ⟶ Bromine + Potassium Chlorine

$$2KBr + Cl_2 \longrightarrow Br_2 + 2KCl$$

Iodine is also liberated from potassium iodide when it reacts with chlorine.

Potassium Iodide + Chlorine ⟶ Potassium Chloride + Iodine

$$2Kl + Cl_2 \longrightarrow 2KCl + I_2$$

Sodium Nitrate is produced when sodium nitrite reacts with chlorine in the presence of water.

Sodium Nitrite + Chlorine + Water ⟶ Sodium Nitrate + Hydrogen Chloride

$$NaNO_2 + Cl_2 + H_2O \longrightarrow NaNO_3 + 2HCl$$

Sodium Sulphite is also converted to sodium sulphate when it reacts with chlorine.

Sodium Sulphite + Chlorine + Water ⟶ Sodium Sulphate + Hydrogen Chlorine

$$Na_2SO_3 + Cl_2 + H_2O \longrightarrow Na_2SO_4 + 2HCl$$

Here from the above reactions it is clear that chlorine is an highly oxidizing agent.

6. Bleaching Action of Chlorine: Chlorine due to power of oxygen addiction has a tremendous power to bleach. In presence of moisture can decolourize any natural matter.

Exp: Introduce a red coloured flower moistened with water in the chlorine for few min. The colour slowly disappears proving the bleaching action of chlorine.

It also oxidizes the organic matter and kills the germs and is therefore, used for disinfectants and for sterilization of drinking water.

For this technique the chlorine is used for the preparation of Bleaching powder.

BLEACHING POWDER

It is also known as Calcium Hypochlorite. The formula is $CaOCl_2$. It is one of the most important compounds of chlorine, rather the salt of hypochlorous acid, but is prepared by the direct action of chlorine and calcium hydroxide. Bleaching powder is an all-purpose disinfectant used for cleaning the environment.

We shall now discuss the industrial preparation of bleaching powder. Bleaching powder is generally prepared by the paper industries for the treatment of pulps; here the process is done by the reaction of calcium hydroxide and chlorine. Chlorine gas is bubbled in the aqueous calcium hydroxide solution and the mixture is continuously stirred. The chlorination is continued until it reaches 30 to 35 g per litre of the solution and the bleaching liquor is then separated from the solution by pumping it out from the reaction tank. A swing pipe, the level of which can be adjusted according to the height of the liquid mixture inside the tank is provided portion of the liquor being drawn in the product. (See Fig. 12.7).

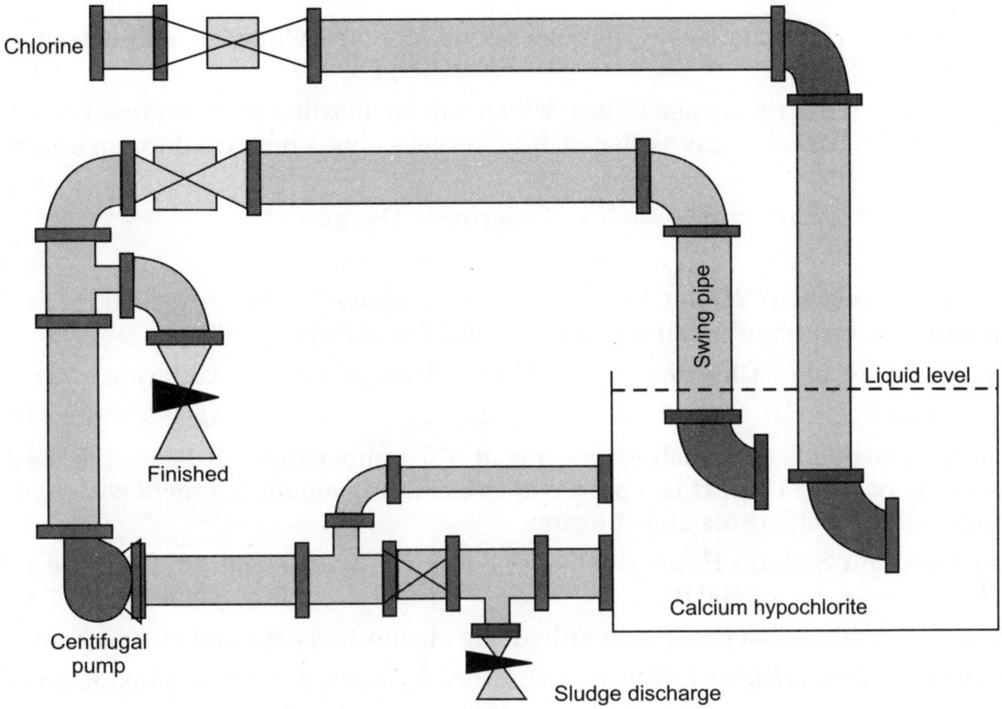

Fig. 12.7. The Industrial Process for Bleaching Powder Manufacture

The composition of calcium hydroxide in percentage is given below:

Calcium Carbonate	98.5%
Calcium Carbonate	2%
Iron Oxide	0.5%
Magnesium Oxide	2%
Water	4.5%

EXPLOSIVE

Bleaching powder is an unstable compound and gradually decomposes when standing and during decomposition it continuously looses chlorine. Bleaching powder should not be exposed in the sunlight because the liberated Chlorine reacts with the hydrogen present in the water to cause an explosion forming hydrogen chloride.

Oxy-Compounds of Chlorine

Chlorine has innumerable compounds among which the oxy-chlorides are going to be discussed.

Dichlorine Oxide (Cl_2O): This compound is prepared by passing chlorine over precipitated mercuric oxide.

$$\text{Chlorine} + \text{Mercuric Oxide} \longrightarrow \text{Mercuric Oxide.Mercuric Chloride} + \text{Dichlorine Oxide}$$
$$2Cl_2 + 2H_2O \longrightarrow HgO.HgCl_2 + Cl_2$$

Dichlorine oxide is an orange liquid, which can be distilled at its boiling point (2.5°C). At higher temperatures the gas explodes. It undergoes photochemical decomposition giving oxygen and chlorine.

$$\text{Dichlorine Oxide} \xrightarrow{\text{Sunlight}} \text{Chlorine} + \text{Oxygen}$$
$$Cl_2O \longrightarrow 2Cl_2 + O_2$$

Chlorine Dioxide (ClO_2): Chlorine dioxide is known as the only dioxide of chlorine. It is prepared by treatment of silver chloride and dry chlorine at 90°C.

$$\text{Silver Oxychloride} + \text{Chlorine} \longrightarrow \text{Chlorine Dioxide} + \text{Silver Chloride} + \text{Oxygen}$$
$$2AgOCl_2 + 2Cl_2 \longrightarrow 2ClO_2 + 2AgCl + O_2$$

Chlorine Dioxide is a reddish yellow gas at 8°C temperature. It is a condensed liquid with a boiling point of 11°C. It is a powerful chlorinating agent. It reacts with hydroxides of metals producing chlorates and chlorites.

$$\text{Chlorine Dioxide} + \text{Sodium Hydroxide} \longrightarrow \text{Sodium Clorite} + \text{Sodium Chlorate} + \text{Water}$$
$$2ClO_2 + 2NaOH \longrightarrow NaCl_2 + NaClO_3 + H_2O$$

It is also considered as the mixed anhydride of chlorous acid and chloric acid.

Dichlorine Hexoxide (Cl_2O_6): It is obtained by mixing chlorine dioxide and ozone.

$$\text{Chlorine Dioxide} + \text{Ozone} \longrightarrow \text{Dichlorine Hexoxide} + \text{Oxygen}$$
$$2ClO_2 + 2O_3 \longrightarrow Cl_2O_6 + 2O_2$$

It is a dark red liquid, which freezes at 4°C. It reacts with hydroxides of metals producing chlorates and perchlorates

$$\text{Dichlorine Hexoxide} + \text{Potassium Hydroxide} \longrightarrow \text{Potassium Chlorate} + \text{Potassium Perchlorate} + \text{Water}$$
$$Cl_2O_6 + 2KOH \longrightarrow KClO_3 + KClO_4 + H_2O$$

Industrial Preparation of Chlorine

Chlorine is produced in industries by the electrolysis of brine (sodium chloride solution). The process is done in the form of electrolytic cells called *diaphragm cells* the procedure is carried out in the *Hooker's cells,* these are typically categorized under diaphragm cells. (see Fig. 12.8). The shape of the cell is cubic. The base consists of a block concentrate from which flat sheets of graphite projects upwards acting as anodes (the positive electrode) is fastened, these positive electrodes are made up of copper. The negative (cathode) electrode consisting of steel fingers are supported horizontally from the side steel frame. The cathode assembly has the sodium hydroxide outlet and also the hydrogen is extracted. There are many pockets made of steel screen having perforations, which has inlet for brine outlet for chlorine gas.

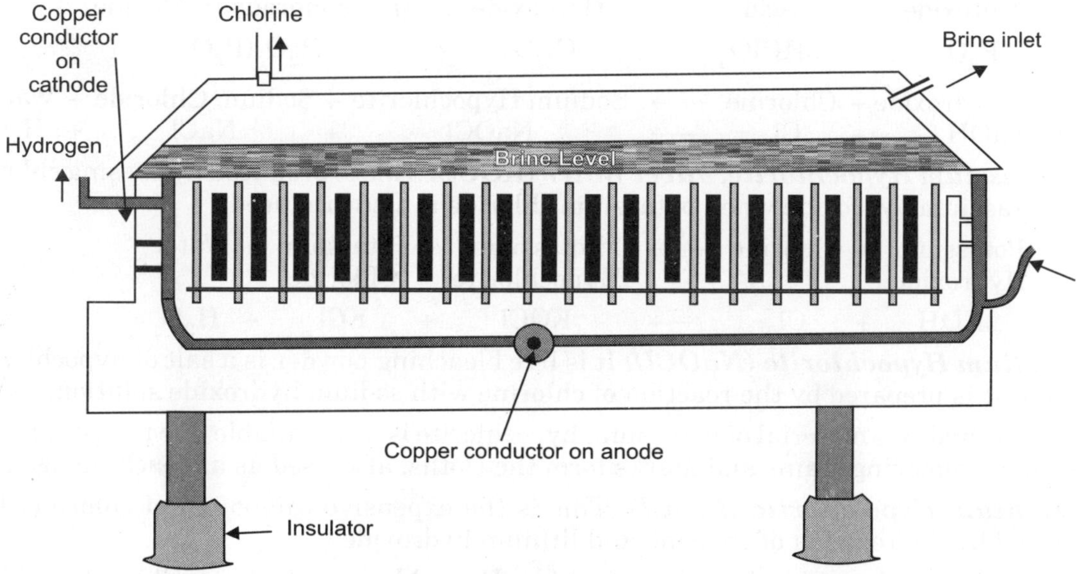

Fig. 12.8. The Hooker Cell (diagrammatic view)

Advantages of Hooker's Cell

1. The shape of the cell is cubic and also the electrodes are kept in the interlaced method, which takes less space.
2. Diaphragm can be removed easily because it is fixed with the lid at bottom.
3. The anode graphite is less consumed because remains submerged in brine.
4. The top cover is easily removed so the internal inspection is easier.
5. The cell is simple in construction, therefore, loss due to heat is less than in other cells.
6. The gap between the anode and the cathode assembly is enough for circulation of brine, which leads to height current efficiency.

Fig. 12.9. The actual Hooker cell

Other Compounds of Chlorine

Chlorine Heptoxide (Cl_2O_7): When perchloric acid is dehydrated with phosphorous pentoxide at a low temperature chlorine heptoxide is produced.

Phosphorous Pentoxide + Perchloric Acid ⟶ Chlorine Heptoxide + Water (Phosphorous Pentoxide)

$$P_2O_5 + 2HClO_4 \longrightarrow Cl_2O_7 + H_2O\ (P_2O_5)$$

Sodium Hydroxide + Chlorine ⟶ Sodium Hypochlorite + Sodium Chloride + Water

$$2NaOH + Cl_2 \longrightarrow NaOCl + NaCl + H_2O$$

Potassium Hypochlorite, Javel Water (KOCl): This is produced by passing chlorine over potassium hydroxide when potassium chloride is also produced.

Potassium Hydroxide + Chlorine ⟶ Potassium Hypochlorite + Potassium Chloride + Water

$$2KOH + Cl_2 \longrightarrow KOCl + KCl + H_2O$$

Sodium Hypochlorite (NaOCl): It is like bleaching powder is a salt of hypochlorous acid but it is prepared by the reaction of chlorine with sodium hydroxide solution.

The anhydrous material of potassium hypochlorite is not available due to instability. It is used for removing stains and marks form the cloths, also used as a bleaching agent.

Lithium Hypochlorite (LiOCl): This is the expensive compound of chlorine. It is produced by the reaction of chlorine and lithium hydroxide.

Lithium Hydroxide + Chlorine ⟶ Lithium Hypochlorite + Lithium Chloride + Water

$$2LiOH + Cl_2 \longrightarrow LiOCl + LiCl + H_2O$$

As lithium compounds of chloride are expensive they are not available commercially

Acids of Chlroine

Chlorine forms many acids some of them are being discussed.

HYDROCHLORIC ACID, MURIATIC ACID, SPIRIT OF SALTS

Hydrochloric acid was known to ancient alchemists as Muriatic acid (muria = Latin is Brine, because this acid is obtained from sodium chloride). Davy in 1810, first proved that it is a compound of chlorine and hydrogen, rather hydrogen chloride. He prepared this compound by the reaction of hydrogen and chlorine in the presence of sunlight.

EXPLOSIVE

Hydrogen + Chlorine $\xrightarrow{\text{Sunlight}}$ Hydrogen Chloride

$$H_2 + Cl_2 \longrightarrow 2HCl$$

Hydrogen chloride is prepared by the processes stated below:

1. Hydrogen Chloride is prepared by the action of potassium chloride and sulphuric acid.

Potassium Chloride + Sulphuric Acid ⟶ Potassium Sulphate + Hydrochloric Acid

$$2KCl + H_2SO_4 \longrightarrow K_2SO_4 + 2HCl$$

2. Hydrogen Chloride is also by the reaction of chlorine with compounds like sodium nitrite and chlorine.

Laboratory Preparation of Hydrogen Chlorine

Theory: Hydrogen chloride is prepared in the lab by the reaction of sodium chloride and sulphuric acid in a flask and the mixture is heated at 200°C when hydrogen chloride is produced. It is collected in a gas jar by upward displacement of air. The gas is dried by passing it through the two filter jars containing water and sulphuric acid. The gas is not dried by passing it through phosphorous pentoxide (see Fig. 12.10).

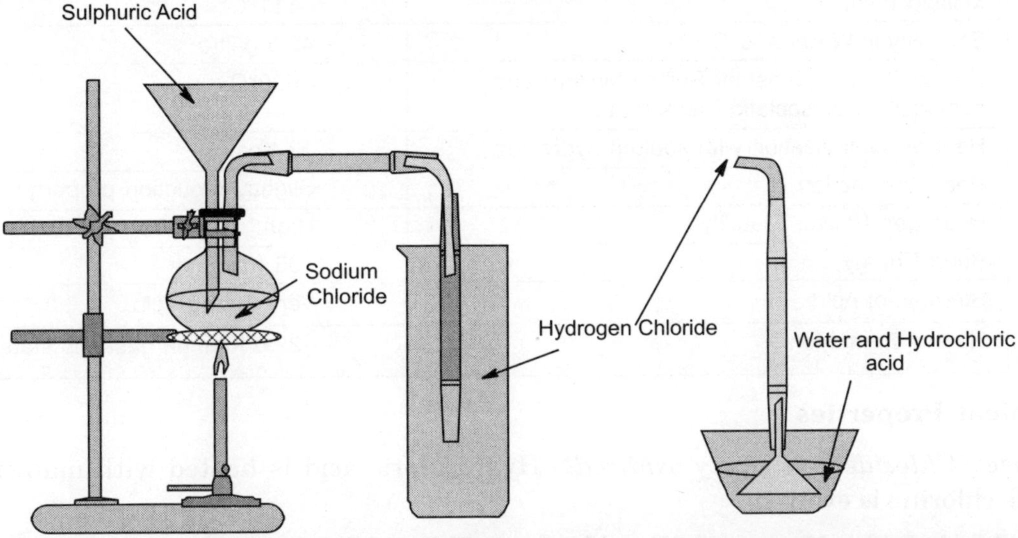

APPARATUS	REAGENTS	TEMPERATURE	PRESSURE
Round bottom flask, thistle funnel, burner, delivery tube etc.	Sodium chloride, Sulphuric acid.	200°C	Normal

Fig. 12.10. The Lab Preparation of Hydrogen Chloride and Hydrochloric Acid

Reaction Equation

Sodium Chloride + Sulphuric Acid ⟶ Hydrogen Chloride (gas) + Sodium Sulphate

$$2NaCl + H_2SO_4 \longrightarrow 2HCl + Na_2SO_4$$

Precaution: *The hydrogen chloride gas should not be dried by passing it through phosphorous pentoxide, because hydrogen chloride gets absorbed slowly. Hydrogen chloride during conversion of hydrochloric acid the delivery tube should be removed first before removing the burner otherwise there is a chance of explosion.*

Conversion of Hydrogen Chloride to Hydrochloric Acid

Exp: A gas jar is full of hydrogen chloride is dipped in the trough of water and the coverlid is removed. The water level rises in the jar and again the lid is covered and the jar is removed, the solution is hydrochloric acid is ready.

PROPERTIES OF HYDROGEN CHLORIDE

Physical Properties

Physical State at 15°C	Gas
Molecular Weight	36.467
Boiling Point	–80°C
Melting Point	–111°C
Solubility in Water at 0°C	42.0 g/litre
Boiling Point of Constant Boiling Mixture and composition dissociation temperature	1500°C
Heat of neutralization with sodium hydroxide	13 Kcal
Reduction Action	Slightly reduction property
Hydrogen Boiling Stability	High
Bond Energy	105 Kcal/mol
Strength of Acid	Very Strong Acid
Heat of Formation	–22.0 Kcal (in gaseous state)

Chemical Properties

Hydrogen Chloride gets easily oxidized: Hydrochloric acid is heated with manganese dioxide chlorine is evolved.

Hydrochloric Acid + Manganese Dioxide ⟶ Chlorine + Manganese Dioxide + Water

$$4HCl + MnO_2 \longrightarrow Cl_2 + MnCl_2 + 2H_2O$$

Hydrogen chloride reacts with potassium permanganate to produce chlorine

Hydrochloric Acid + Potassium Permanganate ⟶ Potassium Chloride + Manganese Dioxide + Chlorine + Water

$$16HCl + 2KMnO_4 \longrightarrow 2KCl + 2MnO_2 + 5Cl_2 + 8H_2O$$

With sodium thiosulphate it liberates sulphur the reaction is slow also termed as 'chemical sunset'.

Sodium Thiosulphate + Hydrogen Chloride ⟶ Sodium Chloride + Sulphur Dioxide + Sulphur + Water

$$Na_2S_2O_3 + 2HCl \longrightarrow 2NaCl + SO_2 + S + H_2O$$

Lead acetate producers lead chloride when reacts with hydrogen chloride

Hydrochloric Acid + Lead Acetate ⟶ Lead Chloride + Acetic Acid

$$2HCl + PbCH_3(COO)_2 \longrightarrow PbCl_2 + CH_3(COOH)_2$$

Water Affinity of Hydrogen Chloride

Exp: In flask some hydrogen chloride is filled and stoppered with a delivery tube. The flask is inverted and kept in a stand. In a beaker some blue litmus solution is taken. The other end of the delivery tube is dipped in the beaker and some litmus solution is violently sucked, this caused a partial vacuum for which more litmus is sucked which rushes in like a fountain and the colour changes from blue to red. (Apparatus is same as the ammonia's fountain experiment, instead of ammonia hydrogen chloride is used and the litmus solution is also altered).

Like sulphuric acid and nitric acid hydrochloric acid also produces hydrogen when it reacts with metals.

$$\text{Hydrochloric Acid} + \text{Zinc} \longrightarrow \text{Zinc Chloride} + \text{Hydrogen}$$
$$2HCl + Zn \longrightarrow ZnCl_2 + H_2$$
$$\text{Hydrochloric Acid} + \text{Iron} \longrightarrow \text{Ferrous Chloride} + \text{Hydrogen}$$
$$2HCl + Zn \longrightarrow FeCl_2 + H_2$$

The metals which stand above hydrogen in the electrochemical series can only displace hydrogen from the dilute acid.

Generally, silver, mercury, gold are not attacked by hydrochloric acid. Copper and lead dissolve in hot and concentrated acid, in presence of air, copper and silver react with the acid very slowly.

$$\text{Hydrochloric Acid} + \text{Copper} \longrightarrow \text{Copper Tetrachloride} + \text{Hydrogen}$$
$$8HCl + 2Cu \longrightarrow 2H_3[CuCl_4] + H_2$$
$$\text{Hydrochloric Acid} + \text{Copper} + \text{Oxygen} \longrightarrow \text{Copper Chloride} + \text{Water}$$
$$4HCl + 2Cu + O_2 \longrightarrow 2CuCl_2 + 2H_2O$$
$$\text{Hydrochloric Acid} + \text{Silver} + \text{Oxygen} \longrightarrow \text{Copper Chloride} + \text{Water}$$
$$4HCl + 4Ag + O_2 \longrightarrow 2CuCl_2 + 2H_2O$$

Hydrogen chloride neither burns in air nor supports combustion. However, burning sodium continues to burn in the gas with a bright yellow flame producing hydrogen and anhydrous sodium chloride.

$$2Na + 2HCl = 2NaCl + H_2.$$

Its aqueous solution is strongly acidic and turns blue litmus red. It is a monobasic acid. In dilute aqueous solution, it completely ionizes into hydrogen and chloride ions arid conducts electricity.

$$HCl \rightleftharpoons H^+ + Cl^-$$
$$H^+ + H_2O = [H_3O]^+$$

In fact, the ion responsible for acidity is not the simple H^+ ion but a hydrated form of it. The salts of this acid are called the chlorides such as KCl, $ZnCl_2$, $AlCl_3$ etc. All metallic chlorides with the exception of $PbCl_2$, AgCl and Hg_2Cl_2 are soluble in water.

Gold or platinum does not react with the acid. Liquid hydrogen chloride does not conduct electricity and has no action upon litmus or metals in the absence of water. However, aluminium dissolves in the liquid.

Hydrochloric acid reacts with metallic oxides and hydroxides yielding salts and water. It liberates carbon dioxide from a carbonate and forms ammonium chloride with ammonia gas or ammonium hydroxide.*

Hydrochloric Acid + Calcium Oxide ⟶ Water + Calcium Chloride

$$2HCl + CaO \longrightarrow H_2O + CaCl_2$$

Hydrochloric acid is readily oxidize, into chlorine. Oxidizing agents like manganese dioxide and lead dioxide. Potassium dichromate converts hot concentrated acid into chlorine. Air or oxygen in presence of copper chloride catalyst oxidizes the acid vapour to chlorine.

Manganese Dioxide + Hydrochloric Acid ⟶ Manganese Chloride + Chlorine + Water

$$MnO_2 + 4HCl \longrightarrow MnCl_2 + Cl_2 + 2H_2O$$

Lead Dioxide + Hydrochloric Acid ⟶ Lead Chloride + Chlorine + Water

$$PbO_2 + 4HCl \longrightarrow PbCl_2 + Cl_2 + 2H_2O$$

Potassium Dichromate + Hyrochloric Acid ⟶ Potassium Chloride + Chlorine + Chromyl Chloride + Water

$$K_2Cr_2O_7 + 14HCl \longrightarrow 2KCl + 3Cl_2 + 2CrCl_2 + 7H_2O$$

Hydrochloric Acid + Oxygen ⟶ Water + Chlorine

$$4HCl + O_2 \longrightarrow 2H_2O + 2Cl_2$$

Potassium permanganate oxidizes the acid at the ordinary temperature.

Potassium Permanganate + Hydrochloric Acid ⟶ Manganese Chloride + Chlorine + Potassium Chloride + Water

$$2KMnO_4 + 4HCl \longrightarrow MnCl_2 + Cl_2 + 2KCl + 2H_2O$$

But hydrochloric acid is not oxidized by concentrated sulphuric acid.

When hydrochloric acid or an aqueous solution of a metallic chloride is added to a solution of lead, silver or mercurous salt, a white precipitate of the chloride of the metal is obtained.

Lead Nitrate + Hydrochloric Acid ⟶ Lead Chloride + Nitric Acid

$$Pb(NO_3)_2 + 2HCl \longrightarrow PbCl_2 + 2HNO_3$$

Lead chloride is soluble boiling water.

---

* When a dilute solution of hydrochloric acid is distilled, the water evaporates and the concentration of the solution increases. But concentrated hydrochloric acid on distillation gives off hydrogen chloride first and resulting, solution becomes more dilute. Any solution of hydrochloric acid on distillation produces ultimately a solution containing 20-20% of HCl by weight. This solution will distil unchanged at a constant temperature (110°C). The acid of this strength is a constant boiling mixture.

$$\text{Silver Nitrate} + \text{Hydrochloric Acid} \longrightarrow \text{Silver Chloride} + \text{Nitric Acid}$$

$$Ag(NO_3)_2 + HCl \longrightarrow AgCl_2 + 2HNO_3$$

$$\text{Mercurous Nitrate} + \text{Hydrochloirc Acid} \longrightarrow \text{Mercurous Chloride} + \text{Nitric Acid}$$

$$Hg_2(NO_3)_2 + HCl \longrightarrow Hg_2Cl_2 + 2HNO_3$$

A mixture of concentrated hydrochloric acid and concentrated nitric acid in the ratio of 3:1 by volume is called *Aqua Regia (kingly water)*. Which dissolves the noble metals like gold, platinum etc.

Concentrated hydrochloric acid on electrolysis liberates hydrogen at the cathode and chlorine at the anode.

Experiments Illustrating the Important Properties or Hydrogen Chloride

1. *Hydrogen chloride is neither combustible nor a supporter of combustion.* When a lighted taper is introduced into a jar of hydrogen chloride, the taper is extinguished and the gas does not burn.

2. *It is highly soluble in water and the aqueous solution reacts acidic.* The solubility in water and acidity of the solution of hydrogen chloride can be proved by the fountain experiment. The experimental procedure has been described while illustrating the similar properties of sulphur dioxide. In this case, the flask is only filled with hydrogen chloride instead of sulphur dioxide.

3. *(a) It reacts with ammonia gas giving ammonium chloride.* A glass rod moistened with concentrated ammonia solution or liquor ammonia is brought near a glass jar containing hydrogen chloride when dense white fumes of ammonium chloride is obtained.

$$NH_3 + HCl = NH_4Cl.$$

(b) A small amount of calcium carbonate taken in a test tube is treated with dilute hydrochloric acid when a colourless, odourless gas is evolved. That the gas is carbon dioxide is proved by the fact that it turns clear limewater milky. Thus, the experiment illustrates that hydrochloric acid liberates carbon dioxide from a carbonate.

(c) The metals such as zinc, magnesium etc. dissolve in dilute hydrochloric acid with the evolution of hydrogen. A little dilute hydrochloric acid is taken in a test tube and a piece of granulated zinc is added to it. The acid and the metal react immediately liberating a colourless, odourless gas. The gas is shown to be hydrogen by its inflammability with a blue flame.

4. *It is oxidized to chorine by oxidizing agents.* Hydrogen chloride is passed through an inlet tube attached to one end of a combustion tube containing heated manganese dioxide. A greenish yellow gas is found to escape from the other end of the tube. The issuing gas is chlorine as it turns a piece of paper soaked in potassium iodide and starch solutions deep blue. Hydrogen chloride may also be converted into chlorine if crystals of potassium dichromate are used in the combustion tube in place of manganese dioxide.

Industrial Preparation of Hydrochloric Acid

Synthetic Method: In this method as shown in Fig. 12.11 the hydrochloric acid is prepared by the direct reaction of hydrogen and chlorine.

Hydrogen + Chlorine = Hydrogen Chloride

$$H_2 + Cl_2 = 2HCl$$

Hydrogen and chlorine after been dried are taken to a tower, similar to oxy-hydrogen blowpipe, which is fitted in a silica brick combustion chamber. A slight excess of hydrogen is desirable so that the conversion of chlorine to hydrogen chloride will be easier. The resulting hydrogen chloride is cooled and then absorbed in an absorption chamber, the solution being constantly agitated and re-circulated in the tower until its condition is saturated. The ejection system shown in the fig works on the same idea as an ordinary filter pump. The unabsorbed hydrogen from the top of the absorption tower is towered to the bummer.

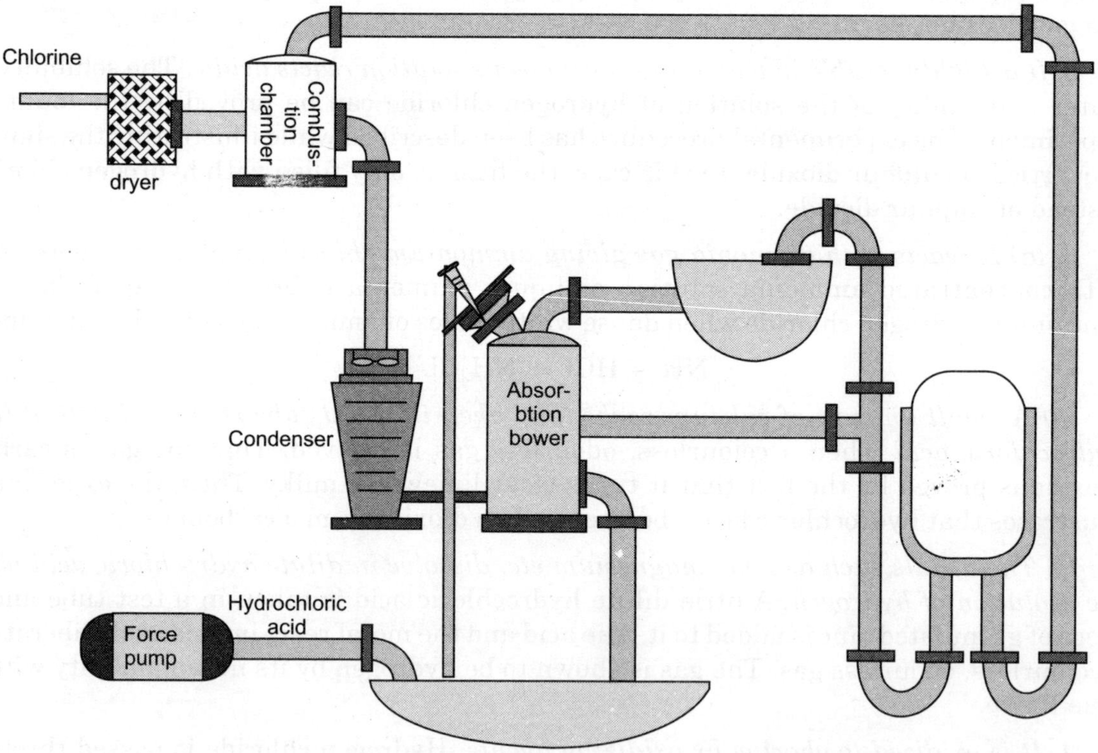

Fig. 12.11. The Industrial Process of Hydrochloric Acid Preparation

Composition of Hydrochloric Acid

The composition of hydrogen chloride is determined by the method of electrical analysis of hydrochloric acid; this experiment is conducted in an apparatus called Hofmann's Voltameter.

The apparatus is shown in the Fig 12.12, consists of three glass vessels. The central one is used as reservoir for the acid solution and the two side limbs are used for main electrolysis, they are connected to two stopcocks and two carbon electrodes are introduced. The apparatus is filled with concentrated hydrochloric acid and then the electric current is passed, it was noticed that hydrogen and chlorine gas has separated in equal volumes. From this we can say that the formula of hydrogen chloride is as $(HCl)_n$ (from Avogadro's hypothesis). Vapour density of hydrogen chloride is 18.25 and the molecular weight is 36.5, therefore the formula must be HCl.

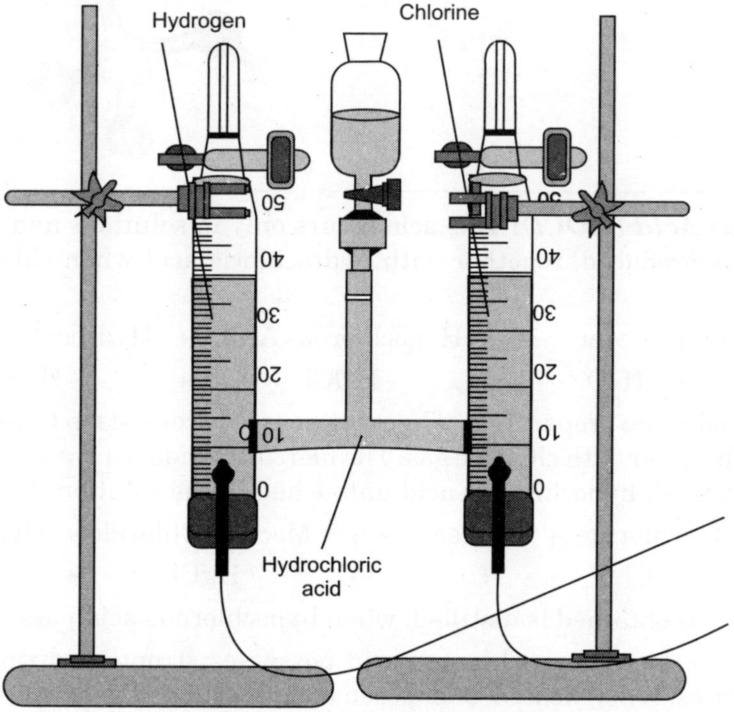

Fig. 12.12. The Hofmann's Voltameter

OXYACIDS OF CHLORINE

Hypochlorous Acid	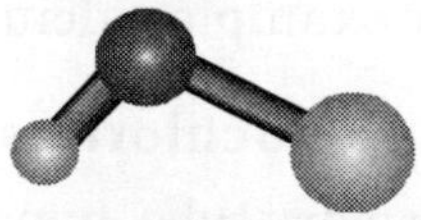
Chlorous Acid	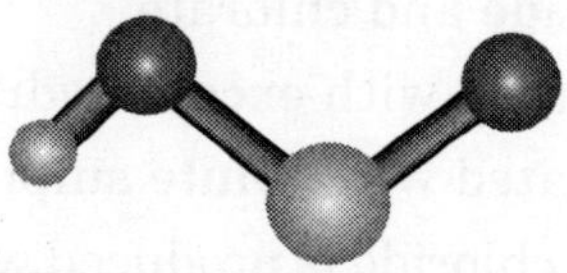

Chloric Acid	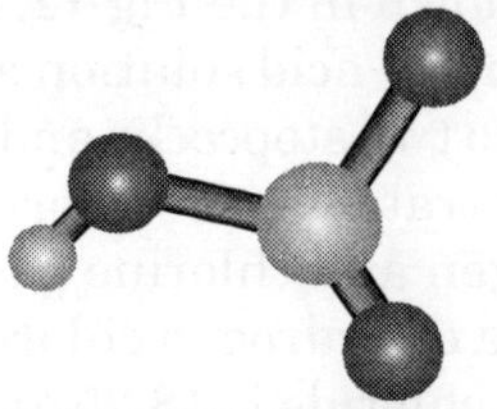
Perchloric Acid	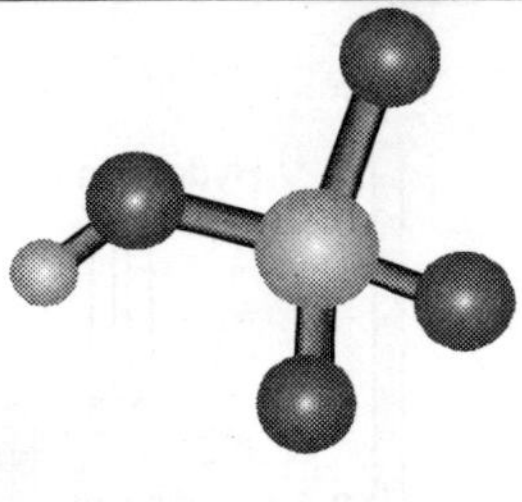

Hypochlorous Acid (HOCl): This acid occurs only in solutions and has not yet been isolated pure. It is produced, together with hydrochloric acid when chlorine reacts with water.

$$\text{Chlorine} + \text{Water} \longrightarrow \text{Hypochlorous Acid} + \text{Hydrochloric Acid}$$
$$Cl_2 + H_2O \longrightarrow HOCl + HCl$$

The best method of the preparation of hypochlorous acid consists in treating a suspension of mercuric oxide in water with chlorine gas. The mercuric oxide neutralizes the hydrochloric acid by leaves the weak hypochlorous acid untouched in the solution.

$$\text{Mercuric Oxide} + \text{Chlorine} + \text{Water} \longrightarrow \text{Mecuric Chloride} + \text{Hypochlorous Acid}$$
$$HgO + Cl_2 + H_2O \longrightarrow HgCl_2 + 2HOCl$$

The solution thus obtained is distilled, when hypochlorous acid passes over.

An aqueous solution of hypochlorous acid possesses strong oxidizing bleaching and germicidal properties. When heated, or exposed to light, the acid decomposes into hydrogen chloride and oxygen.

$$\text{Hypochlorous Acid} \longrightarrow \text{Hydrochloric Acid} + \text{Oxygen}$$
$$2HOCl \longrightarrow 2HCl + O_2$$

The salts of hypochlorous acid, called the hypochlorites, are of much greater importance than the acid itself, for example calcium hypochlorite, better known as bleaching powder.

Analytical Tests for Hypochlorites and Hypochlorous Acid

1. When heated in a dry tube, hypochlorites undergo disportionation yielding a mixture of a chloride and chlorate.
2. When heated with excess hydrochloric acid it produces chlorine.
3. When heated with dilute sulphuric acid hypochlorous acid is produced.
4. Mercuric chloride is produced when hypochlorous acid reacts with metallic mercury.

Chlorous Acid (HClO$_2$): This acid is formed when there is a reaction between chlorine dioxide and water.

Chlorine Dioxide + Water $\longrightarrow$ Chlorous Acid + Chloric Acid

$$2ClO_2 + H_2O \longrightarrow HClO_2 + HClO_3$$

Chlorites are formed when chlorous acid reacts with different hydroxides

Potassium Hydroxide + Chlorine $\longrightarrow$ Potassium Hypochlorite + Potassium Chlorate + Water

$$2NaOH + Cl_2 \longrightarrow KOCl + NaCl + H_2O$$

Sodium hypochlorite is also prepared by the action of sodium hydroxide and chlorine

Sodium Hydroxide + Chlorine $\longrightarrow$ Sodium Hypochloriet + Sodium Chloride + Water

$$2NaOH + Cl_2 \longrightarrow NaOCl + NaCl + H_2O$$

Chloric acid (HClO$_3$): Chloric acid is prepared by the action of dilute sulphuric acid in barium chlorate, here dilute acid is used because concentrated acid will decompose the chloric acid.

Barium Chlorate + Sulphuric Acid $\longrightarrow$ Chloric Acid + Barium Sulphate

$$Ba(ClO_3)_2 + H_2SO_4 \longrightarrow 2HClO_3 + BaSO_4$$

After filtering off the barium sulphate, the solution is concentrated in vacuum to a 40% chloric acid content. Beyond this point decomposition of acid occurs.

Properties of Chloric Acid: Chloric acid reacts only in aqueous solution which is colourless, when the solution is exposed to light the acid undergoes decomposition into hypochlorous acid and nascent oxygen. The acid possesses the strong oxidizing power, combustible substances like wood, paper, immediately catches fire when comes in contact with the acid. The aqueous solution of the acid has been used for oxidizing iodine to iodic acid.

Chloric acid is a strong monobasic acid; the degree of ionization of N solution of chloric is 18°C being 79%

Compounds of Chloric Acid

Chlorates are the compounds of chloric acid; these compounds are produced when chlorine reacts with hydroxide of elements, such as sodium hydroxide.

Sodium Hydroxide + Chlorine $\longrightarrow$ Sodium Chlorate + Sodium Chlorate + Water

$$6NaOH + 3Cl_2 \longrightarrow NaClO_3 + 5NaCl + 3H_2O$$

The chlorate salts are colourless and are quite stable at ordinary temperature and are freely soluble in water. Some of the important reactions are stated below.

Action of Heat: When heated dry, they undergo decomposition, which occurs in two steps. Firstly the chlorate is broken down to chloride and perchlorate.

Potassium Chlorate $\xrightarrow{300°C}$ Potassium Perchlorate + Potassium Chloride + Oxygen

$$2KClO_3 \longrightarrow KClO_4 + KCl + O_2$$

Potassium Perchlorate $\xrightarrow{610°C}$ Potassium Chloride + Oxygen

$$KClO_4 \longrightarrow KCl + O_2$$

Secondly, the decomposition of perchlorate with the evolution of oxygen takes place at 610°C.

With Dilute Sulphuric Acid: When chlorate compounds are heated with sulphuric acid chlorine dioxide is produced.

Potassium Chlorate + Sulphuric Acid ⟶ Potassium Bisulphate + Perchloric Acid + Chlorine Dioxide + Water

$$3KClO_3 + 3H_2SO_4 \longrightarrow 3KHSO_4 + HClO_4 + 2ClO_2 + H_2O$$

Chloric acid is also produced when sulphuric acid reacts with potassium chlorate.

Potassium Chlorate + Sulphuric Acid ⟶ Potassium Bisulphate + Chloric Acid

$$3KClO_3 + 3H_2SO_4 \longrightarrow 3KHSO_4 + HClO_3$$

When heated with dilute hydrochloric acid: A mixture of chlorine dioxide and chlorine is produced.

Potassium Chlorate + Hydrochloric Acid ⟶ Potassium Chloride + Chlorine Dioxide + Chlorine + Water

$$3KClO_3 + 3H_2SO_4 \longrightarrow 3KHSO_4 + 2ClO_2 + Cl_2 + H_2O$$

Structure of Chlorate Ion

In the crystalline chlorates the ClO_3^- ion forms a rather flat pyramid with chlorine at the apex Cl – O bond length is 1.46Å and OClO angle is about 110°. The structure of the chlorate ion is shown in the below:

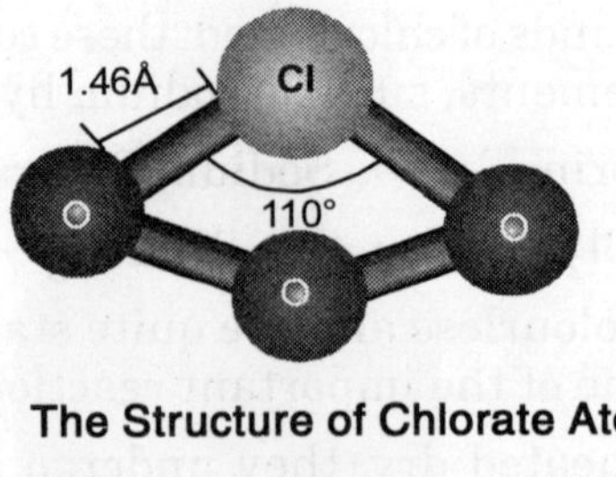

The Structure of Chlorate Atom

Industrial Preparation of Potassium Chloride

In industry potassium chlorate is prepared by the electrolysis of potassium chloride. In a partition less cell consisting of a steel tank, which also acts as cathode. The anodes are the graphite and are arranged very close together (see Fig. 12.13). The temperature is maintained at 40°C by circulating worm water through cooling coils. The chlorine is set free at the anode reacts with the hydroxide formed at the cathode to produce a potassium hypochlorite and chlorate.

Potassium Hypochlorite	+	Chlorine	$\longrightarrow$	Potassium Chloride	+	Potassium Hydroxide	+	Water
2KOH	+	Cl_2	$\longrightarrow$	KCl	+	KOCl	+	H_2O

Potassium Hypochlorite	$\longrightarrow$	Potassium Chlorate	+	Potassium Chloride
3KOCl	$\longrightarrow$	$KClO_3$	+	2KCl

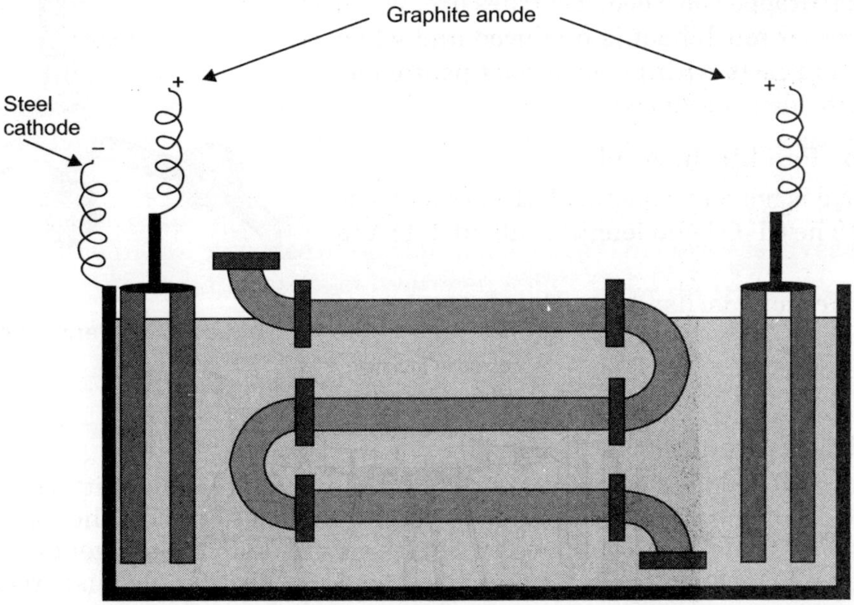

Fig. 12.13. The Industrial Preparation of Potassium Chlorate

The cell liquor is transferred to steel tanks is heated with steam to 90°C to complete the conversion of hypochlorite to chlorate. On allowing the hot liquid to cool, potassium chlorate crystallizes out, the mother liquor after the addition of more potassium chloride, is sent back to the electrolytic cell.

Use of Potassium Chlorate: (1) Large quantities of potassium chlorates are employed in the match industry. (2) In the manufacture of fire works. (3) In the manufacture of potassium perchlorate.

Perchloric Acid ($HClO_4$): Anhydrous perchloric acid is prepared by distilling under reduced pressure a mixture of potassium perchlorate and concentrated sulphuric acid.

Potassium Perchlorate	+	Sulphurc Acid	$\longrightarrow$	Potassium Sulphate	+	Perchloric Acid
$KClO_4$	+	H_2SO_4	$\longrightarrow$	$KHSO_4$	+	$HClO_4$

In the later case, sodium chloride, being sparingly soluble in concentrated hydrochloric acid, gets precipitated almost completely. The excess of hydrochloric acid, is removed by heating the filtrate near to boiling point.

Properties of Perchloric Acid

Anhydrous perchloric acid is a colourless, volatile liquid which fumes in moist air, it is very unstable and decomposes with explosion on heating and some times merely on standing for few days. It can however, be distilled under pressure at about 20°C. Its freezing point is –112°C. Anhydrous perchloric acid is highly dangerous substance, it bums the skin, when dropped on wood, it catches fire. When mixed with water much heat is produced and when perchloric acid reacts with phosphorous pentoxide, chlorine heptoxide is produced.

Structure of Perchloric Acid

Perchloric acid is not a true per acid, the structure is tetrahedral. The Cl–O bond length is about 1.44Å in $KClO_4$.

Perchloric acid is used for the detection of potassium.

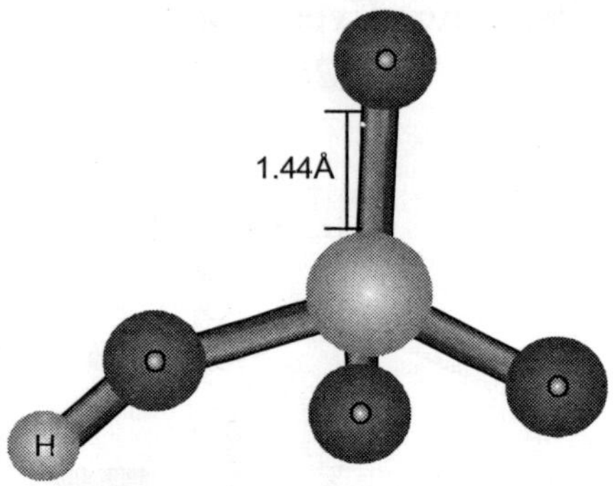

Structure of Perchloric Acid

BROMINE

Hisory and Occurrence

Balard discovered bromine in the year 1825 in the residual liquor (Bittern) left after the separation of common salt from seawater by yellow colonization; from the coloured solution he isolated a dark-reddish brown liquid with a powerful and very pungent smell. This was recognized by him as a new element and also named as Bromos, the disagreeable odour.

Bromine though not a common element occurs in combination in some abundance. As bromides of calcium, magnesium. It also occurs in seawater upto 0.065% on an average.

Preparation of Bromine

Bromine is prepared from its bromide compounds; there are different methods, which are going to be discussed below:

1. From Sea Water

1. Sea water is at first pumped to the plant where it acidified before treating it with chlorine, after the treatment bromine is liberated. The amount of chlorine is being equivalent to the bromine content in the seawater.
2. The liberated bromine is blown out of the brine solution by means of an air current. This is done in a series of towers working on a countercurrent principle.
3. The air carrying the bromine vapour is passed through absorption tower down which a solution of sodium carbonate is sprayed. This dissolves out the bromine in accordance with the following equations.

Sodium Carbonate + Bromine ⟶ Sodium Bromide + Sodium Bormate + Carbon Dioxide

$$3Na_2CO_3 + 3Br_2 \longrightarrow 5NaBr + NaBrO_3 + 3CO_2$$

4. The bromide bromate liquor from the absorption tower is acidifies with the sulphuric acid and distilled to recover to bromine.

Sodium Bromide + Sodium Bromate + Sulphuric Acid ⟶ Sodium Sulphate + Bromine + Water

$$5NaBr + NaBrO_3 + 3H_2SO_4 \longrightarrow 3Na_2SO_4 + 3Br_2 + 3H_2O$$

It is important that the water from which bromine has been extracted, is discharged many miles away from the intake. This is done by locating the intake directly in the ocean and discharging the effluent into a river, which carries the same far away.

2. By passing Chlorine through Potassium Bromide

Potassium Bromide + Chlorine ⟶ Bromine + Potassium Chloride

$$2KBr + Cl_2 \longrightarrow Br_2 + 2KCl$$

Laboratory Preparation of Bromine

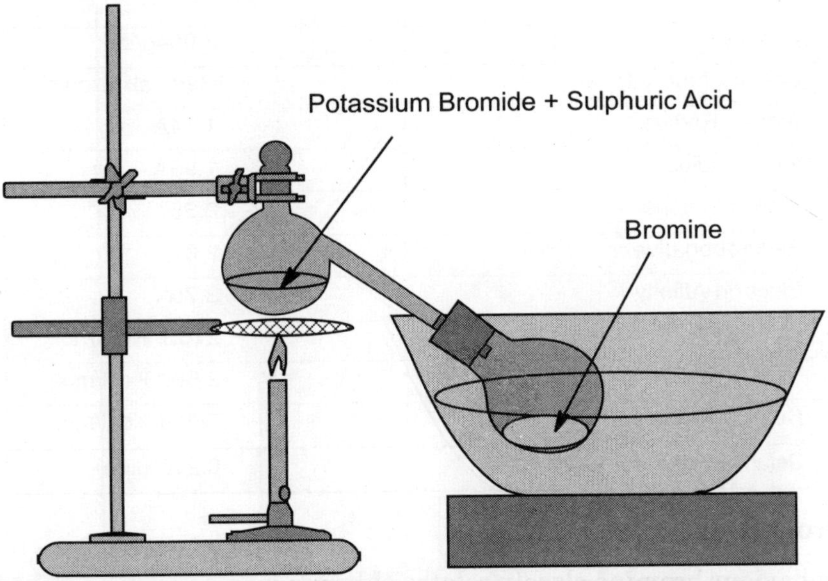

Apparatus	Reagents	Temperature	Pressure	Catalyst
Retort, burner, flask, stand etc.	Sulphuric Acid, Bromide, manganese	Nearly 200°C	Normal	Manganese dioxide

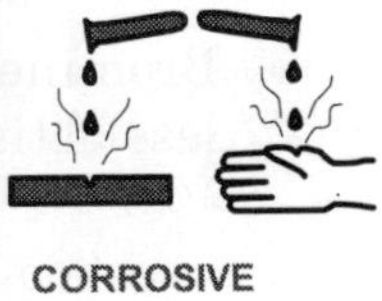

Fig. 12.14. The Lab Preparation of Bromine

Theory: Bromine is prepared by the reaction of potassium bromide and sulphuric acid in the presence of manganese dioxide. The mixture is heated in a retort; then bromine is evolved which is collected in cool reservoir in the form of a red liquid (See Fig. 12.14).

Reaction Equation

Potassium Bromide	+	Sulphuric Acid	+	Manganese Dioxide	⟶	Bromine	+	Potassium Sulphate	+	Manganese Sulphate	+	Water
2KBr	+	$3H_2SO_4$	+	MnO_2	⟶	Br	+	$2KHSO_4$	+	$MnSO_4$	+	$2H_2O$

Precaution: *Bromine is a fuming liquid; should be carefully handled. It causes severe wounds when comes in contact with skin, also causes injuries when comes in contact with eyes.*

Physical Properties

Atomic Number	35
Atomic Weight	789.904
Physical State	Liquid
Boiling Point	58.7°C
Freezing Point	–7.2°C
Density	2.984g/cc
Colour of the Vapour	Raddish-Brown
Atomic Radius	1.14Å
Ionic Radius	1.95Å
Atomic Volume	2.35
Electronegative	2.8
Electron Affinity	3.75
First Ionization Energy	276.1 Kcal/moi
Heat of Fusion	2.580 Kcal/mol
Heat of Vapourization	7.418 Kcal/mol
Solubility in water at 20°C	0.210 (litre)

Chemical Properties

In chemical behaviour bromine closely relates chlorine :

1. Bromine combines with hydrogen forming hydrogen bromide but the reaction is less active than the reaction between hydrogen and chlorine

   Hydrogen + Bromine ⟶ Hydrogen Bromide

   $$H_2 + Br_2 \longrightarrow 2HBr$$

2. Bromine also combines antimony and arsenic

Antimony	+	Bromine	$\longrightarrow$	Antimoy Bromide
2Sb	+	Br_2	$\longrightarrow$	$2SbBr_2$
Arsenic	+	Bromine	$\longrightarrow$	Arsenic Tribromide
2As	+	Br_2	$\longrightarrow$	2HBr

3. Bromine reacts with metals like potassium sodium forming their corresponding bromides.

Sodium	+	Bromine	$\longrightarrow$	Sodium Bromide
2Na	+	Br_2	$\longrightarrow$	2NaBr
Potassium	+	Bromine	$\longrightarrow$	Potassium Bromide
2K	+	Br_2	$\longrightarrow$	2KBr

4. White phosphorous explodes in bromine to form phosphorous Tribromide and phosphorous pentabromide.

Phoshorous	+	Bromine	$\longrightarrow$	Phosphorous Tribromide
2P	+	$5Br_2$	$\longrightarrow$	$2PBr_3$
Phosphorous	+	Bromine	$\longrightarrow$	Phosphorous Pentabromide
2P	+	$5Br_2$	$\longrightarrow$	$2PBr_5$

5. Action of water: Bromine water mixture when exposed to light it slowly decomposes to produce oxygen.

Bromine	+	Water	$\longrightarrow$	Hydrogen Bromide	+	Hypobromous Acid
$5Br_2$	+	H_2O	$\longrightarrow$	HBr	+	HBrO

6. Bromine Water converts sulphurous acid to sulphuric acid, sodium thiosulphate to sodium sulphate

Sulphurous Acid	+	Bromine	+	Water	$\longrightarrow$	Sulphuric Acid	+	Hydrogen Bromide
H_2SO_3	+	Br_2	+	H_2O	$\longrightarrow$	H_2SO_4	+	2HBr

Bromine like chlorine also possesses the bleaching power. It can bleach a moist litmus paper and also any vegetable or organic matter.

7. Action of Hydroxides: Potassium hydroxide reacts with bromine producing both bromide and bromate

Potassium Hydroxide	+	Bromine	$\longrightarrow$	Potassium Bromide	+	Potassium Bromate	+	Water
2KOH	+	Br_2	$\longrightarrow$	KBr	+	KBrO	+	H_2O

In hot state a bromide and bromate are produced.

Bromine reacts with benzene to form hydrogen bromide in the presence of iron powder as a catalyst.

Benzene	+	Bromine	$\xrightarrow{\text{Iron fillings}}$	Benzene Dibromide	+	Hydrogen Bromide
C_6H_6	+	Br_2	$\xrightarrow{\text{Fe}}$	C_6H_5Br	+	HBr

Industrial Preparation of Bromine

Before preparation of bromine a huge amount of Camallite $KCl.MgCl_2\ 6H_2O$ liquor is first prepared, this is known as mother liquor. After extraction of chlorine this mother liquor is heated and allowed to drop gently down a tower (see Fig. 12.15) with earthen were balls, through this tower dry chlorine gas passed from the top to bottom of the tower. Bromine is produced in this reaction, which comes out of the tower through the outlet at the top.

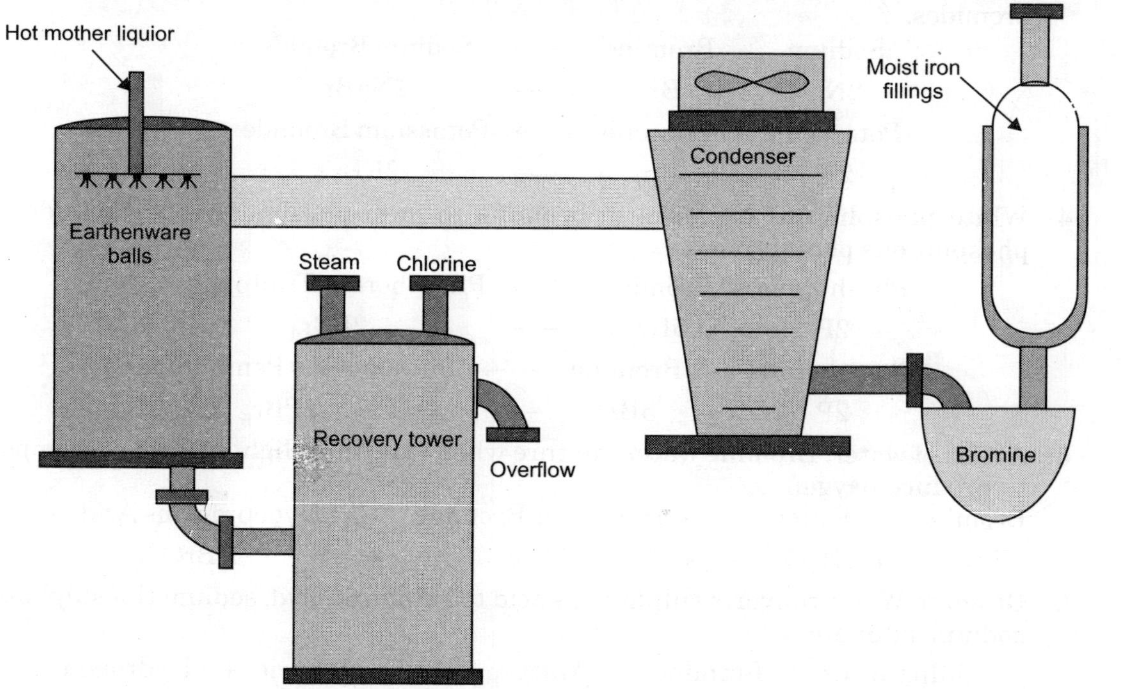

Fig. 12.15. The Industrial Preparation of Bromine

Reaction Equations

$$\text{Potassium Bromide} + \text{Chlorine} \longrightarrow \text{Potassium Chloride} + \text{Bromine}$$
$$2KBr + Cl_2 \longrightarrow 2KCl + Br_2$$
$$\text{Magnesium Bromide} + \text{Chlorine} \longrightarrow \text{Magnesium Chloride} + \text{Bromide}$$
$$MgBr_2 + Cl_2 \longrightarrow MgCl_2 + Br_2$$

Bromine vapours passing out through a condenser where the gas condense to tower packed with iron fillings where the gas is absorbed and yields ferosoferric bromide.

$$\text{Iron} + \text{Bromine} \longrightarrow \text{Ferosoferic Bromide}$$
$$3Fe + Br_2 \longrightarrow Fe_2Br_2$$

Purification of Bromine: Bromine is obtained by this process contains chlorine, moisture and small traces of iodine as impurities. It is purified by shaking it with cold concentrated sulphuric acid (to remove moisture) and then distilling over potassium bromide (to remove the chlorine) and then left over iodine is treated with copper salt where cupric iodide is precipitated. The down coming hot mother liquor carries along with it some bromine in solution, which is collected in the tank to recover the bromine vapours. The dissolved bromine and chlorine are blown out by steam. They pass up the tank and are mixed up with the main current of chlorine.

Uses of Bromine

1. Bromine is used for the preparation of dyestuff, disinfectant, bromine compounds etc.
2. Bromine is also used for the preparation of sedatives as sodium and potassium bromides.
3. Silver bromide is used for photographic plates and films.
4. It is also used for the preparation of ethylene bromide, an antiknock compound used in automobile engines.
5. It is also used for the preparations tear gas e.g. Xylyl Bromide and Bromoacetone.
6. It is used for the preparation synthesis organic compounds.

Tests of Bromine

(*a*) Bromine gives pungent smelling vapour, dark red in colour, which does not turn ferrous solution black (not like nitrogen dioxide).

(*b*) It turns starch paper yellow.

(*c*) It gets dissolved in the carbon disulphide, and turns it orange.

Compounds of Bromine

We shall discuss the compounds of bromine specially the oxides of bromine and other compounds.

Dibromine Oxide (Br_2O): It is prepared by the action of bromine and atomic oxygen.

$$\text{Bromine} + \text{Oxygen} \longrightarrow \text{Dibromine Oxide}$$

$$Br_2 + O \longrightarrow Br_2O$$

It is also prepared by the reaction of bromine and mercuric oxide.

$$\underset{}{\text{Bromine}} + \underset{\text{Oxide}}{\text{Mercuric}} \longrightarrow \underset{\text{Oxide}}{\text{Mercuric}}\ \underset{\text{Bromide}}{\text{Mercuric}} + \underset{\text{Oxide}}{\text{Dibromine}}$$

$$Br_2 + 2HgO \longrightarrow HgO.\ HgBr + Be_2O$$

Dibromine oxide is a dark brown liquid with a freezing point of -18°C

It is also known as the anhydrous of hypobromous acid, HBrO

Bromine Dioxide (BrO_2): It is produced by creating an electrical spark in the mixture of oxygen-bromine, at a temperature of –112°C and at a very low pressure.

Bromine dioxide is only stable at a temperature of –30°C and it is a yellow solid, which dissolve in sodium hydroxide producing bromide and bromates.

Bromine Trioxide (BrO_3) : It is prepared by the action of a slow electric discharge on a mixture of oxygen and bromine between -12°C to -15°C

Bromine	+	Ozone	⟶	Bromine Trioxide
Br_2		$2O_3$	⟶	$2BrO_3$

This compound of bromine exists only below -70°C.

Hydrogen Bromine, Hydrobromic Acid, (HBr)

Preparation

I. Bromine reacts directly with hydrogen but only at a temperature of 2(and in the presence of a suitable catalyst such as platinum.

$$\text{Hydrogen} + \text{Bromine} \xrightarrow{\text{Platinum}} \text{Hydrogen Bromide}$$

$$H_2 + Br_2 \xrightarrow{Pt} 2HBr$$

II. Hydrogen bromide can also be prepared by the action of hydrogen sulphide gas and bromine

Hydrogen Sulphide +	Bromine	⟶	Sulphur Dibromide	+	Hydrogen Bromide
$2H_2S$ +	Br_2	⟶	4HBr	+	$S2Br_2$

III. Hydrogen bromide is also produced when sulphur dioxide is bubbled through a mixture of bromine water and ice. The sulphur dioxide is bubbled until the solution becomes straw colour , indicating that most of the free bromine has disappeared the mixture is distilled and hydrogen bromide is obtained. This method is called *Scott's Method.*

Sulphur Dioxide +	Bromine Water	⟶	Hydrogen Bromide	+	Sulphuric Acid
SO_2 +	Br_2H_2O	⟶	2HBr	+	$H2SO_4$

Laboratory Preparation of Hydrogen Bromide

Theory: In laboratory hydrogen bromide is prepared by dropping bromine on red phosphorous and water.

Reaction Equation

Bromine	+	Phosphorous	⟶	Phosphorous Tribromide		
$2Br_3$		2P	⟶	$2PBr_3$		
Phosphorous Tribromide	+	Water	⟶	Hypophosphorous Acid	+	Hydrogen Bromide
$2PBr_3$	+	$2H_2O$	⟶	H_3PO_2	+	HBr

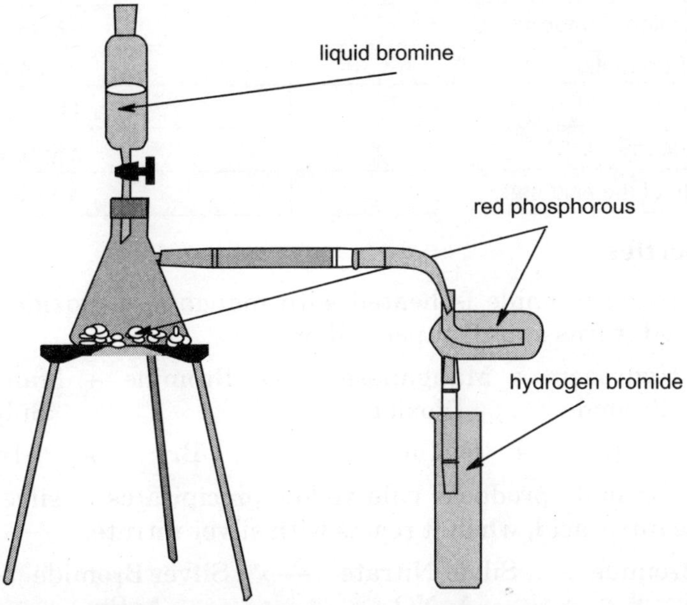

Apparatus	Reagents	Temperature	Pressutre
Conical flask, U tube, dropping funnel, delivery tube etc.	Bromine, Red Phosphorous	Normal	Normal

Fig. 12.16. The Lab Preparation of Hydrogen Bromide

Precaution: Bromine and red phosphorous mixture should be handled with a great care, heating should be done if the reaction does not take place, then only a mild heat should be applied.

Properties of Hydrogen Bromide

Physical Properties: Hydrogen bromide is a colourless gas, fumes in moist air, in weight it is heavier than air, has an irritating smell. The gas is not much soluble in water, produces hydrobromous acid, when dissolved in water. The detailed physical properties are listed below:

Molecular Weight	80.928
Boiling Point	–67°C
Freezing Point	–86°C
Solubility in water at 0°C	49.0g/litre
Boiling Point Contact Boiling Mixture	126

Percentage of Composition	47%
Dissociation Temperature	–8.4Kcal
Heat of Formation	13.84 Kcal
Stability	Not so high
Bond Energy	86 Kcal/mol
Strength of the acid state	Very Strong

Chemical Properties

(I) When hydrogen bromide is heated with manganese dioxide and sulphuric acid, bromine is produced (turns starch paper yellow).

Sulphuric Acid + Hydrogen Bromide + Manganese Dioxide ⟶ Bromine + Manganese Sulphate + Water

$$H_2SO_4 + 2HBr + MnO_2 \longrightarrow Br_2 + MnSO_4 + H_2O$$

(II) Hydrogen bromide produces pale yellow precipitates of silver bromide, which is insoluble in dilute nitric acid, when it reacts with silver nitrate.

Hydrogen Bromide + Silver Nitrate ⟶ Silver Bromide + Nitric Acid

$$2Hbr + AgNO_3 \longrightarrow AgBr + HNO_3$$

(III) Lead dibromate is produced when lead acetate reacts with hydrogen bromide.

Led Acetate + Hydrogen Bromide ⟶ Lead Dibromate + Acetic Acid

$$2PbCH_3(COO)_2 \quad 2HBr \quad Pb_2Br_2 \quad 2Ch_3(COOH)_2$$

(IV) When hydrogen bromide gas is bubbled through chlorine water bromine is produced

Chlorine Water Hydrogen Bromide ⟶ Bromine Hydrogen Chloride

$$Cl_2 \quad 2HBr \longrightarrow Br_2 \quad 2HCl$$

(V) Hydrogen Bromide liberates sulphur dioxide and also bromine is seperated when it reacts with sulphuric acid.

Hydrogen Bromide + Sulphuric Acid ⟶ Sulphur Dioxide + Bromine + Water

$$2HB + H_2SO_4 \longrightarrow SO_2 + Br_2 + 2H_2O$$

(VI) Hydrogen Bromide is a photoreactive substance, in the presence of sunlight and atmosphere oxygen it is decomposed to bromine and water.

Hydrogen Bromide + Oxygen ⟶ Bromine + Water

$$HBr + O_2 \longrightarrow 2Br_2 + 2H_2O$$

(VII) Hydrogen bromide liberates bromine when it reacts with manganese dioxide, lead peroxide, potassium dichromate and nitric acid.

Potassium Permanganate + Hydrogen Bromide ⟶ Potassium Bromide + Manganese Bromide + Water + Bromine

$$2KMnO_4 + 16HBr \longrightarrow 2KBr + 2MnBr_2 + 8H_2O + 5Br_2$$

Lead Peroxide + Hydrogen Bromide ⟶ Lead Dibromide + Water + Bromine

$PbO_2 + 4HBr \longrightarrow PbBr_2 + 2H_2O + Br_2$

Potassium Dichromate + Hydrogen Bromide ⟶ Potassium Bromide + Chromyl Bromide + Water + Bromine

$K_2Cr_2O_7 + 14HBr \longrightarrow 2KBr + 2CrBr_3 + 7H_2O + 3Br_2$

(VIII) With nitric acid it foffi1s aqua regia just like hydrogen chloride.

Industrial Preparation of Hydrogen Bromide

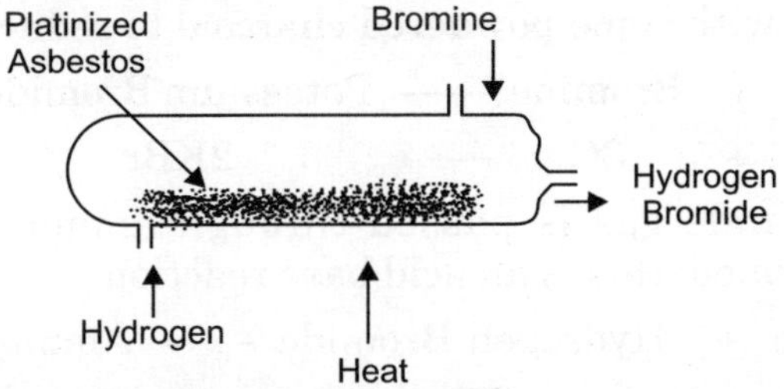

Fig. 12.17. The Industrial Preparation of Hydrogen Bromide

Inter-compounds of Bromine

We shall now discuss the inter-compounds that is the formation of compounds when bromine reacts with other elements and compounds.

1. Compounds of Bromine and Nitrogen

Monobromoamine (NH_2Br): When bromine is dropped in the mixture of eather and ammonia, monobromoamine is produced.

Ammonia + Bromine ⟶ Monobromoamine + Ammonium Bromide

$2NH_3 + Br_2 \longrightarrow NH_2Br + NH_4Br$

Ammonium bromide is then filtered out from the solution and free ammonia is absorbed by the calcium chloride and also the water is frozen, living behind the ethereal solution of monobromoamine. The solution is pale straw in colour, the compound is unstable and also not yet isolated. It reacts with Grignard's reagent to produce primary amines, nitrogen and ammonia.

Dibromoamine ($NHBr_2$): This is also prepared by the above process as in the preparation of monobromoamine, but with excess of bromine

Ammonia + Bromine ⟶ Dibromoamine + Ammonium Bromide

$3NH_3 + 2Br_2 \longrightarrow NHBr_2 + 2NH_4Br$

Dibromoamine is also unstable compound of bromine and nitrogen at elevated temperature it is decomposed to bromine and nitrogen.

Dibromoamine ⟶ Nitrogen + Hydrogen + Bromine

$2NHBr_2 \longrightarrow N_2 + H_2 + 2Br_2$

When bromine reacts with potassium hydroxide potassium bromide is produced.

Potassium Hydroxide + Bromine ⟶ Potassium Bromide + Water + Potassium Bromate

$$2KBrO_3 + 3Br_2 \longrightarrow 5KBr + 3H_2O + KBrO_3$$

Bromides are salts of hydrobromic acid (hydrogen bromide) and bromates are the compounds of hydrobromous acid. But practically these compounds are prepared by the direct reaction of bromine and other compounds.

2. Bromates and Bromides

The resulting solution contains both bromide and bromate, which is then evaporated to dryness and then ignited with some powdered charcoal to reduce the bromate.

Potassium Hydroxide + Bromine ⟶ Potassium Bromide + Carbon Dioxide

$$2KBrO_3 + 3C \longrightarrow 2KBr + 3CO_2$$

When hydrogen bromide gas is passed through ammonium hydroxide solution, ammonium bromide is formed; this is an acid base reaction

Ammonium Hydroxide + Hydrogen Bromide ⟶ Ammonium Bromide + Water

$$NH_4OH + HBr \longrightarrow NH_4Br + H_2O$$

Barium bromate is produced when bromine is added in slight excess to hot concentrated barium hydroxide.

Barrium Hydroxide + Bromine ⟶ Barium Bromide + Water

$$6Ba(OH)_2 + HBr \longrightarrow NH_4Br + H_2O$$

The bromide is soluble and remains in solution. If barium bromate is digested with sulphuric acid and the excess removed by barium hydroxide the filtered solution contains bromic acid.

3. *Bromine and Chlorine Compounds*

Bromine reacts with chlorine *forming* chlorides

Bromine Chloride (BrCl): When chlorine gas is bubbled through liquid bromine at a temperature below 5°C, bromine chloride is formed.

Bromine + Chlorine ⇌ Bromine Chloride

It is reddish yellow liquid at a temperature of 10°C it decomposes into chlorine and bromine.

Bromine Trifluoride (BrF₃): It is prepared by the reaction of bromine vapour and fluorine in the presence of nitrogen

Bromine + Fluorine ⟶ Bromine Trifluoride

$$Br_2 + Cl_2 \longrightarrow 2BrF_3$$

It is also prepared by the action of chlorine trifluoride on bromine at 10°C

Bromine + Chlorine Trifluoride ⟶ Bromine Trifluoride

$$Br_2 + ClF_3 \longrightarrow 2BrF_3$$

Bromine trifluoride is also produced when bromine monofluoride undergoes disproportionation.

Bromine Monofluoride ⟶ Bromine Trifluoride + Bromine

$$3BrF \longrightarrow BrF_3 + Br_2$$

Bromine trifluoride is a colourless liquid with a freezing point of 8.8°C and a boiling point of 127°C. Bromine trifluoride has a high electrical conductivity. This compound is also a good fluorinating agent. It converts several metals, their oxides and chlorides into fluorides e.g sodium.

Sodium + Bromine Trifluoride ⟶ Sodium Fluoride + Bromine

$$6Na + 2BrF_3 \longrightarrow 6NaF + Br_2$$

It dissolves in some of the metal fluorides to yield tetrafluorobromites as in potassium fluoride!

Potassium Fluoride + Bromine Trifluoride ⟶ Potassium Tetrafluorobromite

$$KF + BrF_3 \longrightarrow KBrF_4$$

Or, Hexafluorobromate as in antimony trifluoride.

Antimony Trifluoride + Bromine Trifluoride ⟶ Antimony Hexafluorobromate

$$SbF_3 + 4BrF_3 \longrightarrow SbBrF_6$$

Bromine Monofluoride (BrF): It is formed by the action of bromine trifluoride and bromine.

Bromine Trifluoride + Bromine ⟶ Bromine Monofluoride

$$4BrF_3 + Br_2 \longrightarrow 2BrF_5$$

It is reddish-brown liquid, with a boiling point of 23°C. It is unstable and decomposes readily into bromine and fluorine.

Bromine Pentafluoride (BrF_5): It is prepared by the action of bromine and excess of fluorine.

Bromine + Fluorine ⟶ Bromine Pentafluoride

$$Br_2 + 5F_2 \longrightarrow 2BrF_5$$

Bromine pentafluoride reacts violently and therefore is generally used by diluting with nitrogen whenever required it is also used for organic synthesis

> Bromine pentafluoride is generally diluted by bubbling nitrogen through

Acids of Bromine

Hypobromous Acid (HOBr): This acid like hypochlorous acid occurs only in solutions. An aqueous solution of hypobromous acid can be obtained by shaking bromine water with silver oxide.

Bromine + Water + Silver Oxide ⟶ Hypobromous Acid + Silver Bromide

$$2Br_2 + H_2O + Ag_2O \longrightarrow 2HOBr + 2AgBr$$

The solution is purified by fractional distillation; the bromine passes over first and rejected. Hypobromous acid is pale yellow in colour. They are unstable and cannot be concentrated without decomposition.

Hypobromous solution produces some salts called 'hypobromites' in them sodium hypobromite is discussed only, it is the most common hypobromites used.

Sodium Hypobromites (NaOBr): This compound is very much similar to sodium hypochlorite; it is prepared by the bromine to a cold solution of sodium hydroxide.

Sodium Hydroxide + Bromine ⟶ Sodium Hypobromite + Sodium Bromide + Water

$$2NaOH + Br_2 \longrightarrow NaOBr + NaBr + H_2O$$

As in the case of hypochlorite, rise of temperature causes the conversion of the hypobromite into bromate.

Sodium Hypobromite ⟶ Sodium Bromide + Sodium Bromate

$$2NaOBr \longrightarrow 2NaBr + NaBrO_3$$

The solutions of alkali hypobromites are stronger oxidizing agents than those of the hypochlorites.

A solution of sodium hypochlorite is used in organic synthesis to oxidize amines and amides to nitrogen. The estimation of urea by means of this reagent is a well-known example.

Urea + Sodium Hypobromite ⟶ Carbon Dioxide + Nitrogen + Water + Sodium Bromide

$$CO(NH_2)_2 + 2NaOBr \longrightarrow CO_2 + N_2 + H_2O + 2NaBr$$

Bromic Acid ($HBrO_3$): Like chloric acid this be obtained pure, as its solutions decompose heating.

An aqueous solution of bromic acid is prepared by adding a calculated amount of sulphuric acid to a solution of barium bromate.

Barium Bromate + Sulphuric Acid ⟶ Barrium Sulphate + Bromic Acid

$$Ba(BrO_3) + H_2SO_4 \longrightarrow BaSO_4 + 2HBrO_3$$

Solution of bromic acid possesses stronger oxidizing property than those of chloric acid or iodine any organic compounds are oxidized to carbon dioxide and water by this reagent.

IODINE

Introduction and Occurrence

Bernard Courtoris, a Frenchman, discovered iodine about the year 1811, who was engaged in the production of potassium nitrate from nitre-beeds for Napoleon's Army. He found his copper vessels were corroded by liquors due to the presence of impurity in the soda ash, which he has derived from the seaweed. By adding sulphuric acid to the liquors, this impurity could be isolated as black powder, which yielded a violet vapour when heated and formed a brilliant black crystal looking mass when condensed. He sent this material to J.L. Gay Lussac who recognized it as a new element. Iodine was named as such because of its vapour (Greek *Iodides* = Violet). Gay Lussac prepared many compounds and established his result in the year 1813. Later in August 1819, Dr Coindet a physician of Geneva discovered that tincture of iodine is remedy for goiter and affected some remarkable cures. The use of iodine as tincture was first used by Buchmann in 1828, and its first application in the treatment of battle during American Civil War.

Iodine is the one halogen that occurs in the nature in free state. It also occurs in the form of compounds example, potassium iodide, sodium iodide etc.

Extraction of Iodine

Iodine is obtained from the brines (sodium chloride solutions} used in oil wells. At first the brine solution is made oil free and also all the other impurities are removed. Then silver nitrate is added the iodine present in the brine solution gets deposited as silver iodide, which is filtered and treated with scrap iron to form metallic silver and a solution of ferrous iodide. This ferrous iodide is treated with chlorine to separate iodine.

Iodine is also extracted from Calche as sodium iodate and is allowed to accumulate in the mother liquor forming the crystals of sodium nitrate, this accumulation of crystals are allowed until the concentration about 6 grams per litre has been obtained. Part is then drawn off and treated with exact quantity of sodium bisulphate solution, which converts all the iodates to iodides. This mixture is now acidulated with sulphuric acid resulting the formation of sulphur dioxide, which is then treated with mother liquor to liberate iodine.

Sodium Iodide + Sodium Iodate + Sulphuric Acid ⟶ Iodine + Sodium Sulphate + Water

$$5NaI + NaIO_3 + 3H_2SO_4 \longrightarrow 3I_2 + 3Na_2SO_4 + 3H_2O$$

Laboratory Preparation of Iodine

Theory : In laboratory is prepared by heating potassium iodide with sulphuric acid in the presence of manganese dioxide. The vapourized iodine is collected in a cool reservoir, where iodine is collected as solidified crystals. (See Fig. 12.18 for experimental set up).

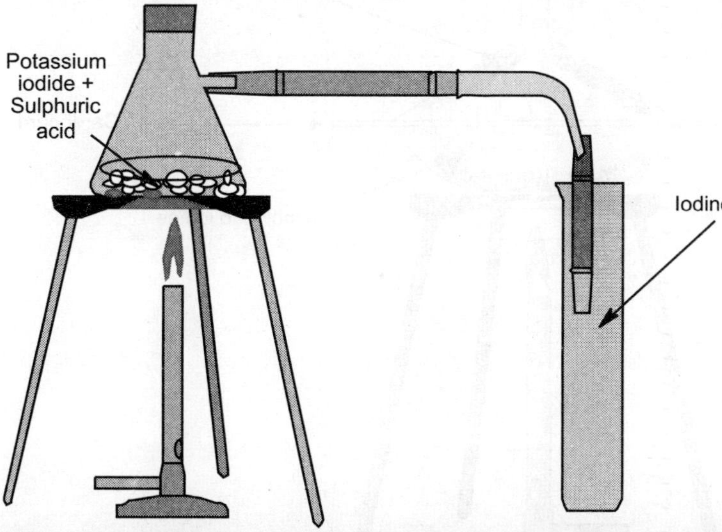

Apparatus	Reagent	Temperature	Pressure
Conical flask, delivery, tube, reservoir, burner	Potassium iodide, sulphuric acid	150-160°C	Normal

Fig. 12.18. The Lab Preparation of Iodine

Reaction Equation

Potassium Iodide	+ Sulphuric Acid	+ Manganese Dioxide	$\longrightarrow$	Iodine +	Sodium Bisulphate	+ Manganese + Sulphate	Water
2KI	+ $3H_2SO_4$ +	MnO2	$\longrightarrow$	I_2 +	$2KHSO_4$	+ $MnSO_4$ +	$2H_2O$

Precaution: *Iodine produces very dense fumes can cause an uneasy atmosphere therefore there should be no leakage and no negligence in handling.*

Properties of Iodine

Physical Properties : (1) Iodine is a steel Grey solid which sometimes on heating gives a beautiful violet coloured vapours. The rhombic crystalline plates have a metallic lusture. (2) The vapours have an irritating smell resembling chlorine to some extent. (3) It is slightly soluble in water, but much more soluble in potassium or sodium to some extent.

Sublimation of Iodine

Experiment 1: This method is applied in the purification of iodine. The word sublimation means *to transform one solid by evaporation and then obtaining the same solid state after*

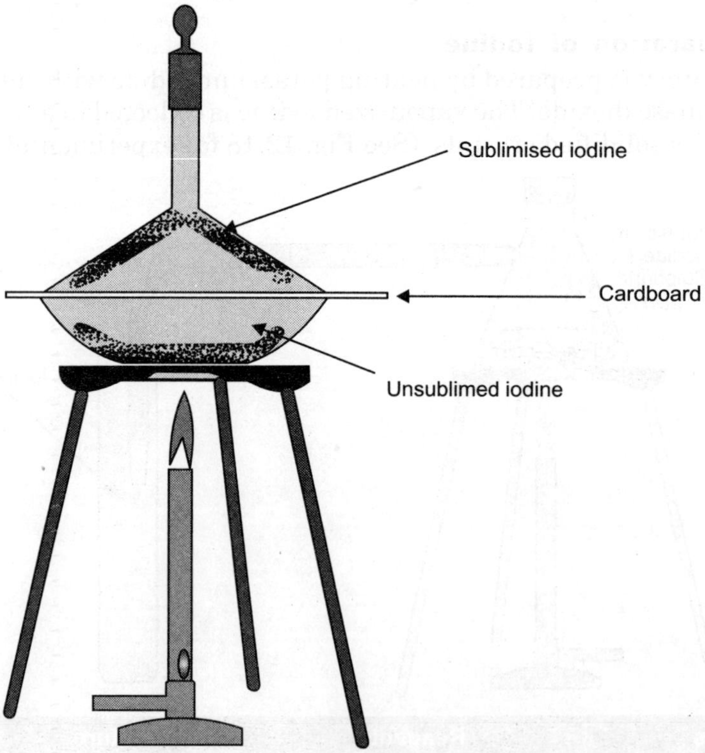

Fig. 12.19. The Sublimation of Iodine

condensation without passing through the liquid state. The process is done by heating iodine crystals in an evaporating dish covered by perforated cardboard and the top of the board a funnel is placed in the inverted position. When the iodine is heated it vapourizes and passes through the cardboard; the vapours of impurities being heavier cannot pass. The iodine crystal condenses at the cold sides (See Fig. 12.19).

Iodine also gets dissolved in carbon disulphide and chloroform giving violet solutions, and with other solvents like eather and alcohol it forms brown colours.

Experiment 2 : Take some water in a test tube and add carbon disulphide, eather now add iodin it will be seen that iodine has dissolved in three layers of solvents forming three distinct colours.

Physical Properties : A list of physical properties are shown in the table below :

Atomic Number	53
Atomic Weight	126.905
Atomic Radius	133pm (pico metre)
Ionic Radius	220pm
Density at 20°C	4.43g/cc
Melting Point	113.6°C
Boiling Point	184°C
Specific Heat	
At Solid State	0.316J/g^{-1} (at 114°C-118°C)
At Liquid State	0.219 J/g^{-1} (at 20°C)
Heat of Vapourization at 184°C	164.4J/g^{-1}
Di-electric Constant at 118°C	11.8
Solubility in 100 g of water at 20°C	0.0290 g

Chemical Properties : Iodine resembles chlorine and bromine but is less reactive than them. Some of the chemical properties are stated below:

1. *Affinity for Hydrogen :* Iodine has less affinity for hydrogen than other halogens. It combines with hydrogen in the presence of a catalyst such as platinum forming hydrogen iodide.

$$\text{Hydrogen} + \text{Iodine} \longrightarrow \text{Hydrogen Iodide}$$
$$H_2 + I_2 \longrightarrow 2HI$$

2. *Oxidizing and Disinfecting Action of Iodine :* Like other halogens iodine is an oxidizing agent, though much less powerful for example.

(*a*) It oxidizes hydrogen sulphide to sulphur

$$\text{Hydrogen sulphide} + \text{Iodine} \longrightarrow \text{Sulphur} + \text{Hydrogen Iodide}$$
$$H_2S + I_2 \longrightarrow S + 2HI$$

(*b*) Sulphur Dioxide is also oxidized to sulphuric acid.

$$\text{Sulphur Dioxide} + \text{Iodine} + \text{Water} \longrightarrow \text{Sulphuric Acid} + \text{Hydrogen Iodide}$$
$$SO_2 + I_2 + 2H_2O \longrightarrow H_2SO_4 + 2HI$$

(*c*) It oxidizes sodium sulphite to sodium sulphate and also sodium arsenite to sodium arsenate

Sodium Sulphite	+	Iodine	+	Water	$\longrightarrow$	Sodium Sulphate	+	Hydrogen Iodide
Na_2So_3	+	I_2	+	H_2O	$\longrightarrow$	Na_2SO_4	+	2HI
Sodium Arsenite	+	Iodine	+	Water	$\longrightarrow$	Sodium Arsenate	+	Hydrogen Iodide
Na_3AsO_3	+	I_2	+	H_2O	$\longrightarrow$	Na_3AsO_4	+	2HI

(*d*) With sodium thiosulphate iodine changes its brown colour to colourless liquid due to the formation of sodium tetrathionate

Sodium Thiosulphate	+	Iodine	$\longrightarrow$	Sodium Tetrationate	+	Sodium Iodide
$2Na_2S_2O_3$	+	I_2	$\longrightarrow$	$Na_2S_4O_6$	+	2NaI

3. Action with Alkalies : When a cold dilute solution of sodium hydroxide is added to iodine, hypoiodous acid and sodium iodide.

Sodium Hydroxide	+	Iodine	$\longrightarrow$	Sodium Iodide	+	Hypoiodous Acid
NaOH	+	I_2	$\longrightarrow$	NaI	+	HIO

Iodine reacts with liquid ammonia producing an explosive called nitrogen triodide.

Ammonia	+	Iodine	$\longrightarrow$	Nitrogen Tri-iodide	+	Hydrogen Iodide
$2NH_3$	+	$3I_2$	$\longrightarrow$	$NI_3.NH_3$	+	3HI

When this nitrogen tri iodide is gently pressurized it decomposes to nitrogen, iodine and ammonium iodide with mild explosion.

Nitrogen Tri-iodide	$\longrightarrow$	Nitrogen	+	Iodine	+	Ammonium Iodide
$NI_3.NH_3$	$\longrightarrow$	$5N_2$	+	$9I_2$	+	$6NH_4I$

Thermal Dissociation : Dissociation of iodide at 700°C and is complete at 1700°C

$$\text{Iodine} \xrightleftharpoons{700-1700^\circ} \text{Iodine}$$

$$I_2 \rightleftharpoons 2I$$

Compounds of Iodine

Iodine reacts with other elements and also with other compounds to form iodine compounds.

Iodine Dioxide (IO_2) : This compound of iodine is obtained by heating iodic acid with concentrated sulphuric acid till fumes of iodine are evolved, cooling and decomposing the resulting basic sulphate of iodine dioxide is produced.

Iodic Acid	+	Sulphuric Acid	$\longrightarrow$	Ioodyl Polysoulphte	+	Water
$2HIO_3$	+	H_2SO_4	$\longrightarrow$	$(IO_2)SO_4$	+	$2H_2O$
Iodyl Polysulphate	+	Water	$\longrightarrow$	Iodine Dioxide	+	Sulphuric Acid
$2(IO)_2SO_4$	+	$2H_2O$	$\longrightarrow$	$2IO_2$	+	H_2SO_4

The oxide is a yellow granular powder, which is only slightly soluble in cold water, when heated with water it dissolves producing iodic acid.

$$\text{Iodine Dioxide} + \text{Water} \longrightarrow \text{Iodic Acid} + \text{Iodine}$$
$$10IO_2 + 4H_2O \longrightarrow 8HIO_3 + I_2$$

For this reason it is considered as the basic iodate or iodyl iodate. $(IO)IO_3$.

When heated to 132°C iodine dioxide decomposes to iodine pentoxide and iodine.

$$\text{Iodine Dioxide} \longrightarrow \text{Iodine Pentoxide} + \text{Iodine}$$
$$10IO_2 \longrightarrow 4I_2O_5 + I_2$$

Iodine Oxide (I_4O_9): This oxide of iodine is obtained as yellow solid by the action of ozonized oxygen and iodine of tripositive iodine $I(IO)_3$

$$\text{Iodine} + \text{Ozone} \longrightarrow \text{Iodine Oxide}$$
$$I_2 + 3O_2 \longrightarrow I_4O_9$$

Iodine Pentoxide, (Iodyloxy) Iodane Dioxide (I_2O_5): It is the best known oxide of iodine, also regarded as the *anhydride of iodic acid.* It is prepared by heating iodic acid to 245°C. It is a white deliquescent solid which then heated readily decomposes to iodine and oxygen. It is insoluble in organic solvents but dissolves in water yielding iodic acid. So it is called iodic anhydride.

$$\text{Iodine Pentoxide} + \text{Water} \longrightarrow \text{Iodic Acid}$$
$$I_2O_5 + H_2O \longrightarrow 2HIO_3$$

Like iodic acid this pentoxide is also possesses oxidizing properties. It is used in the estimation of carbon monoxide in a gaseous mixture; carbon monoxide is converted to carbon dioxide, which is then absorbed in limewater.

$$\text{Carbon Monoxide} + \text{Iodine Pentoxide} \longrightarrow \text{Carbon Dioxide} + \text{Iodine}$$
$$5CO + I_2O_5 \longrightarrow 5CO_2 + I_2$$

Iodine Peroxide (IO_4) : Iodine peroxide is said to be produced by reaction of sodium perchlorate on iodine.

$$\text{Sodium Perchlorate} + \text{Iodine} \longrightarrow \text{Sodium Iodide} + \text{Chlorine Peroxide}$$
$$2NaCIO_4 + I_2 \longrightarrow 2IO_4 + 2CIO_4$$
$$\text{Chlorine Peroxide} + \text{Iodine} \longrightarrow \text{Iodine Peroxide} + \text{Chlorine}$$
$$2CIO_4 + I_2 \longrightarrow 2IO_4 + CI_2$$

It decomposes rapidly into elements, and its existence is rather doubtful.

Iodine Monochloride (ICl): It is formed by passing chlorine over solid iodine at temperature below 0°C.

$$\text{Iodine} + \text{Chlorine} \longrightarrow \text{Iodine Monochloride}$$
$$I_2 + Cl_2 \longrightarrow 2ICl$$

It is a dark red liquid but solidifies on standing. The solid is capable of existing in two forms:

(*a*) *A stable form is obtained* is red needle like crystals on cooling the liquid rapidly. It melts at 26.9°C.

(*b*) *A semi-stable form is obtained* as black solid by cooling the liquid slowly at –10°C. It melts at 14°C.

It is hydrolyzed with disproportionation as shown in the equation below:

Iodine Monochloride + Water ⟶ Iodine + Hydrogen Chloride + Iodic Acid

$$2ICl + 2H_2O \longrightarrow 2I_2 + 5HCl + HIO_3$$

Iodine monochloride is used for the estimation of the iodine number of oils and fats, which is actually the estimation of unsaturation of oil and fats in the form of Wij's Solution.

Iodine Bromide (IBr): It is formed by the direct combinations of the two elements. It is a crystalline solid, resembling iodine in colour and shape. Its vapour dissociates to a small extent on heating.

Iodine + Bromine ⟶ Iodine Bromide

$$I_2 + Br_2 \longrightarrow 2IBr$$

Both iodine monochloride and iodine monobromide are structurally electronic conductors, when in molted state. This proves that they are electrovalent or ionic compounds.

Iodine Trichloride (ICl_3): It is formed by the action of excess of chlorine on iodine or iodine monochloride at 100°C.

Iodine + Chlorine ⟶ Iodine Trichloride

$$I_2 + 3Cl_2 \longrightarrow 2ICl_3$$

It is also prepared when iodine pentoxide is heated with hydrochloric acid.

Iodine Pentoxide + Hydrochloric Acid ⟶ Iodine Trichloride + Water + Chlorine

$$I_2O_5 + 10HCl \longrightarrow 2ICl_3 + 5H_2O + 2Cl_2$$

It is a yellow solid, which fumes readily. It dissociates on heating at 68°C giving chlorine and iodine monochloride. It is also prepared when iodine pentoxide is heated with hydrochloric acid. It is a yellow solid, which fumes readily. It dissociates on heating at 68°C giving chlorine and iodine monochloride.

Iodine Trichloride ⟶ Iodine Monochloride + Chlorine

$$2ICl_3 \longrightarrow ICl + Cl_2$$

It is used in medicines.

Iodine Pentafluoride (IF_5): It is formed by the direct reaction of the elements. Iodine and fluorine, using obviously excess of fluorine.

Iodine + Fluorine ⟶ Iodine Pentafluoride

$$I_2 + 5F_2 \longrightarrow 2IF_5$$

It is also prepared by the reaction of heating iodine pentoxide with fluorine.

Iodine Peroxide + Fluorine ⟶ Iodine Pentafluoride

$$2I_2O_5 + 10F_2 \longrightarrow 2IF_5$$

It is also formed by heating iodine with silver fluoride.

Iodine + Silver Fluoride ⟶ Iodine Pentafluoride + Silver

$$I_2 + 10AgF \longrightarrow 2IF_5 + 10Ag$$

Iodine pentafluoride is a colourless liquid with a boiling point of 98°C. It is also a good conductor of electricity. It also reacts with iodine pentoxide producing iodine oxyfluroide.

Iodine Pentoxide + Iodine Pentafluoride ⟶ Iodine Oxyfluoride

$$I_2O_5 + 3IF_5 \longrightarrow IOF_3$$

Iodine Oxyfluoride is a white solid. It decomposes when heated at 110°C giving black iodine pentafluoride.

Iodine Oxyfluoride ⟶ Iodine Pentafluoride + Iodine Dioxyfluoride

$$2IOF_3 \longrightarrow IF_5 + IO_2F$$

Iodine Heptafluoride (IF_7): It is prepared by the action of iodine pentafluoride and excess of fluorine at a temperature of 300°C.

Iodine Pentafluoride + Fluorine ⟶ Iodine Hexafluoride

$$IF_5 + F2 \longrightarrow IF_7$$

It is a gas at ordinary temperature. It resembles chlorine trifluoride, bromine pentafluoride and also being highly reactive.

Iodine Perchlorate [$I(ClO_4)_3$]: This is obtained by the action of anhydride perchloric acid on ozonized iodine.

Iodine + Perchloric Acid + Ozone ⟶ Iodine Perchlorate + Water

$$I_2 \quad 6HClO_4 \quad O_3 \quad 2I(ClO_4)_3 + 3H_2O$$

Acids of Iodine

Hypoiodous Acid (HOI): This is prepared by dissolving iodine in cold dilute sodium hydroxide.

Iodine + Sodium Hydroxide ⟶ Sodium Iodide + Hypoiodous Acid

$$I_2 + NaOH \longrightarrow NaI + HOI$$

The freshly prepared solution has oxidizing and bleaching properties. It bleaches indigo solution; hydrogen peroxide produces oxygen when reacts with hypoiodous acid, manganous sulphate is converted to manganic sulphate with alcohol it produces iodoform (CHI_3), and even carbonic acid. (Soda water) liberates iodine from the solution. Hypoiodous acid solution, when continuously heated it looses all the properties mentioned above and consists of only iodate and iodide. It is also produced if finely powdered iodine is shaken with mercuric oxide and water. It is an extremely weak acid.

Iodic Acid (HIO_3) : This is the most important acid of iodine. It is prepared by the reaction iodine pentoxide and water.

Iodine	+	Water	+	Chlorine	⟶	Iodic Acid	+	Hydrogen Chloride
I_2	+	$6H_2O$	+	$5Cl_2$	⟶	$2HIO_3$	+	$10HCl$

Iodic acid is also prepared by passing chlorine in the suspension of iodine and water mixture.

Here, hydrochloric acid is removed by adding silver nitrate when silver chloride is precipitated.

Iodic Acid may also be prepared by the action of iodine on chloric acid.

Chloric Acid	+	Iodine	⟶	Iodic Acid	+	Chlorine
$2HClO_3$	+	I_2	⟶	$2HIO_3$	+	Cl_2

Iodic acid is very much soluble in water. The solution at first becomes reddish in colour and then becomes deep red. It changes litmus paper. Iodic acid is insoluble in alcohol. When this is heated with charcoal, sulphur, phosphorous or any organic matter, the acid reacts like an oxidizing agent.

Iodic Acid +	Sulphur	+	Water	⟶	Iodine	+	Sulphuric Acid
$2HIO_3$ +	5S	+	$4H_2O$	⟶	I_2	+	$5H_2SO_4$

Iodic Acid +	Hydrogen Sulphide	⟶	Iodine +	Water	+	Sulphur
$2HIO_3$ +	$5H_2S$	⟶	+		+	

Hydrogen Iodide	+	Iodic Acid	⟶	Iodine	+	Water
5HI	+	HIO_3	⟶	$3I_2$	+	$3H_2O$

Iodic acid melts at 110°C to form a solution and a solid hydrate of I_2O_5, the $3I_2O_5.H_2O$ stable to 196°C but then fusing to form Iodine pentoxide.

Iodates and iodides : These are theoretically the compounds of iodic acid but they are prepared by the reaction of iodine and metallic hydroxides.

(*a*) When hot concentrated sodium hydroxide is made to react with iodine sodium iodide and sodium iodate is produced.

Iodine	+ Sodium Hydroxide	⟶	Sodium Iodate	+	Sodium Iodide	+	Water
$3I_2$ +	6NaOH	⟶	$NaIO_3$	+	5NaI	+	$3H_2O$

(*b*) Potassium Iodate is also produced when potassium chlorate reacts with iodine

Potassium Chlorate	+	Iodine	⟶	Potassium Iodate	+	Chlorine
$2KClO_3$	+	I_2	⟶	$2KIO_3$	+	Cl_2

(*c*) When potassium iodate reacts with barium chloride barium iodate is produced.

Potassium Iodate	+ Barium Chloride	⟶	Barium Iodate	+ Potassium Chloride
$2KIO_3$ +	$BaCl_2$	⟶	$Ba(IO_3)_2$ +	2KCl

Iodic acid is reproduced when barium iodate reacts with sulphuric acid.

Barium Iodate + Sulphuric Acid ⟶ Barium Sulphate + Iodic Acid

$$Ba(IO_3)_2 + H_2SO4 \longrightarrow BaSO_4 + 2HIO_3$$

Iodic acid, although monobasic, forms normal salts and two acid salts:

Normal Potassium Iodate Acid	KIO_3
Potassium Iodate	KIO_3HIO_3 or $KH(IO_3)_2$
Diacid Potassium Iodate	$KIO_3.2HIO_3$ or $KH_2(IO_3)_2$

The normal iodates are slightly soluble in water on heating they decomposes (A) into Iodide and oxygen.

Potassium Iodate ⟶ Potassium Iodide + Oxygen

$$2KIO_3 \longrightarrow 2KI + 3O_2$$

(B) Into Oxide + Iodine + Oxygen e.g., Calcium Iodate

Calcium Iodate ⟶ Iodine + Oxygen + Calcium Iodide

$$2Ca(IO_3)_2 \longrightarrow 2CaO + 2I_2 + 5O_2$$

Iodates are detected by the blue colour, due to the liberation of iodine when sulphurous acid and starch paste is added to the iodate solution. The formula of the iodic acid and the iodine pentoxide are shown in the Fig. below:

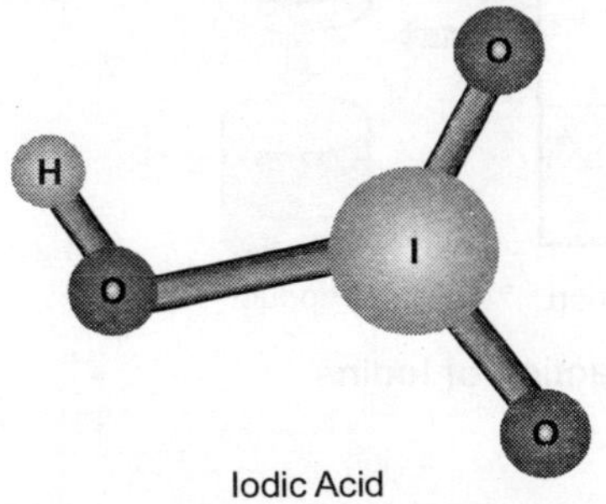

Iodic Acid

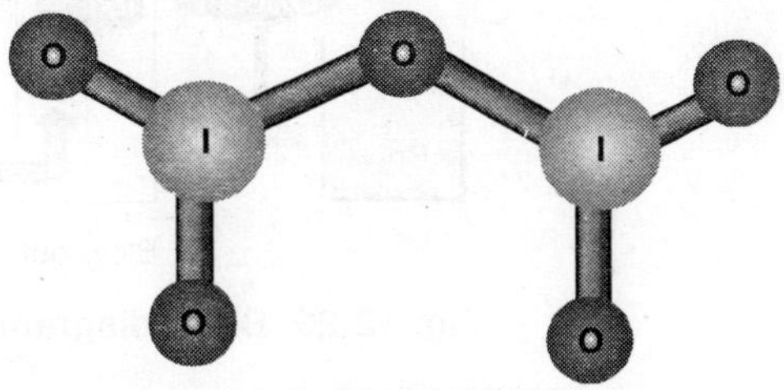

Iodine Pentoxide

Periodic Acid (HIO_4): This was discovered in the year 1833 by Ammermüller and Magnus. Two types of periodic acids are known such as metaperiodic acid (HIO_4) and paraperiodic acid ($HIO_4.2H_2O$). The perchlorates are obtained by the oxidation of iodates, thus when chlorine is passed into a boiling solution of sodium hydroxide containing iodine, disodium paraperiodate is precipitated as a crystalline product. When sodium mesoperiodate is treated with silver chloride silver periodate is produced. When chlorine is passed through silver mesoperiodate. It produces periodic acid.

Silver Mesoperiodate + Chlorine + Water ⟶ Silver Chloride + Paraperiodic Acid + Oxygen

$$4Ag_5IO_5 + 10Cl_2 + 10H_2O \longrightarrow 20AgCl + 4H_5IO_6 + 3O_2$$

The crystals of paraperiodic acid melts with decomposition on heating.

Paraperiodic Acid ⟶ Iodic Acid + Water + Oxygen

$$4H_5IO_6 \longrightarrow 2HIO_3 + 4H_2O + O_2$$

But below 100°C in a vacuum they loose water forming dimesoperiodic acid ($H_4I_2O_9$)

Paraperiodic Acid	$\longrightarrow$	Dimesoperiodic Acid	+	Water		
$2H_5IO_6$	$\longrightarrow$	$H_4I_2O_9$	+	$3H_2O$		
Dimesoperiodic Acid	$\longrightarrow$	Periodic Acid	+	Hydrogen Iodide	+	Oxygen
$H_4I_2O_9$	$\longrightarrow$	$2HIO_4$	+	2HI	+	$5O_2$

Industrial Preparation of Iodine

Iodine is manufactured in industries form seaweeds. Seaweeds are collected then dried and burnt in shallow pits. The resulting ash contains about 1% of iodine in the form of iodides. It is mixed with hot water; the insoluble matter is allowed to settle and the clear liquid is evaporated and the crystalline salt is taken out. The mother liquor containing the iodides is then distilled with manganese dioxide and sulphuric acid from cast iron retorts (see Fig. 12.20). The iodine set free is then collected in stone were receivers called *Aludels* (Fig. 1220A).

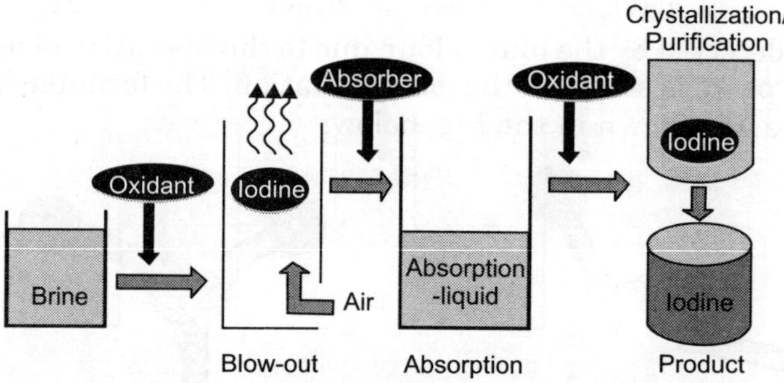

Fig. 12.20. Block diagram of the Extraction of Iodine

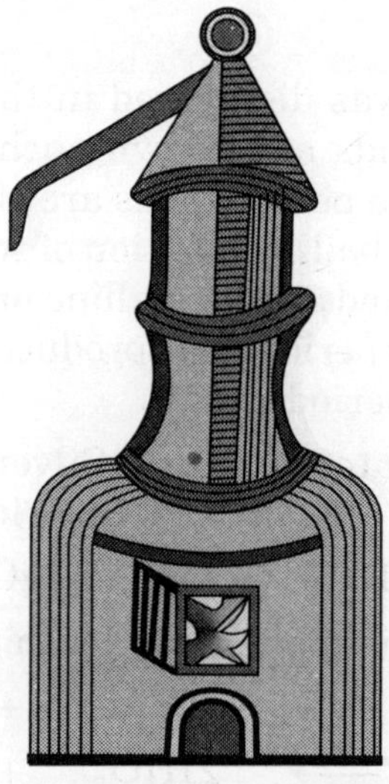

Fig. 12.20A. The Aludel

Hydrogen Iodide, Hydroiodic Acid, Iodized Hydrogen (HI)

Hydrogen iodide is like hydrogen bromide, cannot be prepared by the action of sulphuric acid any iodide, because of its reducing power. The usual methods for the preparation of hydrogen iodide are as follows :

(1) Gaseous hydrogen iodide is best prepared by dropping water on a mixture of red phosphorous and iodine. At first phosphorous and iodine reacts to form phosphorous triodide which reacts with water producing hydrogen iodide.

$$\text{Phosphorous} + \text{Iodine} + \text{Water} \longrightarrow \text{Hydrogen Iodide} + \text{Phosphorus}$$

$$2P + 3I_2 + 6H_2O \longrightarrow 6HI + 2H_3PO_4$$

The gas is fed from the iodine vapour by passing it through a 'fU' tube containing glass beads smeared with red phosphorous and water. Hydrogen iodide can be dried by passing it through moist red phosphorous or phosphorous pentoxide or calcium iodide. Calcium chloride cannot be used because it produces hydrogen chloride while absorbing moisture from hydrogen iodide.

(2) Another method, though less convenient, is done by direct contact of the elements hydrogen and iodine in the presence of platinum asbestos. This process is regarded as the laboratory and the industrial process of hydrogen iodide (see Fig. 12.21).

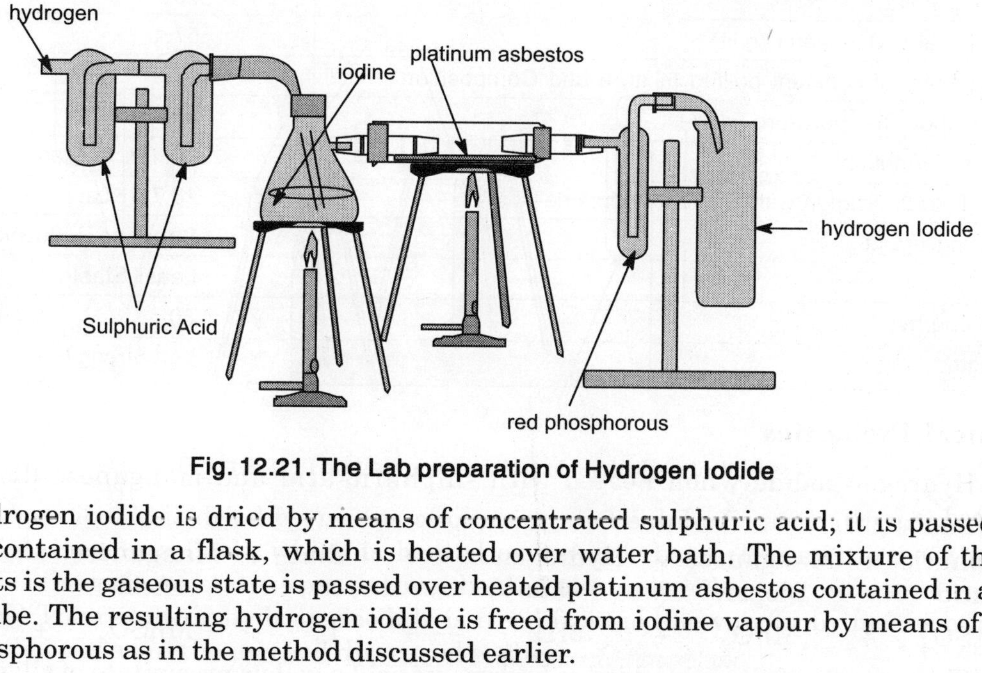

Fig. 12.21. The Lab preparation of Hydrogen Iodide

Hydrogen iodide is dried by means of concentrated sulphuric acid; it is passed over iodine contained in a flask, which is heated over water bath. The mixture of the two elements is the gaseous state is passed over heated platinum asbestos contained in a hard glass tube. The resulting hydrogen iodide is freed from iodine vapour by means of moist red phosphorous as in the method discussed earlier.

$$\text{Hydrogen} + \text{Iodine} \xrightarrow{\text{Platinum}} \text{Hydrogen Iodide}$$

$$H_2 + I_2 \xrightarrow{\text{Pt}} 2HI$$

(3) Aqueous solution of hydrogen iodide is prepared by dissolving the gas in water. It is however best made by the action of hydrogen sulphide upon iodine suspended in water. When the all the iodine has disappeared the sulphur is filtered off and the filtrate is purified by distillation.

Properties of Iodine

Physical Properties

(*a*) Hydrogen iodide is a colourless of pungent smelling gas which fumes strongly in moist air.

(*b*) It is extremely soluble in water; the solution is saturated at 10°C contains of 70% of hydrogen iodide by weight.

(*c*) Hydrochloric acid also forms a constant boiling mixture at 127°C and consists of 57% of hydrogen iodide by weight. The detail physical properties are stated below :

Property	Value
Physical State at 15°C	Gas
Molecular weight	127.905
Boiling Point	–35.5°C
Freezing Point	–50.8°C
Solubility at 0°C in water (g/lit)	57.1
Boiling Point of constant boiling mixture and Composition	127
Dissociation Temperature	180
Heat of Formation	+6.4 Kcal (gas)
Heat of Neutralization with Sodium hydroxide	13.78 Kcal
Reducing agent	Stronger Reducing agent
Stability	Least Stable
Bond Energy	70
Strength	Vert Strength

Chemical Properties

1. Hydrogen iodide when heated with sulphuric acid and manganese dioxide, is produced.

Sulphuric Acid + Manganese Dioxide + Hydrogen Iodide ⟶ Iodine + Manganese Acid + Water

$$H_2SO_4 + MnO_2 + 2HI \longrightarrow I_2 + MnSO_4 + 2H_2O$$

2. When nitrate of silver is added to hydroiodic acid a yellow precipitate of silver iodide is noticed, which is insoluble in nitric acid.

Hydroiodic Acid + Silver Nitrate ⟶ Silver Iodide + Hydrogen Chloride

$$AgNO_3 + AgNO_3 \longrightarrow AgI + HCl$$

3. Hydrogen iodide produces lead iodide when reacting with lead acetate.

Lead Acetate	+	Hydrogen Iodide	⟶	Lead Iodide	+	Acetic Acid
$Pb(CH_3COO)_2$	+	HI	⟶	PbI_2	+	CH_3COOH

4. With chlorine water hydrogen iodide liberates iodine, which produces violet colour with carbon tetrachloride (CCl_4)

Hydrogen Iodide	+	Chlorine	⟶	Hydrogen Chloride	+	Iodine
HI	+	Cl_2	⟶	HCl	+	I_2

5. When heated hydrogen iodide undergoes decomposition at 180°C

Hydrogen Iodide	⟶	Hydrogen	+	Iodine
2HI	⟶	H_2	+	I_2

6 (a) In aqueous solution hydroiodic acid behaves like other acids. It reacts with hydroxide of elements to form salt and water.

Hydrogen Iodide	+	Sodium Hydroxide	⟶	Sodium Iodide	+	Water
2HI	+	NaOH	⟶	NaI	+	H_2O

(b) It also reacts with different metals like zinc, iron producing iodide and hydrogen.

Zinc	+	Hydrogen Iodide	⟶	Zinc Iodide	+	Hydrogen
2Zn	+	2HI	⟶	2ZnI	+	H_2O

(c) It also decomposes carbonate compounds into water and carbon dioxide and water.

Sodium Carbonate	+	Hydrogen Iodide	⟶	Sodium Iodide	+	Carbon Dioxide	+	Water
Na_2CO_3	+	2HI	⟶	2NaI	+	CO_2	+	H_2O

(d) Mercuric chloride is converted to mercuric iodide when it reacts with hydrogen iodide.

Hydrogen Iodide	+	Mercuric Chloride	⟶	Mercuric Iodide	+	Chlorine
HI	+	$HgCl_2$	⟶	HgI	+	Cl_2

Ionic Nature of Iodine and Hydrogen Iodide

1. If potassium iodate is dissolved in a solution of iodine and sulphuric acid, a red-brown precipitation occurs due to the formation of I^{3+} ions.

$$HIO_3 + 7I_2 + 8H_2SO_4 \rightleftharpoons 5I_3 + 3H_3O^+ + 8HSO_4^-$$

2. If potassium iodate is added to iodine in concentrated sulphuric acid consisting of 30% sulphur trioxide by weight a deep blue coloured substance is obtained due to the formation of the ions.

$$HIO_3 + 7I_3^+ + 8H_2SO_4 \rightarrow 7I^+ + 3H_2O + 8HSO_4^-$$

CONFERMATORY OF IODINE BROMINE CHLORINE

Chlorine (Uses)

(i) *As a Bleaching Agent* : About 65% of the total output of chlorine is used for the bleaching of paper and about 20% in the bleaching of cloth. For this purpose chlorine is chiefly used in the form of bleaching powder, but the direct use of liquid chlorine is now greatly supported.

(ii) *As a Disinfectant* : The next important use of chlorine is for disinfecting water, al 10% of the commercial chlorine being used for this purpose. One of the most effective antiseptics used for dressing wounds during World War II known as Carret -Daking solution, which was obtained by passing chlorine into a solution of washing soda.

Other Uses : Chlorine is also used for certain dyestuffs, explosive hypochlorites, chlorates and sulphur monochloride and also muriatic acid.

Confirmatory Test for Chlorine

1. Chlorine turns Starch Iodide Paper Blue.
2. It also precipitates silver chloride when it is treated with silver nitrate.

Bromine (Uses)

The chief uses of bromine at the present times is in the manufacture of ethylene dibromide (a substance used as an anti-knock in the automobile engines). The next important use is in the manufacture of potassium bromide and other bromides. These are used in medicines and also in the preparation of silver bromide used in photographic plates. Large quantities of bromine are also used in the manufacture of organic dyes. A minor use of bromine is an oxidizing agent in chemical analysis.

THE USE AND APPLICATION OF CHLORINE, CHLORINE DIOXIDE, BROMINE AND IODINE FOR DISINFECTION AND POLLUTION TERMINATION

Chlorine was first used in municipal water supply in 1908 in America. Chlorine hydrolyses in water to yield hypochlorous acid. The hydrolysis constant, calculated by Kudesiain 1968, is (3.94-0.021) 10-4 at 25°C and (5.104-0.021) 10-4 at 35°C. The pH value of HOCl is 7.2. The kinetics of hydrolysis of chlorine by temperature jump relaxation technique was found by Eigen and Kustin to be 11.00 sec. The pioneer work in oxidation of organic compounds with chlorine was done by Kudesia and Mukherji. The disinfecting species react with enzymes, which are essential to the metabolic process of living cells. In due course of time, these key substances are inactivated and the cells die. The enzymes are produced in cell plasma and chemical disinfection proceeds first by penetration of the cell walls of bacteria and then reacting with cells enzymes. The 'destruction of living organism is brought by direct chemical combination of chlorine with the protoplasm of the micro-organism. Researcher has established that in certain oxidations of alcohols, aldehydes, ethers and acids with chlorine, the molecular chlorine is 50-200 times more effective than HOCl. From several other experiments it is concluded that, in chlorine water, molecular chlorine is the Actinomycetes and are related to both bacteria and fungi. They give muddy musty odours in water. There activity increases up to 40°C and dies at 45°C. Heavy doses of chlorine for longer time are required to check them. Many types of worms are found in drinking water supplies and water sources. Some of them are red bloodworms, larvae of chironomus flies. Some are grey in colour. The egg hatching process into larvae happens on the walls of water tanks. Remedial measures of cleaning and draining of basins through super chlorination should be adopted once in a month and with special concentration at the time of seasonal changes, mainly onset of rains and at the time of floods and epidemics. Algae are plants of varying shapes and sizes living in water. They contain various pigments like chlorophyll, carotene, and xanthophyll and produce cysts, zygotes and, zoospores. The algae are of various types-green algae, blue green algae, diatoms algae, yellow green algae, brown algae, red algae etc. They get their food from minerals in water, and from nitrogenous compounds and carbon dioxide. They produce undesirable taste and odours. They act as barriers to penetration of oxygen into the water causing suffocation for fish. High concentration of cyanophyceae in can cause poisoning of cattle drinking water. Algae cause no human disease. Dicoplasma are pleuropneumonia like organisms. They do not possess cell walls. They are parasitic and resistant to penicillin. Increasing temperature increases the bactericidal activity of chlorine. The activity is highest when combined with bromine iodine. Small amount of ammonia in water is accelerated by bromide and the rate of bacterial sterilization is increased by chlorine. If too much of organic matter is present in water it will consume high amount of oxidizing agent. Molecular chlorine is an effective oxidizing agent below pH 7 only i.e., slightly acidic conditions. Above pH 7, more time is necessary for killing the germs. Phenols in textile wastewaters consume more chlorine than amines and amino acids. Sulphates in sewerage will need greater doses of chlorine. Sodium nitrite face filitates the destruction of organic matter with low chlorine does. The water contains no interfering substances. All cells of a single species of organism are discreet units equally susceptible to a single species of disinfectant. The rate of disinfection depends on the time of contact, concentration of disinfectant, concentration of organism and temperature of principal effective oxidizing agent. For disinfection 0.5 mg/litre is required to obtain combined available residual chlorine in plain chlorination of cities' surface waters. When chlorine is added to water before any other treatment is given to it, it improves coagulation. It also reduces taste and odours caused by sludge in the sedimentation tank. Sometimes, heavy dose of chlorine is given to water when quick disinfection action is needed because it is necessary that sufficient chlorine should be added to destroy ammonia. A residual can be maintained in the distribution system once its chlorine demand is satisfied and water is not exposed to sunlight. Raw sewerage needs 7 to 23 mg per litre of chlorine, settled sewerage needs 4 to 19 mg per litre, and activated sludge needs 4 to 10 mg per litre of chlorine. When excess of ammonia is present in the water then chloramines are formed in large amount by addition of chlorine to water. These chloramines enter the cell walls and destroy the bacteria. Sometimes, anhydrous ammonia is used for getting chloramines in water, when there is insufficient natural ammonia. In municipal water supply, ammonia is usually added in the ratio one part of ammonia to four parts of chlorine. The ammonium sulphate which is sufficiently pure can be used for this purpose. The break-point chlorination used is to destroy objectionable taste and odours and to eliminate bacteria. It also minimizes biological growth on filters and after-growth in the distribution system. Chlorine destroys bacteria viruses, fungi, actinomycetes, worms, algae, and micro-plasma.

The effect of chlorine was investigated on bacteria, bacteria spores, yeasts and moulds. According to the studies of Kudesia, the minimum dose to which different types of bacteria namely *E. coli*, *S. typhosa* etc. are sensitive to chlorine dose is 0.1 -0.7 ppm. Old spores are sensitive to 50-80 ppm of chlorine. Bacterial enzymes with virulence are sensitive to 20 to 30 ppm where as bacterial spores are sensitive to a high dose of 240 to 300 ppm. Although virucidal action of chlorine has not been extensively investigated, some work has been carried out on polioviruses and entero viruses by Oncllean and Kudesia. Some fungi like mosses, the broyzoa which are really animals, can also be controlled by chlorine disinfection.

The mechanism of action of chlorine on cytoplasmic membrane bound enzymes and intra-cellular enzymes of bacteria is noteworthy. The cytoplasmic membrane is the location for active processes like respiration. So, chlorine acts on respiratory enzymes like succinic dehydrogenises which are linked with cytochrome systems, and in subsequent oxidative phosphorylation reactions where maximum energy is produced. Oxidation of succinic acid yields energy as a result of sub-strain level phosphorylation. Another enzyme important in energy reaction and localized in the cytoplasmic membrane is ATPase. The nitrogen contents of crude exacts is much higher than that of intact cell. Chlorine prevents anaerobic condition in sewerage systems and wastewater treatment works, by reducing decomposition. It destroys hydrogen sulphides in wastewater, oxidizes cyanides in alkaline industrial waste, destroys undesirable growth of algae and other related organisms, checks growth of iron-fixing and slime-forming bacteria in pipelines and sewerages and destroys esychoda (filter flies). It improves the coagulation and reduces BOD of wastewaters.

Disinfection by Chlorine Dioxide

Chlorine dioxide is a versatile oxidizing agent. It is used in U.S.A., West Germany, France, Switzerland, Italy, and Netherlands and to some extent in Belgium for purifying the drinking water. It reduces odours and taste, destroys phenol, and oxidizes iron and manganese. It is a yellow to red gas soluble in water at room temperature. It is liberated by treating a solution of sodium chloride with a strong chlorine solution or with an acidified solution of calcium or sodium hydrochloride. The gas should be collected in coloured bottles at low temperature. Chlorine dioxide is also prepared by the action of concentrated sulphuric acid on potassium chlorate, $KClO_3$ and oxalic acid. The reaction of sodium chlorite produces chlorine dioxide free from chlorine. The ClO_2 molecule remains in the same form over a pH range 4.0 to 8.4. The general doses which are given to water are between 0.2 and 0.3 mg per litre. It is very destructive on phenolic substances in water. The toxicity of ClO_3 on yeast moulds, bacteria, and bacterial spores has been studied by Kudesia. It has been observed that chlorine dioxide is more powerful and effective than chlorine, bromine or iodine for removal of algae from contaminated cooling towers. 0.1 ppm of ClO_2 killed all the *E. coli* present in tap water within five minutes at 25°C. The polioviruses and entero viruses were completely inactivated by exposure to 0.3 ppm of ClO_2 at pH 7 and 25° C. It is not wasted by ammonia. It also destroys hydrogen sulphide in wastewater and does not form tribal methane, like chlorine in drinking water. It reduces concentration of humic acid* in water. It is costly and very reactive and hence needs special precautions for use. The chlorite and the chlorate which are formed in water have been reported to be toxic substances and hence the usage must be restricted to only the developed cities.

Disinfection by Bromine

Bromine is a strong oxidizing agent. The oxidation potential for formation of molecular bromine from ionic bromide is (-) 1.065 Vat 25° C. Similar values for fluorine, chlorine and iodine are 2.05 V, 0.544 V, and -0.28 V respectively. However, oxidation potential cannot be a decisive factor in determining the reaction rate. Bacterial activity of Bromine increases with temperature and the effect is greater than that is reported for chlorine and less than that is reported for iodine. Wyss Stockton investigated the effect of temperature in the range of 5° C to 35° C on bromine activity against *E. coli* and Aerobacter aerogens to arrive at that conclusion.

Johansson demonstrated the superiority of bromamine over chloramines when tested against Bacillus meteins spores at 25° C. Break point reactions with Bromine is reached almost instantaneously in the presence of even a slight excess of Bromine. The very strong bactericidal properties of bromamine may be exploited by the use of combined bromine residual because there is no need to reach the break point in order to achieve strong disinfectant effects. This property of bromine makes it more economical than chlorine. Bromine also acts as a germicidal agent as observed by Wood *et al.* The toxicity of bromine on yeasts, moulds, bacteria and bacterial spores has been studied in detail. Marks Strandskov succeeded in comparing the effect of equivalence of chlorine, bromine and iodine on the killing time of bacteria at pH 7 and 25° C and discovered that order of effectiveness against bacillus mittens is chlorine, iodine and bromine. Bromine is also found to be more powerful and effective than chlorine for the removal of algae from contaminated cooling towers. It has also been proved to be more toxic than chlorine has been to yeasts and algae. According to a survey of Kudesia the ground water of different places in Uttar Pradesh in India contains bromide ions naturally in small amounts. It has been found that there is a reduction in the count of bacteria, which survived after chlorination in the presence of ammonia and distilled water with chlorine dose of 0.06 ppm. It was also observed that the rate of bacterial stabilization by chlorine in waters containing free ammonia is accelerated in the presence of small amount of bromide. The exact role of bromine and chlorine on *E. coli* suspended in tap water was studied by Kudesia. He found that a mixture of halogens is a superior germicide in comparison to any single halogen or halide. Bromamine are more effective bactericidal compounds than chloramines and the bromine compounds in swimming pools do not produce any mucous, skin irritation or itching. Bromine destroys the halogen sulphide in wastewaters; changes cyanide in alkaline industrial wastewaters into cyanates and destroys other undesirable algae and related organisms in water and waste waters. It reduces the BOD of wastewaters but does not

*It is the part of humus of soil.

form the tribromine with nitrogen to give bad odour to water. Residual bromine stays in water for a longer time than chlorine. Bromine is more toxic than chlorine to yeast, algae etc. The bromine attacks sites other than affected by chlorine in bactericidal action and is more effective against spores (Soil bacteria contains spore producers) than chlorine. But, use of bromine has certain disadvantages. Even low concentration (0.5 ppm) may cause irritation of the eyes. The concentration of 7 ppm of bromine in air can cause fatal illness in humans after an exposure of 30 to 60 minutes. Liquid bromine is more harmful than chlorine and rapidly attacks the skin. The bromine vapours cause lung diseases by preventing absorption of oxygen and diminishes the red cells and hæmoglobin.

Disinfection by Iodine

In dilute or with water solution it imparts a yellowish tinge to the water, and a careful control of iodine residual and pH of water and avoid development of aquamarine-tint. As elemental iodine is not soluble in water, at 25° C powdered iodine is scattered over surfaces of pools and rivers and also in drinking water tanks. Potassium iodide cannot be used, as potassium is a plant food and could encourage the growth of algae. Instead, sodium iodide is used over the water surface, an exact amount required to liberate the desired amount of iodine from the iodide is added in the form of a second chemical, may be chlorine, chloramines etc. It does not burn the organic matter as its sister halogens do. However, under specific conditions, it reacts with alcohols, aldehydes, phenols, oxalic acids, hydrazines, aromatic amines etc. At pH 5 almost all iodine is represented in the form of At pH 6 it is hydrolyzed to HIO_3 and 10. Above pH 8, we get HIO_3 and 10. At pH 8 only 12% is present as 12, 88% as HIO, and only 0.005% as 10. Above pH 8 we get HIO_3 as one of the oxidants. Molecular iodine is the principle-oxidizing agent. The germicidal power decreases with increase iodide concentration due to formation of 130. The addition chloride or bromide ions enhance the germicidal power to great extent. There is no available combined iodine in water. Bactericide activity of iodine increases with increasing temperature. The *E. coli* and Aerobacter aerogens are to be affected by bromine and chlorine.

Iodine does not react with organic ammonia or nitrogen. So, there is no wastage of Iodine in side diversions. Due to this fact and all due to its stability to ultraviolet light, iodine is a useful bactericide disinfectant. The effect of iodine on *E. coli* and other vegetative bacteria namely S. typhosa, S. cocci at pH 7 and concentration 0.5 ppm in 1 15, 20 minutes at 25°C has been studied by Kudesia. It has been found in all the three cases that percentage kill of bacteria increases with increase in time and the value is zero at 0.1-ppm iodine. 1 45°C, the killing effect is very high. Several studies indicate that the mixture of iodine with bromine or chlorine can prove an effective boon if used in drinking water instead of chlorine, which cause, several troubles in human bodies. Iodine has several advantages over chlorine. It does not cause irritation to eyes and nose. There is evidence of any undesirable effect on general health or on function, thyroid gland. Iodine demand of water is characteristically lower the chlorine demand, as it is not wasted in side diversion and reversibility of iodides to iodine. Iodine is a weaker oxidizing agent and a large number of terms remain unaffected. Excess amount of iodine impart yellow colour to the drinking water.

Other Uses : Chlorine is also used for certain dyestuffs, explosive hypochlorites, chlorates and sulphur monochloride and also muriatic acid.

Confirmatory Test for Chlorine

(*i*) Chlorine turns Starch Iodide Paper Blue

(*ii*) It also precipitates silver chloride when it is treated with silver nitrate.

Other Uses of Bromine

The chief uses of bromine at the present time is in the manufacture of ethylene dibromide (a substance used as an anti-knock in the automobile engines).

The next important use is in the manufacture of potassium bromide and other bromides. These are used in medicine and also in the preparation of silver bromide used in photographic plates. Large quantities of bromine are also used in the manufacture of organic dyes. A minor use of bromine is an oxidizing agent in chemical analysis.

Confirmatory Test of Bromine

Free bromine may be identified by the following tests :

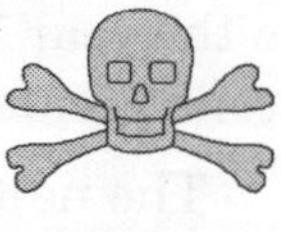

POISON

(*i*) At room temperature bromine is a heavy reddish liquid with a strong chlorine like smell.

(*ii*) To distinguish the vapour of bromine from nitrogen dioxide and chromyl chloride it is dissolved in water, add a few drops of carbon disulphide and shake up. The bromine water solution will produce an orange-yellow colour, nitrogen dioxide will produce a colourless solution, and chromyl chloride will produce a yellow solution.

(*iii*) Also the bromine vapour does not blacken the ferrous sulphate solution (distinction from nitrogen dioxide.

Iodine (Other Uses)

Iodine is used in the manufacture of certain coal tar dyes, iodoform, potassium iodide etc. In laboratory iodine is generally used for volumetric analysis of some substances. Tincture iodine (made by 27ml of each iodine and potassium iodide and distil water and making the solution to 500 ml by using rectifies spirit) it is used for the treatment of wounds, reduction swelling etc.

Confirmatory Tests of Iodine

(*i*) Fine violet colour of the vapour.

(*ii*) The solubility of the element in carbon tetrachloride or carbon disulphide or chloroform producing violet colour.

(*iii*) Formation of deep blue colour in the presence of starch, which on heating disappears.

CHAPTER

13

Phosphorus and Its Compounds

History and Occurrence

In the year 1671-1675, Brun alchemist of Humburg was actually working on Philosopher's stone accidentally discovered phosphorus by distilling the evaporated urine.

The name phosphorus was given because it glows in dark (Greek *Fhos* = light, Latin *fero* = produce).

In the year 1777 Scheele made phosphorus by distilling phosphoric acid with charcoal. Phosphorus is never found in free state in elementary form. It occurs principally as the ores of phosphorus. They are, *Phosphorite,* $Ca_2(PO_4)_2$, *Fluoro a petite* $3Ca_3(PO_4)_2\ CaF_2$, *Chlorapetite.* The phosphorite mainly occurs in U.S.A. and Russia while appetite is abundant in Canada. The phosphorite minerals and bone ash (which consists mainly calcium phosphate) are the two sources from which phosphorus, phosphoric acid and phosphotic fertilizers are obtained in large scale. Bone ash being purer than normal phosphates is therefore used for the preparation of phosphoric acid and monosodium phosphate, while the mineral phosphate is used for the preparation of "super phosphate" and phosphorus.

Phosphorus is an essential element of all living matter; it is also found in many plants and animal-tissues as phospho-proteins, which are complex compound proteins with phosphoric acid derivatives. Examples of phosphoric-proteins are Casein found in bananas and also Vitellin found in eggs.

Phosphorus Cycles

Phosphorus plays an important role in all life process. It is an essential constituent of all living matter. Phosphorus is required in small quantities both by plants and by animals for their proper growth. Plants assimilate phosphorus directly from the soil where it is present in the form of phosphates. If the plants growing on the soils are not removed, they will die and decay. Normally, the phosphorus of the plant is passed on to the animals in the form of food and fodder. But this makes no difference at all. Herbivorous animals also returns there phosphorus to the soil when they die. Thus, the soil if left to it, will conserve its stock of phosphorus. Also this will be returned indefinitely because the soil does not lose it's phosphorus as it looses its nitrogen by the action of denitrifying bacteria.

But this cycle is disturbed by man. He cultivates the land and removes the crops along with their phosphorus consumed by man runs into rivers with sewage or is barred in cemeteries and is no longer available for the plant life. As a result the cultivated soil looses its phosphorus year by wear. This deficiency has to be made up by artificial phosphate manure to the soil regularly. The one of such is Calcium Super phosphae (see Fig. 13.1).

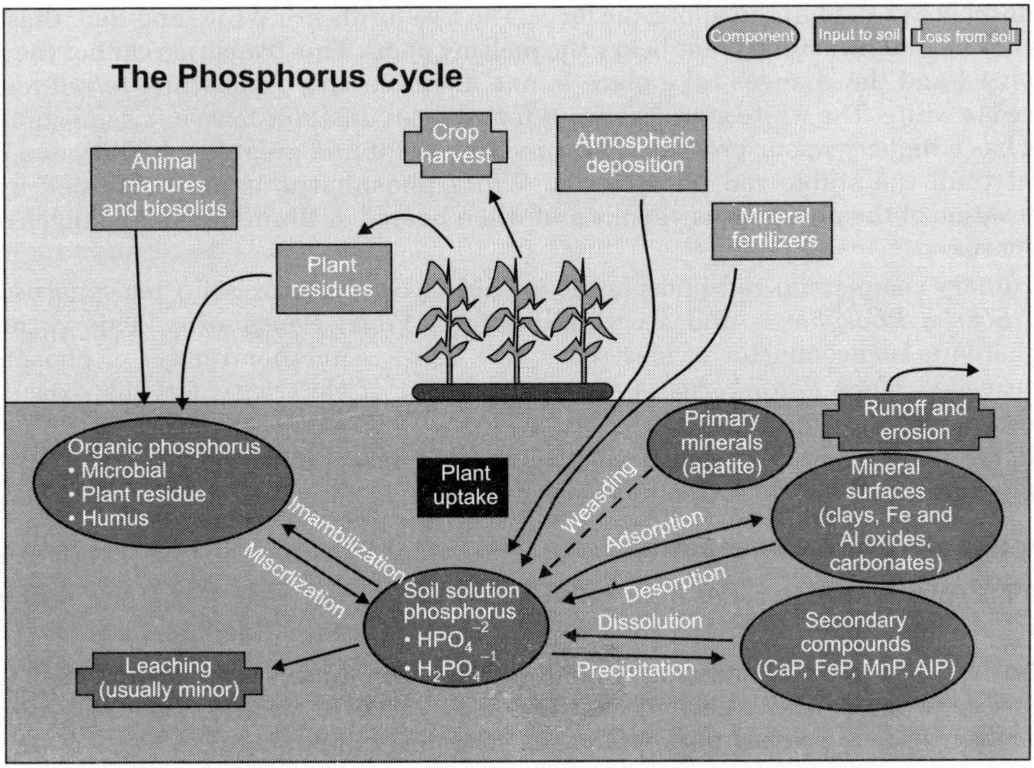

Fig 13.1. The Phosphorus Cycle in Nature

Allotropy of Phosphorus

Some elementary substances like phosphorus and carbon *exist* in more than one form and they are having little bit difference in properties. The different forms are called allotropes and the property is called allotropy. Therefore *allotropy is defined as the property of an element which allows the element to exist in more than one form.* Here we are going to discuss the different allotropes of phosphorus.

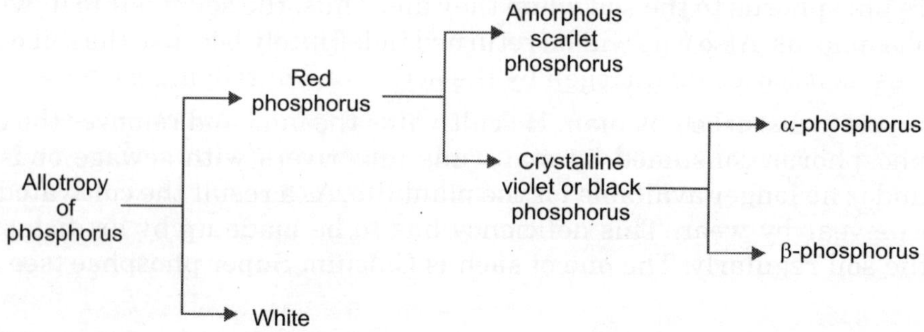

Phosphorus exists in the allotropic form. The two forms are White and Red, these two forms has no real transition point below the melting point. This transition cannot therefore be realized and the changes take place in one direction only from white to red and not form red to white. The white state is always found in an unstable form or a semi stable for, also it has a higher vapour pressure lower melting point and greater solubility in a given solvent than the stable red phosphorus. White phosphorus is only obtained by the condensation of the phosphorus vapour and when heated or illuminated it changes to red phosphorus.

Ordinary commercial red phosphorus is said to be a mixture of a phosphorus form called *Scarlet Phosphorus* and a crystalline form *Violet Phosphorus.* This variety of phosphorus is non-conductor of electricity; but there is another variety of phosphorus present called *Black Phosphorus* is a good conductor of electricity, for this reason it is regarded as *metallic phosphorus.*

All the varieties of red phosphorus are more or less stable and also they are very strongly oxidized in air at the room temperature and do not give of poisonous vapours.

Extraction of White Phosphorus

White phosphorus is extracted from phosphorite rock by an electro thermal process. A mixture of phosphorite, sand and coke is fed to an arc furnace (See Fig. 13.2).

The discharge of an alternating current between the carbon-electrodes at the base of the furnace produces the high temperature (about 3000°C) necessary for the reaction, which is supposed to proceed in the following reactions in two stages.

$$\text{Phosphorite + Silicon Dioxide} \longrightarrow \text{Calcium Silicate + Phosphorus Pentoxide}$$

$$Ca_2(PO_4)_2 \; + \; 3SiO_2\,2 \longrightarrow 3CaSiO_3\,3 \; + \; P_2O_5$$

$$\text{Phosphorus Pentoxide + Coke} \longrightarrow \text{Carbon Monoxide + Phosphorus}$$

$$P_2O_5 \; + \; 3C \longrightarrow 5CO \; + \; 2P$$

The fusible calcium silicate is removed as a long slag through the slag pipe. It is condensed by means of water and then caste in the form of sticks.

The fusible calcium silicate is removed as a slag through the slag pipe. It is condensed by means of water and then caste in the form of sticks.

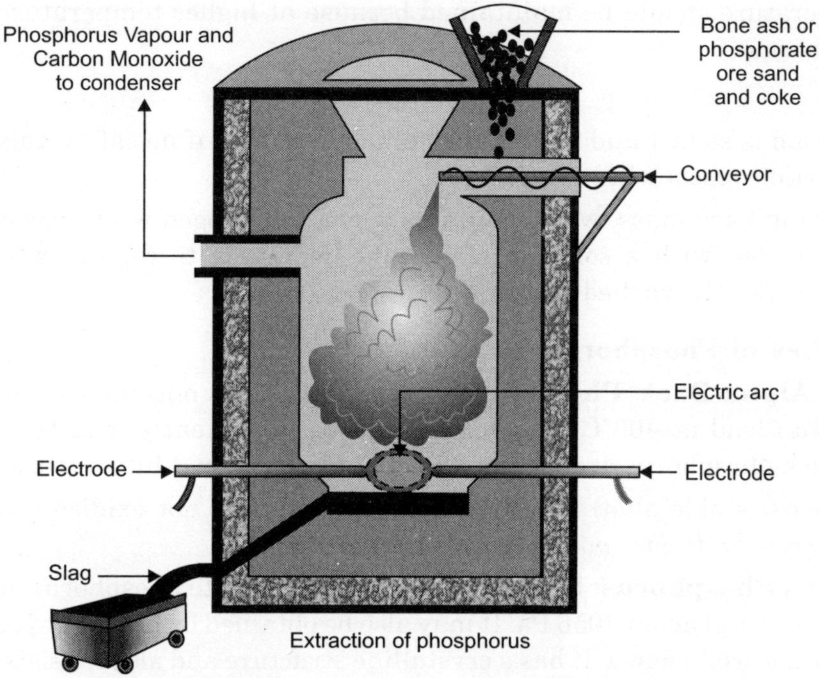

Fig. 13.2. Diagram of the Extraction of White Phosphorus

Abstraction of Red Phosphorus

Red phosphorus is prepared by heating white phosphorus in an iron pot provided with an upright pipe, a safety valve and thermometers T_1 and T_2 (see Fig. 13.3). Pot is heated at a temperature of 245°C; great care must be taken so that temperature does not rise above 260°C.

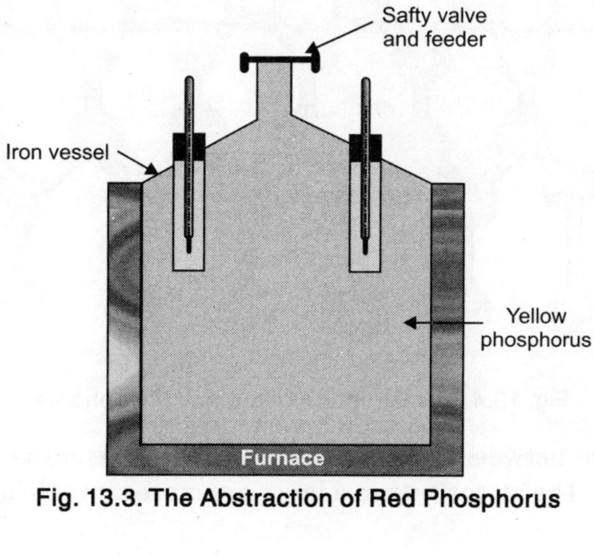

Fig. 13.3. The Abstraction of Red Phosphorus

The temperature should be maintained because at higher temperature the reaction becomes exothermic.

$$\text{P(white)} \longrightarrow \text{P(red)}$$

This reaction is so fast and produces so much heat that if no safety valve is provided the whole reaction vessel will explode.

The resulting hard mass, which contains some, unchanged white powder under water, boiled with a solution of sodium hydroxide to remove white phosphorus and finally washed and dried.

EXPLOSIVE

Other Varieties of Phosphorus

Metallic or Alpha-Back Phosphorus: This variety is obtained by dissolving red phosphorus into lead at 400°C in a sealed tube for sufficiently long time. On cooling, crystals of black phosphorus are obtained lead is dissolved in dilute nitric acid.

It is the most stable allotrope of phosphorus and does not oxidize in air until it is heated very strongly. *It does not conduct electricity.*

Beta-Black Phosphorus: It is obtained by heating white phosphorus at 200°C under a very high pressure of about 4056 Pa. It may also be obtained by heating white phosphorus whose structure is well known. It has a crystalline structure and also consists of corrugated sheets, each phosphorus atom being covalently bonded to the three neighbouring phosphorus atoms as shown in Fig. 13.4.

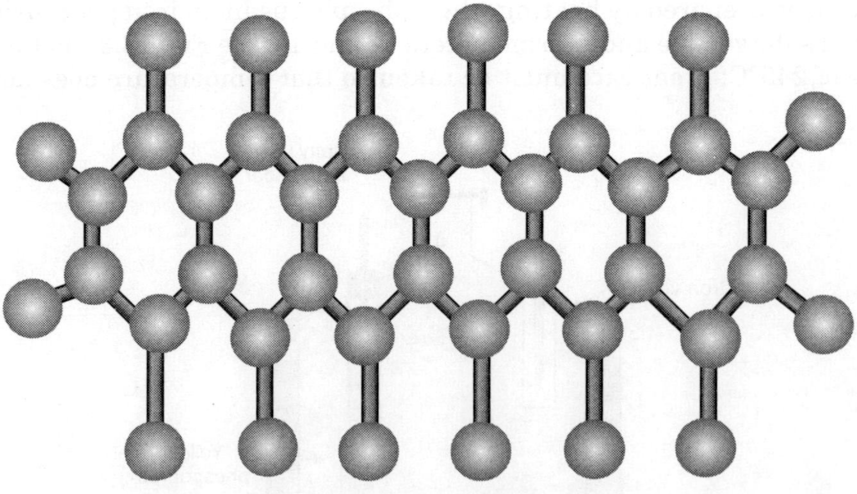

Fig. 13.4. The Structure of b-black Phosphorus

The P – P distance between two nearest phosphorus atom is 2.18 Å. The adjacent layers are 3.68Å apart. The atoms within a layer are more strongly bonded than the atoms in adjacent layers. This gives rise to flag crystals resembling to those of graphite. It has a definite sharp melting point of 580°C. Its specific gravity is 2.69. It is also a Stable form of phosphorus and does not bum in air even upto 400°C. Like graphite it is also a fair conductor of electricity.

Scarlet Phosphorus: This allotrope is derived from an amorphous scarlet powder by boiling a 10% solution of phosphorus in phosphorus tribromide for about 10 hours. Pure scarlet phosphorus can be prepared by heating phosphorus tribromide with mercury at 240°C.

Phosphorus Tribromide + Mercury $\longrightarrow$ Mercuric Bromide + phosphorus

$$2PBr_3 \quad 3Hg \longrightarrow 3HgBr_2 + 2P$$

Scarlet phosphorus resembles the red variety in its physical property and the white phosphorus in its chemical properties. It is however, slowly oxidized in air.

Violet Phosphorus: This variety is obtained by heating white phosphorus with a trace of sodium at 230°C under high pressure; it is also crystalline in structure.

Extremely Poisonous Gas do not Inhale

PROPERTIES OF PHOSPHORUS

Physical

White Phosphorus: This form of phosphorus is obtained when phosphorus vapour is cooled rapidly. When prepared fresh it is colourless, rather transparent waxy solid. On exposure to light it gets suddenly covered with red variety, which adds a yellow colour, hence the name has been given as *yellow phosphorus,* which is sometimes wrongly given to this variety.

White phosphorus has garlic like odour.

White phosphorus remains ignited in the air for a long time and also in chlorine.

Determination of the Ignition Point of Phosphorus

Make the apparatus set up as shown in the Fig. 13.5. Place a dry piece of phosphorus in the boiling tube and then heat the water gently with constant stirring, as the phosphorus catches fire immediately note down the temperature (30°C).

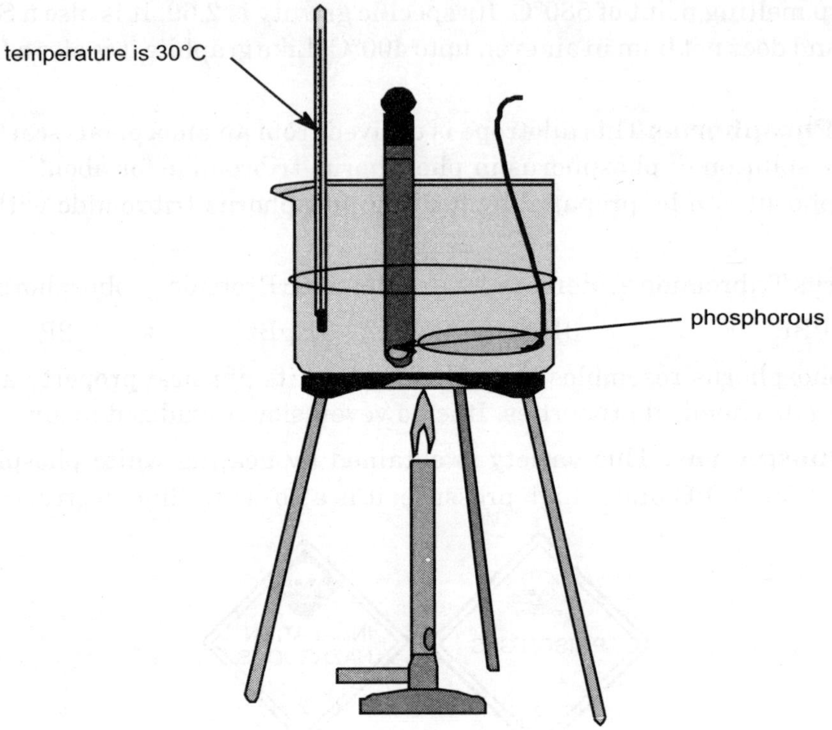

Fig. 13.5. the set up for the Determination of the Ignition point of the White Phosphorous

Physical Properties of Phosphorus

Property	Red	White
Melting Point	500°C	44.15°C
Boiling Point		277°C
Ignition Point	240°C	30°C
Density	2.1 g/cc	1.81 g/cc
Ionic Radius	2.12 Å	
Dielectric Constant	4.096	
Electronegativity	2.1	
Atomic Volume	17.0 cc	
Enthalpy of Vapourization	12.4 kJ/mol at 25°C	
Enthalpy of Fusion	0.66 kJ/mol at Boiling Point	
Latent Heat of Fusion	0.15 kcal/mol	

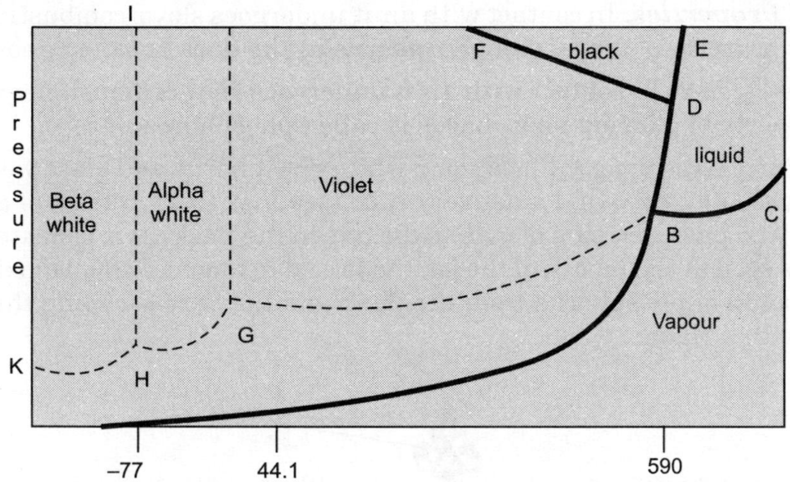

Fig. 13.6. The Phase Diagram of Allotropes of Phosphorus

This is the various equilibria that exists in the phosphorus at different phases are shown in the form of a diagram (see Fig. 13.6). GB, KH and HG are the vapourization curves of violet, beta white and alpha white phosphorus, respectively. While the curves BC is the vapourization curve of the liquid phosphorus, ending abruptly at critical point. The curves DF and HI represent the effect of pressure on the transition point of violet phosphorus into black phosphorus and on that of alpha-phosphorus into beta-phosphorus allotrope respectively. The points B and G are the melting point of the violet and alpha white variety, while H (–77°C) is the transition of alpha and beta forms. The area DEF is the region of black phosphorus.

Formula of Phosphorus

White phosphorus has the molecular formula p 4 both in the solid and in the vapour states at the moderate temperature. The four atoms present in the molecule are arranged at the comers of a tetrahedron as shown below. At higher temperatures (about 700°C) however dissociation to diatomic molecules P_2 occurs.

$$P \rightleftharpoons 2P_2$$

The structure of phosphorus (red) has not yet been completely determined but there is evidence that it is polyatomic and consists of chains of p 4 tetrahedron, linked together possibly in the manner shown in the figure below:

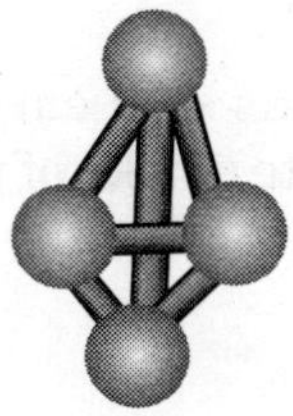

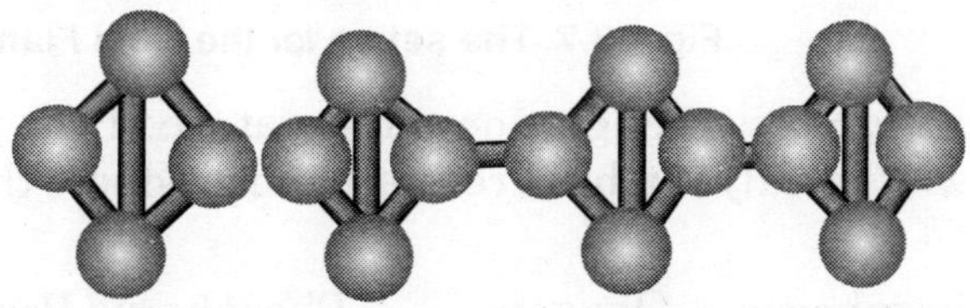

Chemical Properties: In contact with air it undergoes slows combustion and glows in the dark. This property of glowing of phosphorus in the dark is called phosphorescence.

(i) Phosphorescence: In contact with air it undergoes slow combustion and glows in the dark. This property of glowing phosphorus is called phosphorescence.

(ii) Cold Flame Experiment: Place some white phosphorus and glass wool in the round bottom flask with a jet tube and a delivery tube (see Fig. 13.7). Switch of all the lights of the laboratory and pass a stream of carbon dioxide in the flask, as it is heated on the water bath, a flame is seen at the mouth of the jet and is cold to touch. Carbon dioxide is sweeping out all the vapours of phosphorus from the flask and they are escaping through the tube and burning with a flame.

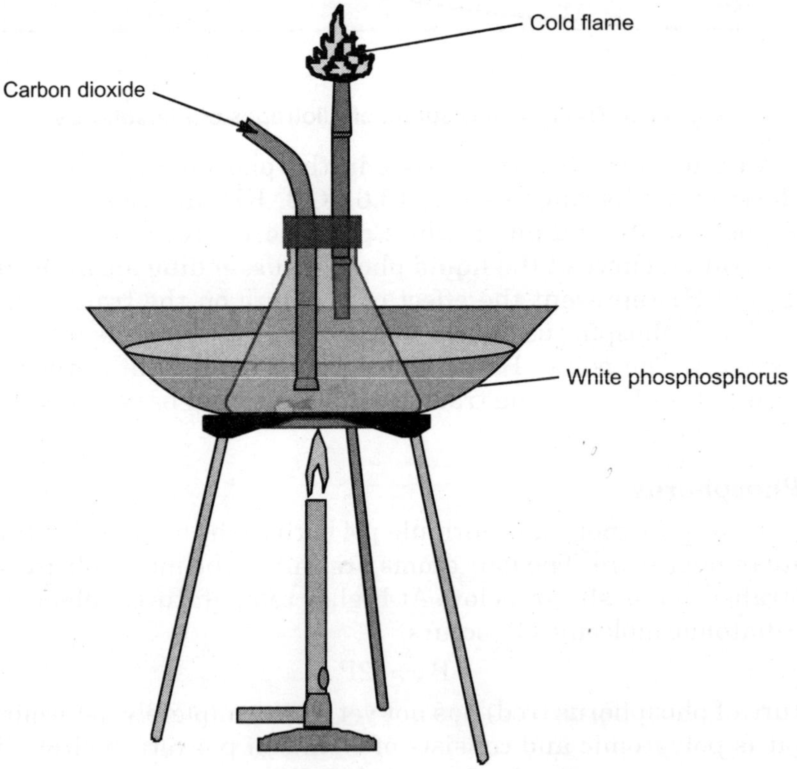

Fig. 13.7. The set up for the Cold Flame Experiment

(iii) Burning: It is the ignition temperature of the phosphorus as said earlier, is very low, (30°C) so it rapidly catches fire in air and produces the dense white smoke of phosphorus pentoxide.

$$\text{Phosphorus} + \text{Oxygen} \longrightarrow \text{Phosphorus Pentoxide}$$

$$2P + 5O_2 \longrightarrow 2P_2O_5$$

White phosphorus is extremely poisonous, must be handled with care and always stored under water, and never touched it with finger. To handle white phosphorus tweezers should be used. By chance if phosphorus falls on hand it should be washed with copper sulphate solution.

Exp.: Dissolve some white phosphorus in carbon disulphide solution (2 g of phosphorus and 10g of carbon disulphide) then dip a filter paper it and through it on the floor. After few seconds it will catch fire because the carbon disulphide being volatile gets evaporated and phosphorus burns in coming contact with air.

(iv) With Sodium Hydroxide: Phosphorus is dissolved in sodium hydroxide solution; on boiling it in an inert atmosphere phosphine is produced.

Sodium Hydroxide + Phosphorus + Water ⟶ Phoaphene + Sodium Hypophophite

$$NaOH + 2P_2 + H_2O \longrightarrow PH_3 + 3NaH_2PO_2$$

(v) With Halogen: White phosphorus ignites spontaneously in chlorine and reacts violently with the other halogens like bromine and iodine. Cold and concentrated nitric acid is converted to phosphoric acid when reacts with phosphorus.

Nitric Acid + Phosphorus ⟶ Phosphoric Acid + Water

$$20HNO_3 + 2P_2 \longrightarrow 4H_3PO_4 + H_2O$$

Phosphorus also reacts with chlorine producing trichloride and then pentachloride. White phosphorus reacts more violently than red phosphorus, for instance, white phosphorus spontaneously catches fire in chlorine producing phosphorus trichloride (PCl_3) and pentachloride (PCl_5), while red phosphorus catches fire in chlorine only on heating. It also reacts with bromine and iodine.

Phosphorus + Chlorine ⟶ Phosphorus Trichloride

$$2P_2 \quad 6Cl_2 \quad 4PCl_3$$

Phosphorus + Chlorine ⟶ Phosphorous Pnetachloride

$$2P_2 \quad 10Cl_2 \quad 4PCl_5$$

Phosphorus + Bromine ⟶ Phosphorous Tribromide

$$2P_2 + 6Br_2 \longrightarrow 4PBr_3$$

Phosphorus + Iodine ⟶ Phosphorous Trichloride

$$2P_2 \quad 6I_2 \quad 4PI_3$$

(vi) Reaction with Sulphur: Phosphorus combines with sulphur through an explosion forming a number of sulphides such as phosphorus disulphide (P_4S_2) phosphorus pentasulphide (P_4S_5) and phosphorus heptasulphide (P_4S_7) and also phosphorus trisulphide (P_4S_3), phosphorus trisulphide is used for 'strike anywhere matches'.

(vii) Reaction with Metals: Phosphorus combines with number of metals forming phosphides, for example: with magnesium and calcium it forms calcium phosphide and magnesium phosphide.

Phosphorus + Calcium $\longrightarrow$ Calcium Phosphide

$2P_2$ + $6Ca$ $2Ca_3P_2$

Phosphorus + Magnesium $\longrightarrow$ Phosphorous Pentachloride

$2P_2$ $6Mg$ $2Mg_3P_2$

Sodium and Potassium Reacts Violently in Producing Flame

Phosphorus + Sodium $\longrightarrow$ Trisodium Tetraphosphate

$2P_2$ + $3Na$ $\longrightarrow$ Na_3P_4

Phosphorus + Potassium $\longrightarrow$ Tripotassium Tetraphosphate

$2P_2$ $3K$ K_3P_4

(viii) Reducing Action: Since the white phosphorus is easily oxidized in air, it can reduce many chemical compounds. It reduces nitric acid to nitrogen dioxide and sulphuric acid to sulphur trioxide. It is also reduces the solution of copper gold to and silver salts to the corresponding metals.

Thus, a colloidal solution of gold may be prepared by reducing a solution of gold chloride with phosphorus dissolved in eather.

(ix) Reaction with other compounds: White phosphorus with copper sulphate producing metallic copper and phosphoric acid.

Silver nitrate also reacts with phosphorus forming silver phosphide.

Comparison between Red and White Phosphorus

Property	White	Red
1. Colour	Almost Colourless	Reddish violet
2. Odour	Garlic Smell	Odourless
3. Physical State (S.T.P.)	Crystalline solid	Crystalline
4. Specific Gravity	1.82	2.1
5. Stability	Metastable at ordinary temperature	Stable
6. Melting Point	44.1°C	597°C
7. Boiling Point	280°C	Very High
8. Solubility in water	Almost Insoluble	Insoluble
9. Electrical Conductivity	Very Feeble Conductor	Feeble
10. Solubility of Carbon Disulphide	Soluble	Insoluble
11. Physiological Effect	Poisonous	Non-poisonous
12. Chemical Activity	Very active	Less Active
13. Ignition Temperature	30°C	260°C

This is the annexure of the phosphorus, which is named as Reaction of Phosphorus. This annexure consists of different reactions of phosphorus, which will be helpful for the students who are studying for different competitive examinations and also doing research work in chemistry about phosphorus. The reactions are as follows:

$$\text{Phosphorus} + \text{Barium Hydroxide} + \text{Water} \longrightarrow \text{Barium Hypophosphite} + \text{Phosphene}$$
$$4P_2 + 3Ba(OH)_2 + 6H_2O \longrightarrow 3Ba(H_2PO_2)_2 + 2PH_3$$

$$\text{Phosphorus} + \text{Nitric Acid} \longrightarrow \text{Phosphoric Acid} + \text{Nitrogen Dioxide} + \text{Water}$$
$$P + 5HNO_3 \longrightarrow H_3PO_4 + 5NO_2 + H_2O$$

$$\text{Phosphorus} + \text{Ammonium Chloride} \longrightarrow \text{Phosphorus} + \text{Hydrogen} + \text{Hydrogen Chloride}$$
$$P_2 + 2NH_4Cl \longrightarrow 2PNH_2 + 2HCl + H_2$$

$$\text{Phosphours} + \text{Sodium} \longrightarrow \text{Sodium Phosphide}$$
$$P_2 + 2Na \longrightarrow 2NaP$$

$$\text{Calcium} + \text{Phosphorus} \longrightarrow \text{Calcium Phosphide}$$
$$3Ca + P_2 \longrightarrow Ca_3P_2$$

$$\text{Phosphorus} + \text{Potassium Cyanide} \longrightarrow \text{Phosphorus} + \text{Potassium}$$
$$P_2 + KCN \longrightarrow P_2CN + N$$

Use of Phosphorus in Match Industry

Phosphorus is used in match industry in large scale. Simbal wood is used for matchsticks. Generally the matchsticks are divided into two categories. (I) Safety Matches, which are generally used in domestic purpose. (II) Lucifer Matches or Strike anywhere Matches, or Friction Matches, these are not used for domestic purpose being unsafe.

Lucifer Matches: These matches prepared by giving three successive dips in different chemical substances. The description for preparation is stated below:

(a) First Dip: The matchsticks are firstly dipped in an emulsion of a combustible material of molten sulphur or liquid paraffin to about two inches length.

(b) Second Dip: At the extreme tip the match sticks are dipped in a paste of following substances:

1. A readily combustible substance like antimony trisulphide.
2. A low kindling material, which starts combustion. White phosphorus was previously used but as it is forbidden red phosphorus is being widely used.
3. An oxidizing agent like potassium chlorate, potassium nitrate or red lead is used for required oxygen supply.
4. An inert substance like glass powder or chalk, which increases the friction also, decreases the sensitiveness is used.

(c) Third Dip: These are next dipped in molten wax to provide a thin protective coating to the matchsticks.

To prevent the after-glow the matchsticks are previously impregnated with borax. Matchsticks are boiled with a concentrated borax solution. This puts off the fire immediately after they are blown out without glowing for any appreciable.

Lighting Procedure of the Lucifer Matchsticks

The sides of the box are coated with sand glue mixture. When the stick is rubbed against the surface the low kindling material phosphorus trisulphide in this case, catches fire due to heat supplied by friction. Easily combustible substance (antimony trisulphide) is next lightened. Sulphur is next to burn and finally the matchstick catches fire. Oxidizing agents supply oxygen necessary oxygen for combination.

Safety Matches: The composition of the safety matches differs from that of from the tip. This is present in the composition pasted on the side of box. The paste applied to the side of the box is detained by mixing red phosphorus or phosphorus trisulphide with sand or glass powder (to cause friction) and glue (adhesive).

Lighting of a safety match: To light the safety match it is rubbed against this special surface. A trace of phosphorus is detached and ignited locally by heat due to friction. This tiny flash ignites antimony trisulphide and sulphur and the matchstick in turn, is lighted as in Lucifer matches.

Phosphorus is used largely in fertilizer industry for the manufacture of much common and also useful fertilizer, they are being discussed.

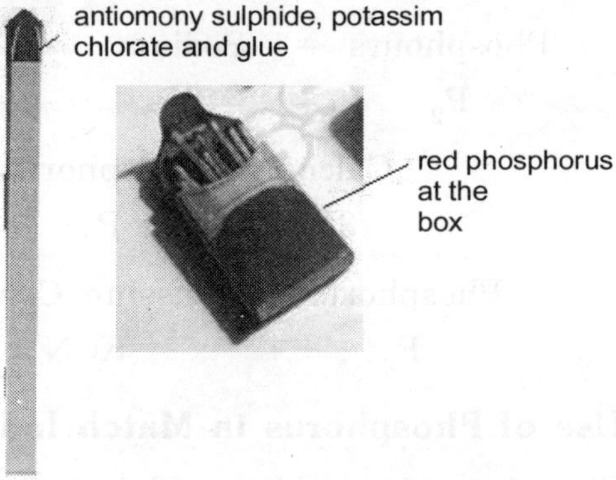

Match box and a safety match

Phosphorus in Fertilizer Industry

Calcium Super phosphate [$Ca(H_2PO_4)$]: This is manufactured by treating the insoluble phosphate rock with a calculated quantity of sulphuric acid (commercial) when we get the more soluble calcium soluble calcium dihydrogen phosphate (super phosphate) and calcium sulphate.

$$\begin{array}{lcccc} \text{Apetite} & + \text{ Sulphuric Acid} & \longrightarrow & \text{Calcium Sulphate} + & \text{Calcium Super phosphate} \\ Ca_3(PO_4)_2 + & 2H_2SO_4 & \longrightarrow & 2CaSO_4 \quad + & Ca(H_2PO_4)_2 \end{array}$$

A sufficient amount of water is then added to the calcium sulphate to convert it into calcium sulphate dihydrate ($CaSO_4.2H_2O$) also known as Gypsum. The mixture of calcium dihydrogen phosphate and gypsum is then utilized as the 'super phosphate of lime' and used as the fertilizer.

The powder rock phosphate (apetite) mixed with a calculated amount of sulphuric acid is placed in the cast iron mixture pot, which is fitted with a mechanical stirrer and can handle about 4 tonnes of charge at a time. After mixing for about 5 minutes, the mass is dumped into one of the brick-lined dens. Successive charges from the mixture continue in the den for about 24 hours. The temperature rises to about 100°C with the result that most of the water is driven off and the resulting product is a solid mass. It is allowed to remain in the den for a few days till the mass is dried up completely. It is dug out powder and put into market.

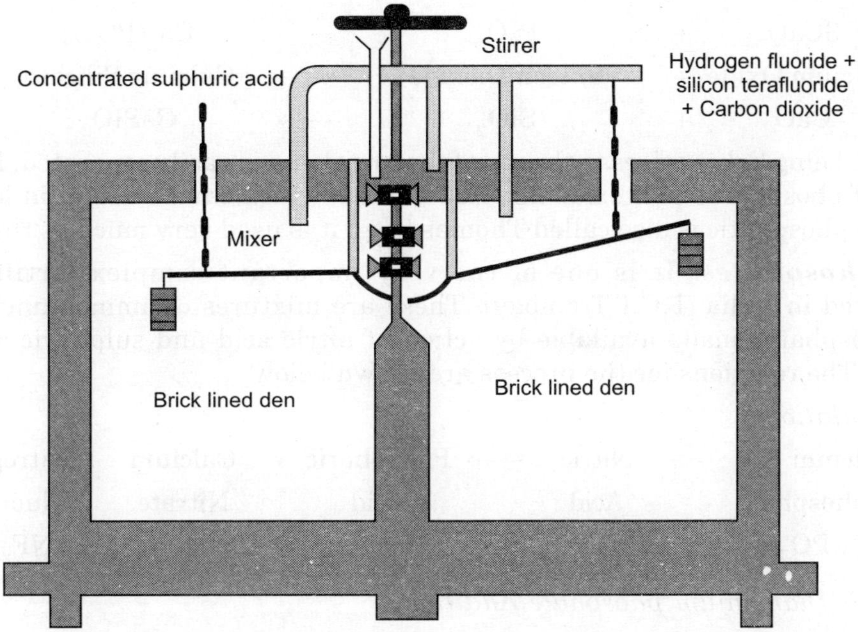

When one den is full, the charge form the mixture is dumped to the second. The calcium fluoride present in the apetite is attacked sulphuric acid, which produced hydrofluoric acid, which further attacks the silica mater present in the mass, producing the silicon tetrafluoride.

Some carbon dioxide and sulphur dioxides are also formed by the action of sulphuric acid on organic impurities. Therefore, the gases issuing out from the dens and the mixture are scrubbled with water in one or two towers. The resulting solution of hydrofluoric acid is neutralized by sodium carbonate to get sodium fluoride. Alternatively the solution is treated with sand to get hydroflurosilicic acid (H_2SiF_6), which is neutralized, by sodium carbonate or magnesium to get sodium silicofluoride (Na_2SiF_6) or magnesium silicofluoride ($MgSiF_6$). These are important byproduct of this process. Among them sodium silicofluoride is used for the preservatives of the cement about 18% of phosphorus pentoxide is considered as the good phospahtic fertilizers. However, care should be taken in storage as it as a tendency of hard lumps.

> Byproducts are sodium silicofluoride or magnesium silicofluoride.

Phosphatic Slag: It is the byproduct of the steel industry. When the ore is burnt with limestone in the presence of air, calcium appears as a slag. The following reactions occur:

$$\text{Phosphorus} + \text{Oxygen} \longrightarrow \text{Phosphorus Pentoxide}$$

$$4P + 5O_2 \longrightarrow 2P_2O_5$$

$$\text{Calcium Oxide} + \text{Phosphorus Pentoxide} \longrightarrow \text{Calcium phosphate}$$

$$3CaO + P_2O_5 \longrightarrow Ca_3(PO_4)_2$$

$$\text{Calcium Oxide} + \text{Silicon Dioxide} \longrightarrow \text{Calcium Silicate}$$

$$CaO + SiO_2 \longrightarrow CaSiO_3$$

The slag being lighter rises to the top of the metal and is easily separated. It consists of 14 to 185 of phosphorus pentoxide and 405 of lime, which can be useful in lowering soil acidity. The phosphatic is also called Thomas Slag; it is used very much as the fertilizers.

Nitrophosphates: It is one of the very important complex fertilizers being manufactured in India (F.C.I Trombay). These are mixtures of ammonium nitrate and various phosphates made available by action of nitric acid and sulphuric acid on rock phosphate. The reactions for the process are shown below:

(i) Acidulation

$$\underset{\text{Fluorophosphate}}{\text{Calcium}} + \underset{\text{Acid}}{\text{Nitric}} \longrightarrow \underset{\text{Acid}}{\text{Phosphoric}} + \underset{\text{Nitrate}}{\text{Calcium}} + \underset{\text{Fluoride}}{\text{Nitrogen}}$$

$$Ca_{10}F_2(PO_4)_6 + 20HNO_3 \longrightarrow 6H_3PO_4 + 10Ca(NO_3)_2 + 2NF_2NF$$

(ii) DAP (Diamonium phosphate Addition)

$$\text{DaP} \longrightarrow \underset{\text{Nitrate}}{\text{Ammonium}} + \underset{\text{Diphosphate}}{\text{Calcium}} + \underset{\text{Fluoride}}{\text{Calcium}} + \underset{\text{Acid}}{\text{Phosphoric}}$$

$$7(NH_4)_4HPO_4 \longrightarrow 14NH_4NO_3 + 3Ca(H_2PO_4)_2 + CaF_2 + H_3PO_4$$

(iii) Ammoniation

$$\text{Oxygen} + \text{Ammonia} \longrightarrow \underset{\text{Nitrate}}{\text{Ammonium}} + \underset{\text{Phosphate}}{\text{Ammonium}} + \underset{\text{Phosphate}}{\text{Calcium Hydrogen}} + \underset{\text{Fluoride}}{\text{Calcium}}$$

$$O + 10NH_3 \longrightarrow 20NH_4NO_3 + 4NH_4H_2PO_4 + 9CaHPO_4 + CaF_2$$

In this new process developed by FCI, Trombay unit, rock is acidulated with nitric acid in a series *of "V"* shaped reactors made of stainless steel. The product thus obtained contains phosphoric acid calcium nitrate. Hydrogen fluoride produced is either removed with the help of silica or by absorbing it in week alkaline-water in the scrubber-absorbed. To this mass is added water-soluble diammonium phosphate (DAP) and the every state to check the *NIP* (Nitrogen-Phosphorus) ratio. The final product consists of ammonium nitrate, dicalcium phosphate and monoammonium phosphate. The resulting product is mixed with potassium salt and spary dried in a spheridiser to get complex fertilizer mixture in the form of uniform, pallets with both citrate and water soluble form of phosphate. These are screened and waxed with coating agent to provide anti-cracking properties.

Compounds of Phosphorus

Phosphorus is chemically active that is why it forms different compounds. We shall discuss the preparation, both laboratory and industrial process of these compounds.

Phosphorus forms hydrides and as well as acids forming their respective phosphides. Phosphorus also reacts with halogens forming halides.

Hydro-compounds of Phosphorus

Phosphorus forms number of hydro-compounds or hydrides, such as Phosphine, diphosphine, etc.

PHOSPHINE

Invention: Phosphine is that hydride of phosphorus which is similar to ammonia. Gengenbre discovered it in the year 1783.

Preparation: In laboratory Phosphine is prepared by boiling white phosphorus with sodium or potassium hydroxide, in the presence of carbon dioxide atmosphere. This is because the gas phosphine will react with atmospheric oxygen, the gas is bubbled in the water and as it comes in contact with oxygen immediately catches fire and produces vortex rings (see Fig. 13.8).

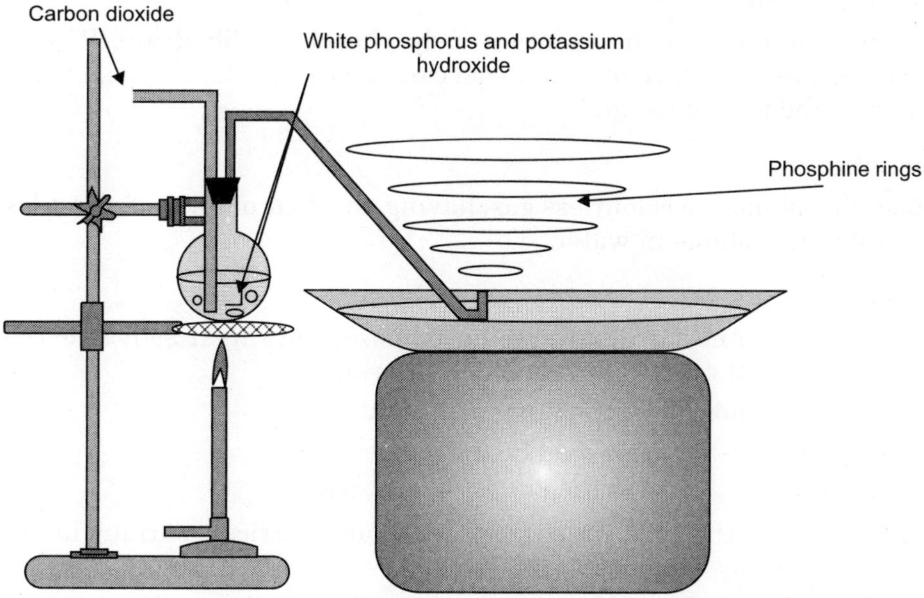

Fig. 13.8. The Lab Preparation of Phosphine

Reaction Equation

Phosphorus + Potassium Hydroxide + Water ⟶ Phosphine + Potassium Hypophosphate

$$2P_2 + 3KOH + H_2O \longrightarrow PH_3 + 3KH_2PO_2$$

Precaution: Phosphine consists of highly inflammable and poisonous diphosphine, therefore, it should be handled with care and the gas should never be inhaled at any conditions directly.

Preparation of Pure Phosphine

1. Pure phosphine is prepared by heating phosphorus iodide with sodium hydroxide.

Phosphonium Iodide + Sodium Hydroxide ⟶ Sodium Iodide + Water

$$PH_4I + NaOH \longrightarrow PH_3 + H_2O$$

2. A convenient method of preparation of phosphine is that hydrolyzing a metallic phosphide with water or dilute hydrochloric acid.

Calcium Phosphide + Water ⟶ Phosphine+ Calcium Hydroxide

$$Ca_3P_2 + 6H_2O \longrightarrow PH_3 + 3Ca(OH)_2$$

3. Phosphine is also prepared by heating phosphorus acid

Phosphorus Acid ⟶ Phosphoric Acid + Phosphine

$$4H_3PO_3 \quad 3H_3PO_4 + PH_3$$

The gas thus obtained is pure.

Phosphine also can be freed from the spontaneously phosphine dihydrate, P2~ by passing through a freezing mixture. Activated charcoal can also be used for the absorption of the liquid hydride from the phosphine gas.

Properties of Phosphine

Physical Properties: Phosphine is a colourless gas, having an odour of rotten fish and is highly poisonous. It is slightly soluble in water.

Chemical Properties

1. *Stability to Heat:* (*i*) Phosphine is completely decomposed into hydrogen and red phosphorus at the temperature of 445°C or by electrical sparking.

(*ii*) It burns in air forming phosphorus pentoxide and water.

Phosphine + Oxygen ⟶ Phosphorus Pentoxide + Water

$$2PH_3 + 4O_2 \quad P_2O_5 + 3H_2O$$

(*iii*) Phosphine and oxygen in the pure state if exposed to an electrical sparking under reduced pressure, violently explodes.

2. *Reaction with Halogens:* (*A*) Phosphine reacts with all the halogens liberating a great amount of heat.

Phosphine + Chlorine ⟶ Phosphorus Pentachloride + Hydrogen Chloride

$$PH_3 + 4Cl_2 \longrightarrow PCl_5 + 3HCl$$

(B) A mixture of phosphine with nitrous oxide when sparked violently explodes.

Phosphine + Nitrous Oxide ⟶ Phosphorous Pentoxide + Water + Nitrogen

$$PH_3 + 8N_2O \longrightarrow P_2O_5 + 3H_2O + 8N_2$$

3. Basic Property of Phosphine: Phosphine is highly basic and reacts with hydrogen halides forming phosphine halides. Thus, it reacts with hydrochloric acid forming phosphonium chloride.

$$\text{Phosphine} + \text{Hydrochloric Acid} \longrightarrow \text{Phosphonium Chloride}$$
$$PH_3 + HCl \longrightarrow PH_4Cl$$

Similarly with hydrobromic acid and hydroiodic acid, the phosphine also reacts as above.

$$\text{Phosphine} + \text{Hydrogen Iodide} \longrightarrow \text{Phosphonium Iodide}$$
$$PH_3 + HI \longrightarrow PH_4I$$
$$\text{Phosphine} + \text{Hydrogen Bromide} \longrightarrow \text{Phosphonium Bromide}$$
$$PH_3 \quad HBr \quad PH_4Br$$

4. As Reducing Agent: Because of the decomposition of phosphine into phosphorus and hydrogen when heated, phosphine acts as a very powerful reducing agent. Thus, it precipitates silver and gold from solutions of their respective salt solutions.

$$\text{Phosphine} + \text{Silver Nitrate} \longrightarrow \text{Silver Nitrate} \cdot \text{Silver Phosphide} + \text{Nitric Acid}$$
$$PH_3 + 3AgNO_3 \longrightarrow 3AgNO_3 \cdot Ag_3P + HNO_3$$
$$\text{Silver Nitrate} \cdot \text{Silver Phosphide} + \text{Water} \longrightarrow \text{Silver} + \text{Nitric Acid} + \text{Phosphorus Acid}$$
$$3AgNO_3 \cdot Ag_3P + H_2O \longrightarrow 6Ag + HNO_3 + H_3PO_3$$

5. Similarity to Ammonia: Like ammonia, with some of the halides, phosphine also combines to form phosphine complexes.

Ammonia and phosphine are both having some similar and dissimilar properties, which are tabulated below:

Ammonia	Phosphine
1. A colourless gas lighter than air.	1. A colourless gas heavier than air.
2. It possesses the pungent smell.	2. The odour of phosphene is like rotten fish
3. It is non-poisonous.	3. It is highly poisonous.
4. Highly soluble in water.	4. It is sparingly soluble in water.
5. It is not a good supporter of combustion; it burns in excess of air or oxygen. $4NH_3 + 3O_2 \longrightarrow 2N_2 + 6H_2O$	5. It is also a bad supporter of combustion. It burns in air forming phosphorous pentachloride. $2PH_3 + 4O_2 \longrightarrow P_2O_5 + 3H_2O$
6. It reacts with chlorine vapour to produce nitrogen trichloride and hydrogen chloride.	6. It spontaneously inflames and burns in chlorine gas forming phosphorous pentachloride.
7. It reacts with acids forming ammonium salts. $NH_3 + HCl \longrightarrow NH_4Cl$	7. It reacts with halogen acids giving phosphonium salts. $PH_3 + HCl \longrightarrow PH_4Cl$ $PH_3 + HBr \longrightarrow PH_4Br$ $PH_3 + HI \longrightarrow PH_4I$

Ammonia	Phosphine
8. It is reducing only at high temperature. It reduces cupric oxides and lead monoxides to metal at very high temperature. $3CuO + 2NH_3 \longrightarrow 3Cu + 3H_2O + N_2$	8. It is a reducing agent it precipitates silver and gold solutions.
9. When ammonia is passed in the solution of the salts of metals the metallic hydroxides, are usually precipitated.	9. Phosphene is converted into its elements phosphorous and hydrogen when at 445°C and also sparked.
10. It can be prepared by heating sodium hydroxide and ammonium chlrode.	10. It can be prepared sodium hydroxide and phosphorus. $2P_2 + 3NaOH + H_2O \longrightarrow PH_3 + 3Na + H_2PO_3$

Composition of Phosphine

Electric sparks are passed through a known volume of phosphorus enclosed over mercury in a eudiometer tube (see Fig. 13.9). The volume of hydrogen obtained is found to be 1 Y2 times the volume of phosphorus taken.

$$\text{Phosphine} \longrightarrow \text{Phosphorous} + \text{Hydrogen}$$
$$1 \text{ Volume} \longrightarrow 1/2 \text{ volume}$$
$$1 \text{ molar} \longrightarrow 1/2 \text{ Molar} = 3 \text{ atoms}$$

With the applications of Avogadro's Hypothesis we learn that one molecule of phosphorus (say, x). Thus the formula of phosphine is P_xH_3 with molecular weight $31x + 3$.

Equating this with molecular weight of phosphine (= 34) determined experimentally we have $31x + 3 = 34$, or x = I. Hence molecular formula is PH_3.

Why the boiling point of PH_3 is lower than that of NH_3? The molecular weight of phosphine (34) is twice that of ammonia (17). So, according to the general rule, the boiling point of PH_3 should be higher than that of NH_3; but, in actuality, PH_3 boils at a much lower temperature than NH_3. This is due to the existence of extensive hydrogen bonding among the molecules of NH_3. A large number of NH_3 molecules become associated by hydrogen bonding, as a result of which, the effective molecular weight of such associated species becomes much higher than that of discrete PH_3 molecules.

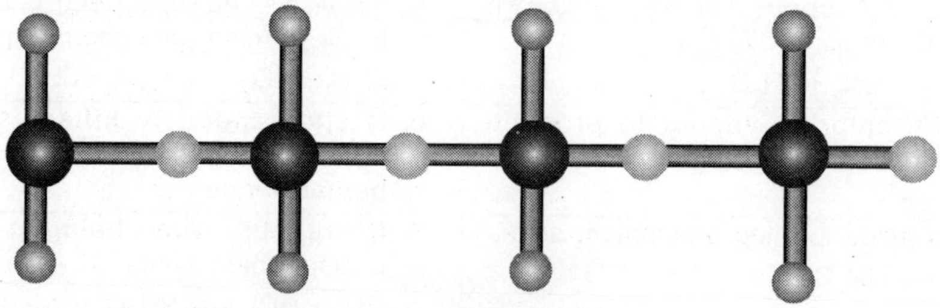

Hydrogen bonding in NH_3 molecule

Due to lower electronegativity of phosphorus, hydrogen bonding is not possible among the NH3 molecules. For this reason, phosphine has a lower boiling point than ammonia.

Ammonia is more basic than phosphine: The more the ability of a compound to donate electron to H^+ of an acid, the more is its basic character. In the NH_3 and PH_3 molecules, both the N and the P atoms contain a lone pair of electrons. Of these two molecules, the one, which can more easily donate this electron pair, will be more basic. As nitrogen is more electronegative than phosphorus, the electron-density on the N-atom .of the NH_3 molecule is more than that on the P-atom of the PH_3 molecule. (This is due to the reason that the N-atom of NH_3 draws the electron pairs of the covalent N-H bonds more towards itself.) As a result, NH_3 molecule can donate electron from its N-atom very easily, but PH_3 molecule cannot do it so easily from its P-atom. For this reason, NH_3 is more basic than PH_3.

Further, due to the easy removal of electron from the N-atom, ammonia molecule has a strong affinity for water to form NH_4OH which dissociates into NH_4 and OH ions. For the phosphine molecule, this affinity is very small and as such, little PH_4OH arid subsequent OH ions are produced. For this reason ammonia is more basic than phosphine. (See Structure of Phosphine).

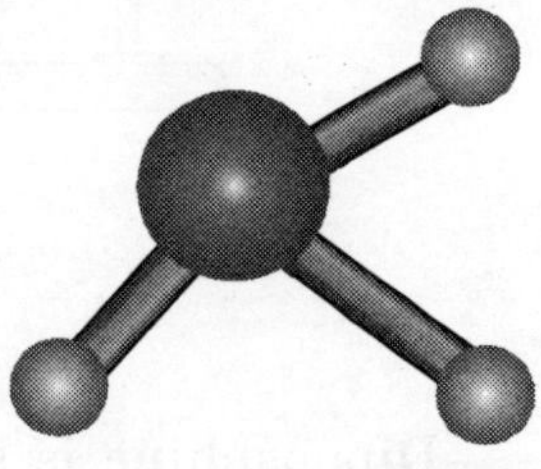

Structure of Phosphine

Uses of Phosphine

1. Phosphine is used in Holm's signal, which is a mixture of calcium carbide and calcium phosphide is taken in a container, which is prepared and thrown into the sea. Phosphine evolved catches fire and lights the evolved acetylene form calcium carbide; the flash light the caused gave signal to the ships.
2. Phosphine also is used in smoke screen, calcium phosphide is used in smoke screens, and phosphine produced catches fire to give the needed smoke.

Detection of Phosphine

(*i*) Its smells like rotten fish.

(*ii*) A black precipitate is also formed when the gas is passed through silver nitrate solution.

Diphosphine, Phosphorus Dihydride (P_2H_4): Diphosphine is formed along with phosphine by the action of water on calcium phosphide.

$$\text{Calcium Phosphide} + \text{Water} \longrightarrow \text{Calcium Hydroxide} + \text{Diphosphine}$$

$$Ca_2P_2 + H_2O \longrightarrow Ca(OH)_2 + P_2H_4$$

It is also prepared by the reaction of sodium hydroxide and white phosphorus and then boiling the mixture. When the dihydride is produced along with phosphine, it is separated by passing the mixture through a freezing mixture; diphosphine is condensed as a yellow liquid (see Fig 13.9).

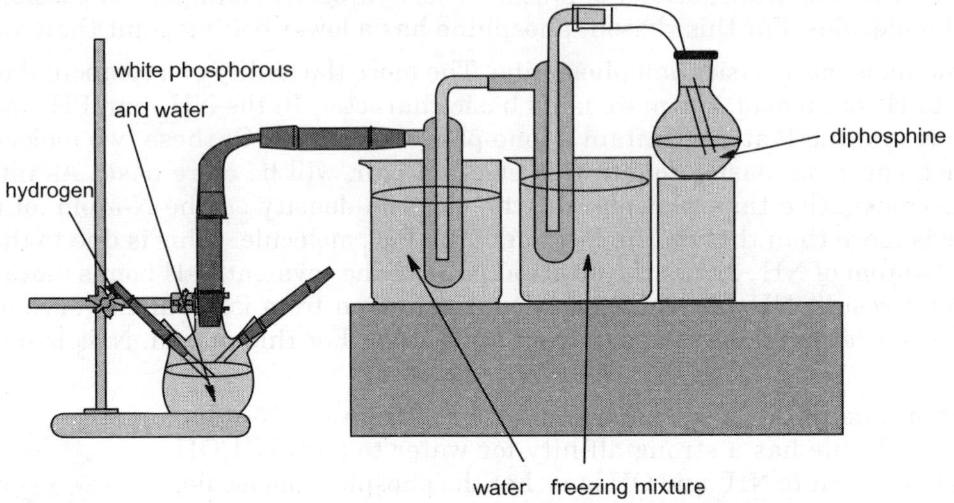

Fig. 13.9. The Lab Preparation of Diphosphine

Diphosphine is spontaneously inflammable and decomposes on standing to form amorphous yellow solids having compositions varying around P2~. These solids are insoluble in common solvents. Diphosphine differs from hydrazine and has no basic character.

Oxides of Phosphorus

Phosphorus forms three oxides:

1. Phosphorus Trioxide P_2O_3 or Phosphorus (*iii*) Oxide
2. Phosphorus Tetraoxide P_2O_4 or Phosphorus (*iv*) Oxide
3. Phosphorus Pentoxide P_2O_5 or Phosphorus (*v*) Oxide

Phosphorus Trioxide (P_2O_3): The formation of phosphorus trioxide takes place by the slow oxidation of phosphorus in air or combustion in a limited supply of air, which was noticed by Sage (1777), but pure phosphorus trioxide was first isolated by Trope and Tutten (1890 - 1891). White phosphorus was burnt is a stream of air in a tube, the phosphorus pentoxide was also formed which was removed by filtration through glass wool and the phosphorus trioxide vapour is condensed by cooling. Sticks of 11 long are placed in a hard glass (see Fig. 13.10) connected with a brass inner tube; 1 inch in diameter and 2 feet in length, of Liebig's Condenser containing water at 60°C. A plug of glass wool is fitted in the brass tube at the furthest end of the phosphorus. Which serves to the filter out solid phosphorus pentoxide while the phosphorus trioxide vapour passing in is condensed in the 'U' tube immersed in ice. A rapid stream of air is passed through the whole apparatus by a water pump connected with a sulphuric acid wash bottle, and the phosphorus ignited. The trioxide condensed in the "U" tube is melted by warming and the liquid runs down in the bottle is kept beneath.

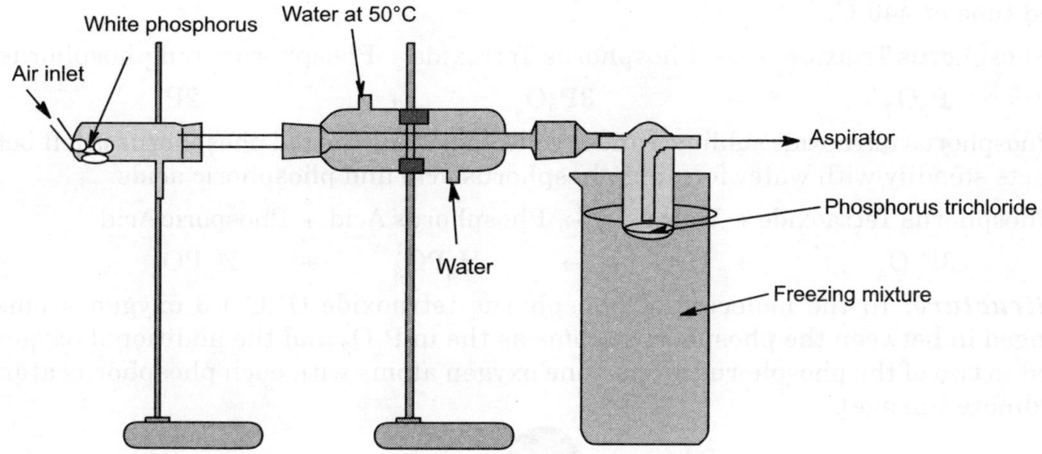

13.10. The Preparation of Phosphorus Trichloride

Properties of Phosphorus Trioxide (Physical)

Pure phosphorus trioxide is a colourless crystalline solid. It is waxy when impure. The boiling point of phosphorus trioxide is 171.3 °C. The melting point is 23.8°C.

Properties of Phosphorus Trioxide (Chemical)

Unless quite pure, phosphorus slowly turns red when exposed to weather because of the presence of impure red phosphorus in it. It can be purified by repeated exposure to light and sublimation. When quite pure it does not glow in air. It condenses in air or oxygen and forms phosphorus pentoxide.

$$\text{Phosphorus Trioxide} + \text{Oxygen} \longrightarrow \text{Phosphorus Pentoxide}$$

$$P_2O_3 + O_2 \longrightarrow P_2O_5$$

It fumes spontaneously in chlorine, forming phosphorus oxychloride.

$$\text{Phosphorus Trioxide} + \text{Chlorine} \longrightarrow \text{Phosphorus Oxychloride} + \text{Metaphosphoryl Chloride}$$

$$P_2O_3 + 2Cl_2 \longrightarrow POCl_3 + POCl$$

Structure of this compound has been studied in the vapour state by the electron diffraction method. The molecule is P_2O_3 and is also structurally related to P 4 molecule by a P-O-P bridge.

The bond angle of the oxygen phosphorus bonds is a single and the other a double bond. The structure of the oxide in the crystalline state is so far not yet been determined.

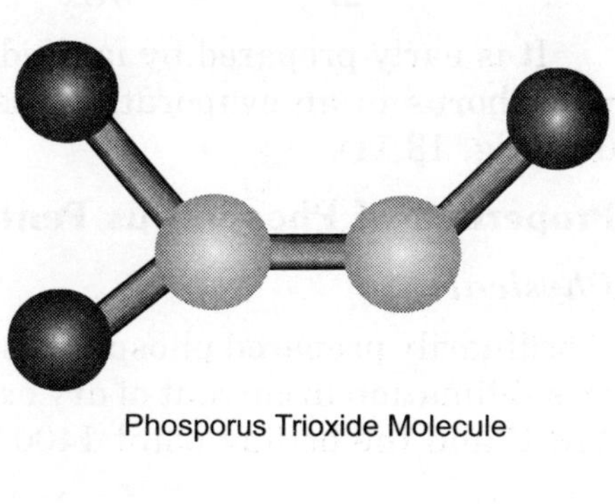

Phosporus Trioxide Molecule

Phosphorus Tetraoxide (P_2O_4): It is obtained by heating phosphorus trioxide in a sealed tube at 440°C.

Phosphorus Trioxide ⟶ Phosphorus Tetroxide + Phosphorus (red phosphorus)

$$P_2O_3 \longrightarrow 3P_2O_4 + 2P$$

Phosphorus tetraoxide sublimes and is collected, while the red phosphorus is left behind. It reacts steadily with water forming phosphorus acid and phosphoric acids.

Phosphorus Tetraoxide + Water ⟶ Phosphorus Acid + Phosporic Acid

$$3P_2O_4 + H_2O \longrightarrow H_3PO_3 + H_3PO_4$$

Structure: In the molecule of phosphorus tetraoxide (P_2O_4) 3 oxygen atoms are arranged in between the phosphorus atoms as the in P_2O_3 and the additional oxygen are linked to two of the phosphorus atoms (one oxygen atoms with each phosphorus atoms by co-ordinate linkage).

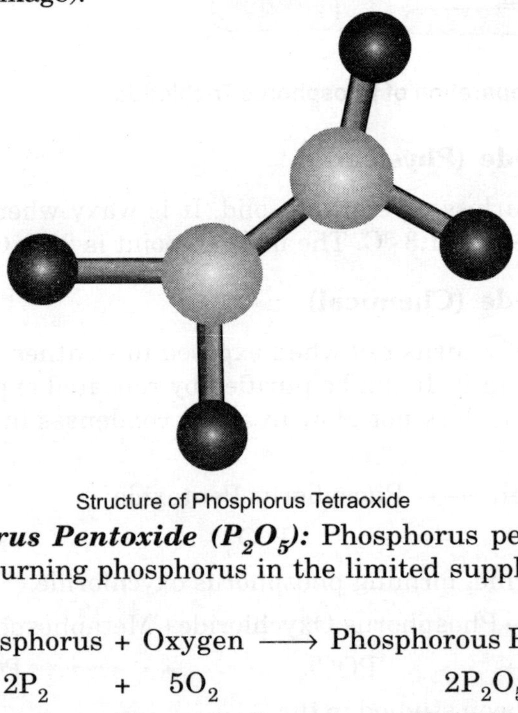

Structure of Phosphorus Tetraoxide

Phosphorus Pentoxide (P_2O_5): Phosphorus pentoxide is obtained by burning phosphorus in the limited supply of air or oxygen.

Phosphorus + Oxygen ⟶ Phosphorous Pentoxide

$$2P_2 + 5O_2 \qquad 2P_2O_5$$

It is early prepared by ignited a very small portion of white phosphorus in an evaporating dish covered by a small bell jar (see Fig. 13.11).

Fig. 13.11. Expremental setup for phosphorus pentoxide

Properties of Phosphorus Pentoxide

Physical

If ordinarily prepared phosphorus pentoxide has a faint odour like garlic, but it is purified by sublimation in current of dry oxygen it is completely odourless. It has a melting point of 418°C and the boiling point 1400°C.

Chemical Properties

(*i*) Phosphorus pentoxide has a great affinity for water. It dissolves in cold water with a hissing sound to form the metaphosphoric acid, HPO_3.

Phosphorus Pentoxide + Water ⟶ Metaphosphoric Acid

$$2P_2O_5 + H_2O \longrightarrow 2HPO_3$$

This gradually changes to orthophosphoric acid is much quicker if hot water is introduced instead of cold.

(*ii*) On account of great affinity towards water, phosphorus pentoxide possesses; it is the most powerful dehydrating agent known. It removes the elements of water from most organic and inorganic compounds. The sulphuric acid, when treated with phosphorus pentoxide trioxide is produced.

Sulphuric Acid + Phosphorus Pentoxide ⟶ Sulphur Trioxide + Metaphosphoric Acid

$$H_2SO_4 + P_2O_5 \longrightarrow SO_3 + 2HPO_3$$

And nitric acid gives nitrogen anhydride (Dinitrogen pentoxide).

Nitric Acid + Phosphorus Pentoxide ⟶ Dinitrogen Pentoxide + Metaphosphoric Acid

$$2HNO_3 + P_2O_5 \longrightarrow N_2O_5 + 2HPO_3$$

Similarly, amides of carboxylic acid give the corresponding nitrates.

Acitamide + Phosphorus Pentoxide ⟶ Acetonitrite + Metaphosphoric Acid

$$CH_3CONH_3 + P_2O_5 \longrightarrow CH_3CH + 2HPO_3$$

For the same reason phosphorus pentoxide chars wood, paper, sugar etc.

Exp: Take two dry test tubes, in one place take some cane sugar and in the other one a mixture of cane sugar and phosphorus pentoxide, Heat the tubes simultaneously. Note that the pure sugar is slightly affected while the mixture is rapidly chard.

When strongly heated with carbon it is reduced to phosphorus

Phosphorus Pentoxide + Carbon ⟶ Phosphorus + Carbon Monoxide

$$P_2O_5 + 5C \longrightarrow 2P_2 + 5CO$$

Structure of Phosphorus Pentoxide

The structure of the phosphorus pentoxide has been determined in the vapour as well as in the crystalline state. The electron detraction study of the vapour shows that the structure is derived from the phosphorus trioxide by adding one extra oxygen atom to each of the two phosphorus atoms so as to complete the as depicted in the structure.

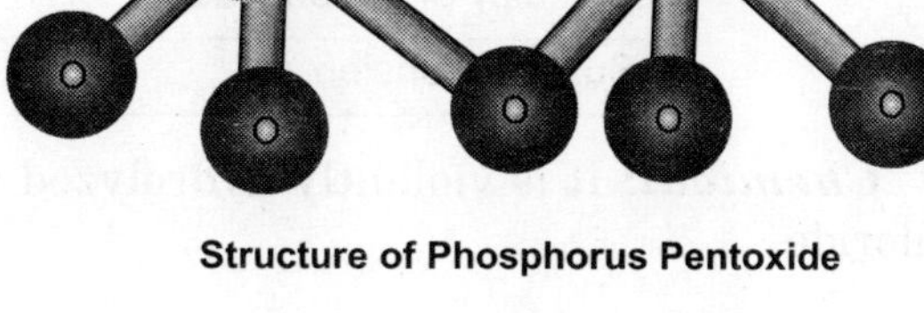

Structure of Phosphorus Pentoxide

Halides of Phosphorus

Phosphorus combines with all the halogens forming halides, which are all having co-valent bonds. Phosphorus chlorides are more important than the fluorides, bromides and iodides, tri and pentachloride are more common but the trichloride of phosphorus is not very stable.

Phosphorus Trichloride (PCl_3): Phosphorus trichloride is prepared by heating white phosphorus in a current of dry chloride. The apparatus is shown in the Fig. 13.12, dry phosphorus is heated on a water bath in the atmosphere of dry chlorine. Phosphorus trichloride formed being volatile is distilled of and collected in the water-cooled reservoir.

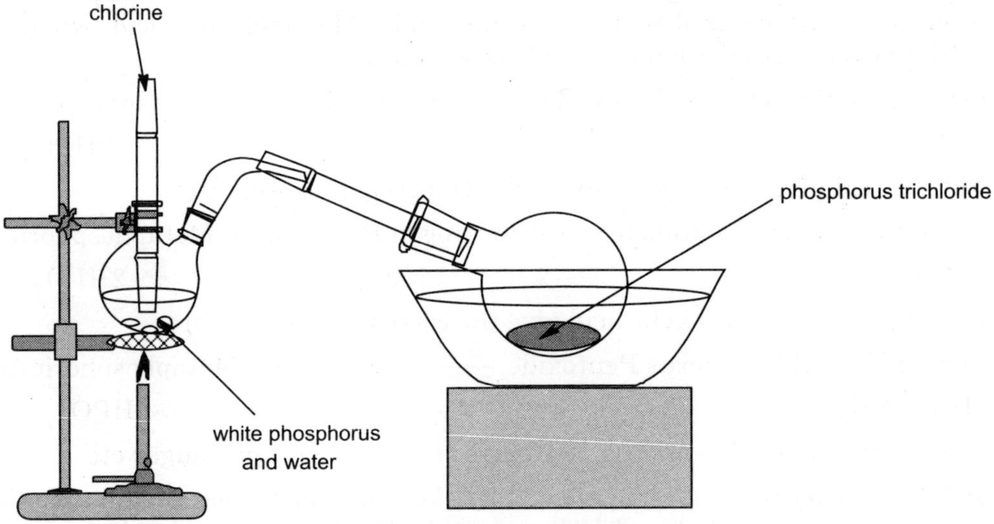

Fig. 13.12. The Lab Preparation of Phosphorus Trichloride

$$\text{Phosphorus} + \text{Chlorine} \longrightarrow \text{Phosphorus Trichloride}$$

$$2P_2 \quad + \quad 6Cl_2 \quad \longrightarrow \quad 4PCl_2$$

Physical Properties

Physical State	Liquid
Colour of the Physical state	White
Melting Point	–91°C
Boiling Point	76°C
Density	1.59 g/ml
Enthalpy of Vapourization	30.5 kJ/mol
Surface Tension	27.9 mN/m(millinewton per metre)

Chemical: It is violently hydrolyzed by water giving phosphorus acid and hydrogen chloride.

Phosphorus Trichloride + Water $\longrightarrow$ Hydrogen Chlrodie + Phosphorus Acid

$$PCl_3 + 3H_2OO \longrightarrow 3HCl_3HCl + H_3PO_3$$

(*ii*) It readily combines with oxygen phosphorus oxychloride.

Phosphorous Trichloride + Oxygen $\longrightarrow$ Phosphorous Oxychloride

$$2PCl_3 + O_2 \longrightarrow 2POCl_3$$

(iii) It reacts with chloride or sulphuryl chloride forming phosphorus pentachloride.

Phosphorous Trichloride + Chlorine $\longrightarrow$ Phosphorous Pentachloride

$$2PCl_3 + Cl_2 \longrightarrow PCl_5$$

(iv) Phosphorus trichloride also reacts with sulphur trioxide violently

Phosphorous Trichloride + Sulphur Trioxide $\longrightarrow$ Phosphorous Oxychloride + Sulphur Dioxide

$$2PCl_3 + SO_3 \longrightarrow POCl_3 + SO_2$$

(v) It is decomposed by hot concentrated sulphuric acid.

Phosphorous Trichloride + Sulphuric Acid $\longrightarrow$ Sulphuryl Hydrochloride + Metaphosphoric Acid + Hydrogen Chloride

$$PCl_3 + 2H_2SO_4 \longrightarrow SO_3HCl + HPO_3 + 2HCl$$

It reacts with organic compounds like acetic acid and alcohol where the hydroxyl ion (–OH) group is replaced by the chloride atom and phosphorus acid is also produced.

$$3CH_3CO\boxed{OH + P}Cl_3 \longrightarrow 3CH_3COl + H_3PO_3$$

$$3C_2H_5\boxed{OH + P}Cl_3 \longrightarrow 3C_2H_5Cl + H_2PO_3$$

Phosphorus Trichloride as Electron Donor

If we look at the electronic structure rather the electric dot formula. We will be able to see that phosphorus has one line pair of electrons, which it can donate other electron deficient species (ions and molecules). In other words it acts as an electrons donor that is a Lewis Base.

$$Cl\!:\!\ddot{\underset{\cdot\cdot}{P}}\!:\!Cl$$

Electron Dot Formula of PCl_3

Industrial Preparation of Phosphorus Trichloride

Phosphorus trichloride is usually prepared by the direct reaction of phosphorus and chlorine. In order to control the heat of the reaction of the elements are premixed and then continuously refluxed. The diagrammatic representation of the plant is shown in the Fig. 13.13. Liquid Phosphorus and chlorine are continuously introduced into the reaction vessel, which is arranged such that a significant portion of the phosphorus trichloride is refluxed.

The remaining phosphorus trichloride is distilled into a pot, where it is traced with additional chlorine is continuously introduced into the reaction vessel, which is arranged such that a significant portion of the phosphorus trichloride is refluxed.

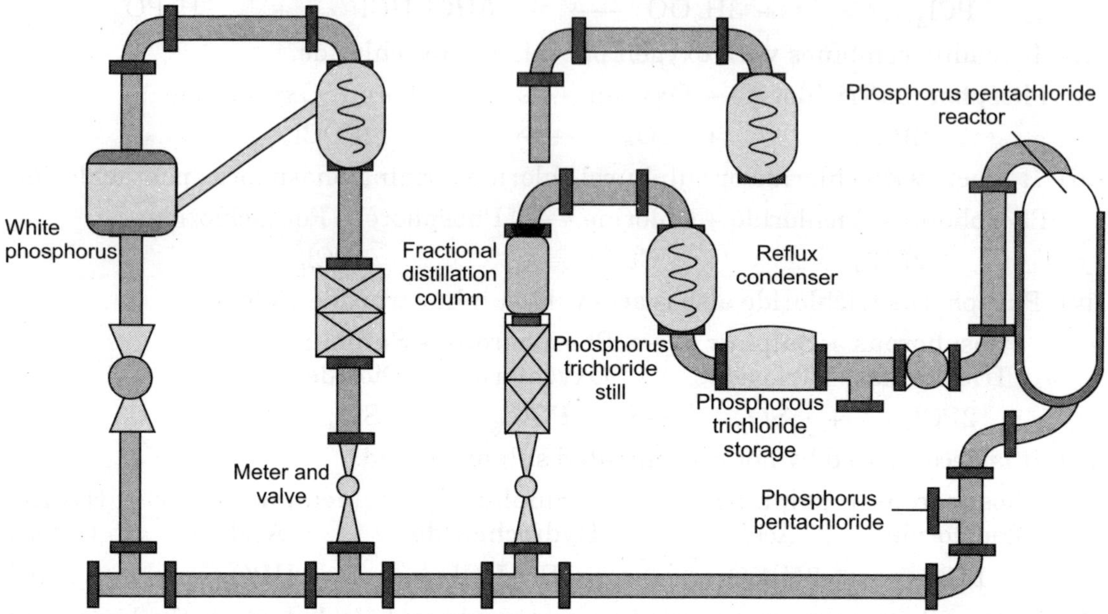

Fig. 13.13. The Industrial Preparation of Phosphorus Pentachloride and Trichloride

The remaining phosphorus trichloride is distilled into a pot, where it is traced with additional chlorine to remove traces of concentrated phosphorus. The resulting crude phosphorus trichloride is then fractionally distilled to free it from small amounts of organic matter consisting of portions of chloride group and also any traces of water carried in with phosphorus. A yield of about 95% of purified phosphorus trichloride is usually obtained with respect to both the phosphorus trichloride is usually dissolved in the carbon trichloride before treating with chlorine. In a water-cooled jacket provided with an efficient stirrer in a tight cover with a reflux condenser, a mixture of about one part of phosphorus trichloride to one part of carbon trichloride is introduced. The chlorine is then allowed to enter the part above the part liquid level and the crystals of phosphorus pentachloride begin to settle in the liquid. When the reaction is completed, the suspension of crystals of phosphorus pentachloride are drawn out of the vessel into a filter, which removes the crystals. Circulating hot water through the jacket of filter dries the crystals and the carbon tetrachloride is returned to the reacting vessel. In the plant shown in the Fig. 13.14, the phosphorus pentachloride is made continuously by running the phosphorus trichloride countercurrent to the chlorine and collecting the resulting phosphorus pentachloride at the bottom of the reaction tower. Lead pumping and head limited steel equipment is usually in the construction of the manufacturing of phosphorus pentachloride

Phosphorus Oxychloride ($POCl_3$): It is obtained by the partial hydrolysis of phosphorus pentachloride with potassium chlorate.

Phosphorus Trichloride + Potassium Chlorate ⟶ Phosphorus Oxychloride + Potassium Chloride

$$3PCl_3 + KClO_3 \longrightarrow 3POCl_3 + KCl$$

It is best prepared by the reaction of oxalic acid and phosphorus pentachloride.

Phosphorus Pentachloride + Oxalic Acid ⟶ Phosphorus Oxychloride + Hydrogen Chloride + Carbon Dioxide + Carbon Monoxide

$$PCl_5 + H_2C_2O_4 \longrightarrow POCl_3 + 2HCl \quad CO_2 \quad CO$$

It is colourless fuming liquid boiling at 106°C, it is readily hydrolyzed into orthophosphoric acid.

Phosphorus Oxychloride + Water ⟶ Orthophosphoric + Hydrogen Chloride

$$POCl_3 + 3H_2O \longrightarrow H_3PO_4 + 3HCl$$

Phosphorus oxychloride also reacts with many organic compounds removing the hydroxyl ion from that that compound.

Phosphorus Oxychloride + Alcohol ⟶ Hypophosphoric Acid + Ethyl Chloride

$$POCl_3 + 3C_2H_5OH \longrightarrow H_3PO_2 + 3C_3H_5Cl$$

It has the advantage over phosphorus pentachloride in being a less chlorinating agent, behaviour that is frequency more desirable.

Industrial Preparation of Phosphorus Oxychloride

Phosphorus oxychloride is manufactured in industry by bubbling chlorine through an insoluble matter of phosphorus pentoxide in phosphorus trichloride and distilling the resulting phosphorus oxychloride from the reaction kettle (reactor).

A measured quantity of raw phosphorus trichloride is charged into the reactor. This phosphorus pentachloride is added in slight excess of the reaction is dissipated by refluxing phosphorus trichloride the rate of chlorine is adjusted to prevent flooring of the reflux condenser. As the reaction approaches completion, then the steam is admitted into the reactor jacket to maintain the reaction temperature. As soon as the chlorine appears in the vapours, the chlorination is stopped and the product sent for distillation. Yield of phosphorus oxychloride is 96% to 98%.

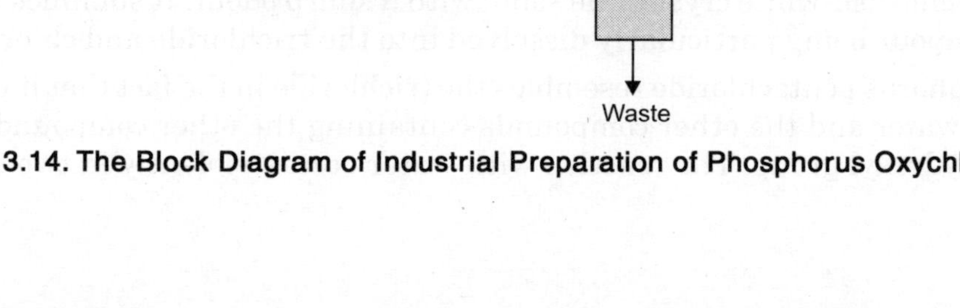

Fig. 13.14. The Block Diagram of Industrial Preparation of Phosphorus Oxychloride

Phosphorus Pentachloride (PCl_5): Phosphorus pentachloride is obtained by the action of dry chlorine used in excess on phosphorus trichloride. The apparatus set up is shown in Fig. 13.15. A wide mouth bottle is fitted with a cork carrying (1) Dropping Funnel, (2) an inlet tube for dry chlorine and (3) an outlet tube. The dropping funnel and the outlet tube are provided with calcium chloride guard tubes to keep the moisture out. Phosphorus trichloride is slowly dropped while a current of dry chlorine is passed through the flask. When a sufficient amount of phosphorus trichloride is collected the apparatus is disconnected and the flask is tightly stoppered.

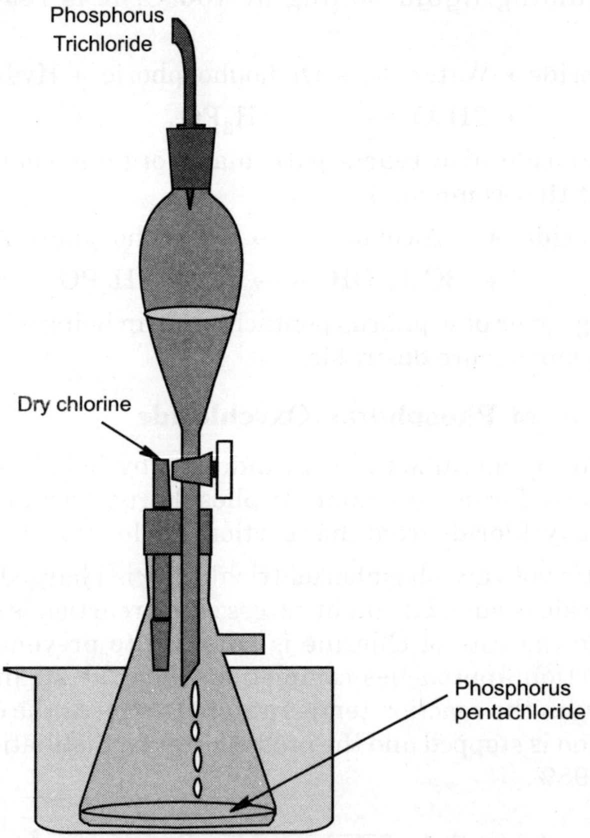

Fig. 13.15. The Lab Preparation of Phosphorus Pentachloride

Properties of Phosphorus Pentachloride

(*i*) It is yellowish-white crystalline solid, with a sharp odour. It sublimes below 100°C, the vapour being particularly dissolved into the trichloride and chloride.

(*ii*) Phosphorus pentachloride resembles the trichloride in the fact that it readily reacts with water and the other compounds containing the other compounds containing the hydroxyl group. The reaction with water occurs violently in two steps.

Phosphorus Trichloride	+	Water	$\longrightarrow$	Phosphorous Oxychloride	+	Hydrogen Chloride
PCl_3	+	H_2O	$\longrightarrow$	$POCl_3$	+	2HCl
Phosphorus Oxychloride	+	Water	$\longrightarrow$	Phosphoric Acid	+	Hydrogen Chloride
$POCl_3$	+	H_2O	$\longrightarrow$	H_3PO_4	+	2HCl

Structure of Phosphorus Pentachloride

In liquid and also in gaseous state PCl_5 has trigonal by pyramidal structure.

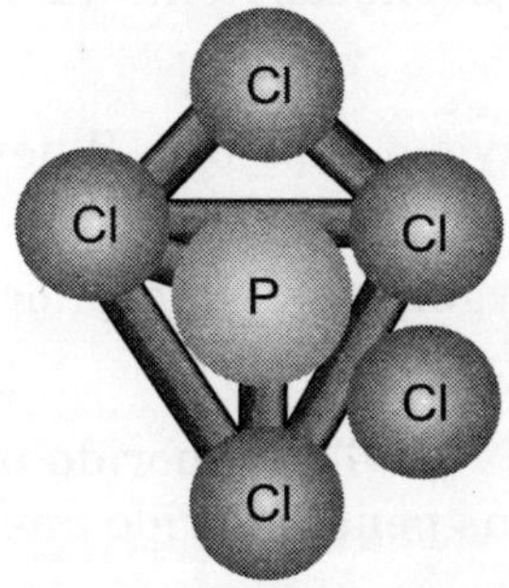

Structure of Phosphorus Pentachloride

Phosphorus Sulphochloride ($PSCl_3$): This compound of phosphorus is easily obtained by the direct addition of sulphur and phosphorus trichloride.

Phosphorus Trichloride + Sulphur $\longrightarrow$ Phosphorous Sulphachloride

$$2PCl_3 + S_2 \longrightarrow 2PSCl_3$$

The temperature of this reaction being around 175°C to 180°C. The reaction proceeds smoothly when the phosphorus trichloride vapour is passed through an inert solvent. The use of sodium hydroxide or barium hydroxide catalyzes the reaction so that it takes place at 115°C. When an excess of sulphur is used, the residue of the sulphur forming the following charge the yield is better than 98% based upon either reaction materials. Phosphorus sulphochloride can also be prepared by reaction of the pentasulphide and pentachloride.

Other Halides of Phosphorus

Besides chloride, phosphorus also reacts with bromine, fluorine and iodine to form respective compounds. They are now being discussed below.

Phosphorus Trifluoride (PF_3) : It was first discovered by Mossan in the year 1884. It is produced by the action of arsenic trifluoride and phosphorus trichloride.

Phosphorus Trichloride	+	Arsenic Triflouride	$\longrightarrow$	Phosphorus Trifluoride	+	Arsenious Trichloride
PCl_3	+	AsF_3	$\longrightarrow$	PF_3	+	$AsCl_3$

It can also be obtained by the reaction phosphorus tribromide and zinc fluoride.

Phosphorus Tribromide + Zinc Trifluoride ⟶ Phosphorus Trifluoride+ Zinc Tribromide

$$5PBr_3 + 3ZnF_3 \longrightarrow 2PF_3 + ZnBr_3$$

The gas does not fume in air and has no action on glass in the cold, it is hydrolyzed by water.

Phosphorous Trifluoride + Water ⟶ Phosphorous Acid + Hydrogen Fluoride

$$2PF_3 + 3H_2O \longrightarrow H_3PO_4 + 3HF$$

And it is also decomposed by sparking.

Phosphorus Trifluoride ⟶ Phosphorus Pentafluoride + Phosphorous

$$2PF_3 \longrightarrow 3PF_5 + 2P$$

A mixture of the vapour with oxygen explodes. When sparked, produces phosphorus oxyfluoride.

Phosphorus Trifluoride + Oxygen ⟶ Phosphorus Oxyfluoride

$$2PF_3 + O_2 \longrightarrow 2POF_3$$

Phosphorus Pentafluoride (PF_5): This fluoride of phosphorus was discovered in 1877. It is produced when phosphorus pentachloride and arsenic trifluoride is mixed in a freezing mixture.

Phosphorus Pentachloride + Arsenic Fluoride ⟶ Phosphorus Pentafluoride + Arsenic Trichloride

$$3PCl_3 + 5AsF_3 \longrightarrow 3PF_5 + 5AsCl_3$$

It is also obtained when phosphorus flurobromide (made by a mixture of bromine and phosphorus trifluoride at –19°C) at 15°C.

Phosphorus Fluorobromide ⟶ Phosphorus Pentafluoride + Phosphorus Pentabromide

$$5PF_3Br_2 \longrightarrow 3PF_5 + 2PBr_5$$

Also prepared by heating a mixture of 25g of phosphorus pentoxide and 55g of powdered calcium fluoride in an iron tube.

Phosphorus Pentoxide + Calcium Fluoride ⟶ Phosphorus Pentafluoride + Calcium Metaphosphate

$$6P_2O_5 + 5CaF_2 \longrightarrow 2PF_5 + 5Ca(PO_3)_2$$

The gas has the normal density corresponding with PF_5, does not attack glass when dry, fumes m air.

Phosphorus Pentafluride + Water ⟶ Phosphorus Oxyfluoride + Hydrogen Fluoride

$$PF_5 + H_2O \longrightarrow POF_3 + 2HF$$

And combines with ammonia gas to form solid $2PF_5NH_3$.

Phosphorus Oxyfluoride (POF_3): It is obtained by heating a mixture of hydrogen fluoride and phosphorus pentoxide.

Phosphorus Pentoxide + Hydrogen Fluoride ⟶ Phosphorus Oxyfluoride + Water

$$P_2O_5 + 6HF \longrightarrow 2POF_3 + 3H_2O$$

The reaction is carried out in a brass tube, it is a colourless gas with a boiling point of –39.8°C and melting point –39.4°C. the gas is collected over mercury. The gas is also prepared by the reaction of zinc fluoride and phosphorus oxychloride.

Phosphorus Oxyfluoride + Zinc Fluoride ⟶ Phosphorus Oxyfluoride + Zinc Chloride

$$2POF_3 + 3ZnF_2 \longrightarrow 2POF_3 + 3ZnCl_2$$

Phosphorus. Dichloride (PCl_2): It is an oily fuming liquid formed by the action of a silent discharge on a mixture of phosphorus trichloride vapour and hydrogen.

Phosphorus Trichloride + Hydrogen ⟶ Phosphorus Dichloride + Hydrogen Chloride

$$PCl_3 + H_2 \longrightarrow 2PCl_2 + 2HCl$$

It is also obtained by the action of zinc arc on phosphorus trichloride.

Phosphorus Trichloride $\xrightarrow{\text{Zinc arc}}$ Phosphorus Dichloride + Zinc Chloride

$$2PCl_3 + (Zn) \longrightarrow PCl_2 + ZnCl$$

Phosphorus Bromide and Iodide

White phosphorus explodes in contact with liquid chlorine or bromine. Liquid bromine when dropped on red phosphorus in a cooled flask reacts with evolution of light and phosphorus tribromide is then produced.

By adding more bromine to this phosphorus pentabromide is produced.

Phosphorus Tribromide + Bromine ⟶ Phosphorus Pentabromide

$$2PBr_3 + 2Br_2 \longrightarrow 2PBr_5$$

This vapour gets dissociated.

Phosphorus Oxybromide ($POBr_3$): It is a crystalline solid with melting point of 56°C and boiling point of 108°C is formed by the action of all little water or of oxalic acid on PBr_3 and by distilling PBr_5 with P_2O_5.

Phosphorous Tribromide + Phosphorus Pentoxide ⟶ Phosphorous Oxybromide

$$PBr_3 + P_2O_5 \longrightarrow 5POBr_3$$

White phosphorus inflames in contact with iodine, phosphorus di-iodide is produced, formula PI_2 and then the phosphorus tri-iodide, PI_3 are formed by mixing solutions of iodine and white phosphorus in carbon disulphide in the correct ration and evaporating, PI_5 and PIO_3 are not definitely known. The probably $PI_2.I_2P$.

Oxy Acids of Phosphorus

Besides compounds phosphorus also produces acids they are being listed below:

Name	Formula	Structure formula	Remark
Hypophosphorus Acid	H_3PO_2	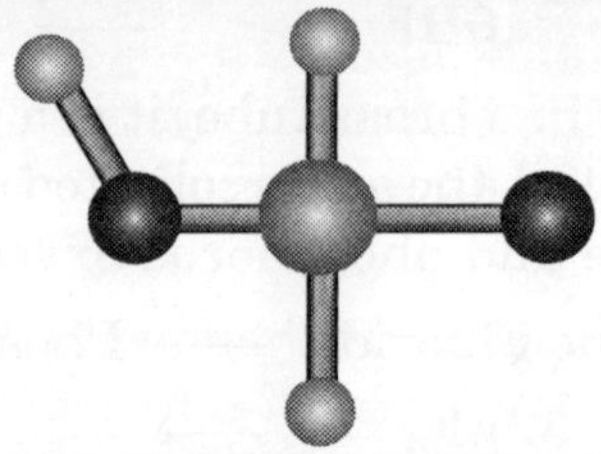	Tetrahedral Monoprotic
Phosphorus Acid	H_3PO_3	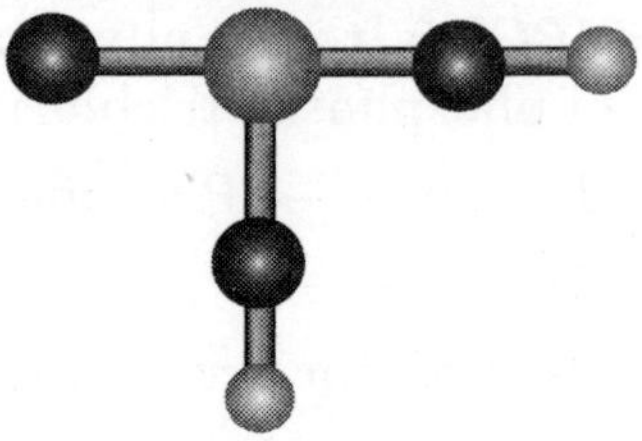	Tetrahedral, Diprotic
Metaphosphorus Acid	HPO_3	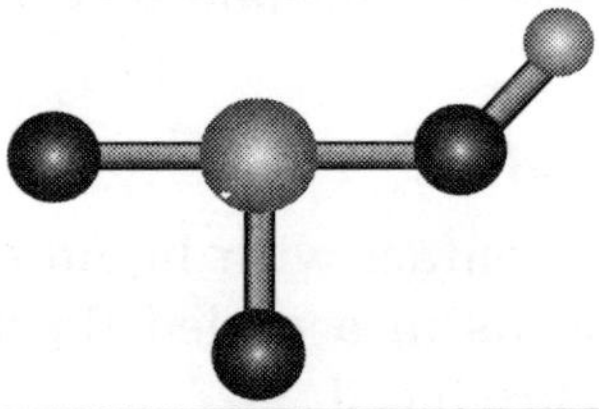	Long Chain, PO_4
Orthophosphoric Acid	H_3PO_4	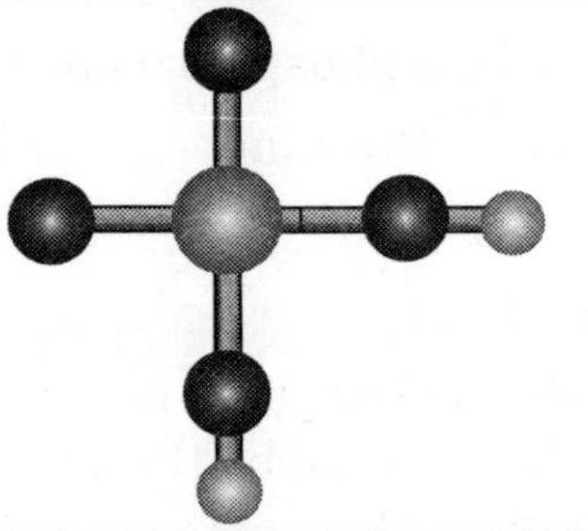	Tetrahedral,highly hydrogen bonded
Pyrophosphoric	$H_4P_2O_7$	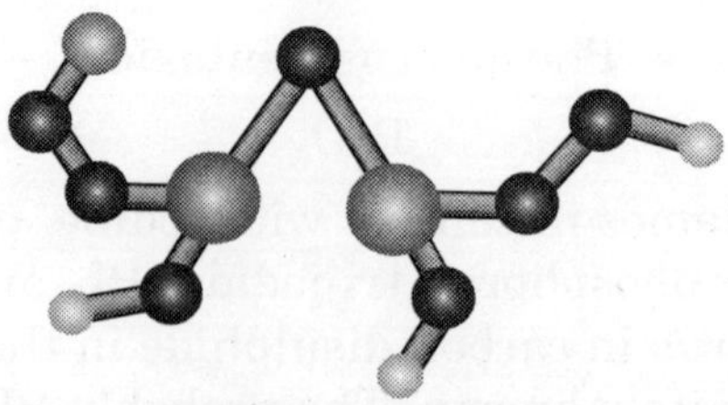	Two PO_4 tetrahedral linked through Oxygen bridge (tetraporic)
Hypophosphoric Acid	$H_4P_2O_6$	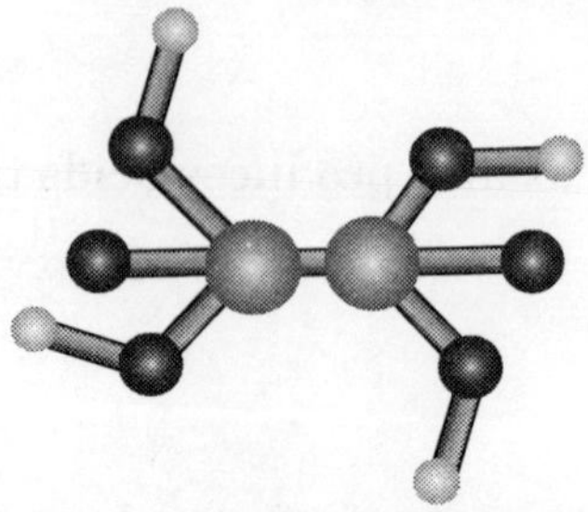	Two PO_3 tetrahedral having P-P bond (traprotic)

We are now going to discuss the acids of phosphorus:

Hypophosphorus Acid (H_3PO_2): When phosphorus is prepared by heating white phosphorus with barium hydroxide solution, barium hypophosphite is formed in the solution.

$$\begin{array}{lllll} \text{Phosphorus} & + \text{ Barium Hydroxide} & + \text{ Water} \longrightarrow & \text{Phosphene} & + \text{ Barium Hypophosphite} \\ 8P & + \quad 3Ba(OH)_2 & + \ 6H_2O \longrightarrow & 2PH_3 & + \quad 3Ba(H_2PO_2)_2 \end{array}$$

Phosphorus is in fact may be said to suffer hydrolysis and disproportion to become phosphene and phosphorus acid.

$$\begin{array}{llll} \text{Phosphorus} & + \text{ Water} \rightleftharpoons & \text{Phosphine} & + \text{ Hypophosphorus Acid} \\ 4P & + \ H_2PO_4 \rightleftharpoons & PH_3 & + \quad H_3PO_2 \end{array}$$

The solution remaining behind in flask is treated with carbon dioxide to remove the excess of barium hydroxide as carbonate. The filtrate from the carbonate is evaporated till the barium hypophosphite crystallizes. The crystals are then made to react with the calculated amount of sulphuric acid.

$$\begin{array}{llll} \text{Barium Hypophophate} & + \text{ Sulphuric Acid} \longrightarrow & \text{Barium Sulphate} & + \text{ Hypophosphorus Acid} \\ Ba(H_2PO_2)_2 & + \quad H_2SO_4 \longrightarrow & BaSO_4 & + \quad 2H_3PO_2 \end{array}$$

The filtrate is evaporated below 130°C to syrup and cooled in a freezing mixture it crystallizes. Hypophosphorus acid from colourless crystallizes producing phosphene at 50°C.

$$\begin{array}{lll} \text{Hypophosphorus Acid} \longrightarrow & \text{Phosphine} & + \text{ Phosphorus Acid} \\ 3H_3PO_2 \longrightarrow & PH_3 & + \quad 3H_3PO_3 \end{array}$$

At higher temperature phosphorus acid and phosphite further disproportionate resulting phosphine and phosphoric acid or phosphate.

$$\begin{array}{lll} \text{Phosphorus Acid} \longrightarrow & \text{Phosphine} & + \text{ Phosphoric Acid} \\ 4H_3PO_3 \longrightarrow & PH_3 & + \quad 3H_3PO_4 \end{array}$$

Hypophosphorus acid and its salts are powerful reducing agents precipitating silver, gold, platinum and mercury from their respective salt solutions. From a salt of copper sulphate phosphorus acid liberates cupric hydride.

$$\begin{array}{llllll} \text{Copper Sulphate} & + \text{ Hypophosphorus Acid} \longrightarrow & \text{Water} & + \text{ Cupric Hydride} & + \text{ Sulphuric Acid} & + \text{ Phosphoric Acid} \\ 4CuSO_4 & + \quad 3H_3PO_2 \longrightarrow & H_2O & + \ 2Cu_2H_2 & + \ 4H_2SO_4 & + \ 3H_3PO_4 \end{array}$$

It also liberates gold from gold chloride.

$$\begin{array}{llllll} \text{Gold Chloride} & + \text{ Hypophosphoric Acid} \longrightarrow & \text{Water} & + \text{ Phosphoric Acid} & + \text{ Hydrogen Chloride} & + \text{ Gold} \\ 2AuCl_3 & + \quad 3H_3PO_2 \longrightarrow & 3H_2O & + \ 3H_3PO_4 & + \ 6HCl & + \ 2Au \end{array}$$

Structure: By structure it was only one hydrogen atom that it is a monobasic acid. The structure of this acid shown below:

Hypophosphorus acid produced salts called hypophosphites. These compounds were popular in medical practice, as it was supposed that they would be easily assimilated by the human body especially the brain and nerve tissue. The belief has now been proved useless because these Compounds pass through the body without any change.

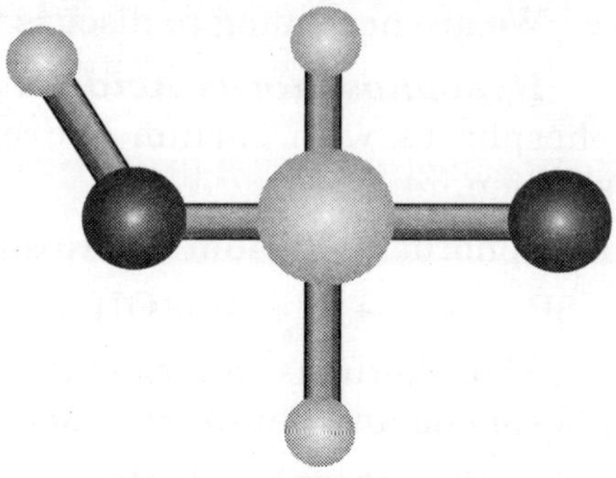

Structure of Hypophosphorus Acid

Preparation of hypophosphites: The first step in the of commercial hypophosphites is the production of calcium hypophosphite $Ca(H_2PO_2)_2$ by the action of hypophosphorus acid milk of lime (calcium hydroxide solution) heated at 55°C.

Calcium Hydroxide + Hypophosphorus Acid ⟶ Calcium Hypophosphite + Water

$$Ca(OH)_2 + 2H_3PO_2 \longrightarrow Ca(H_2PO_2)_2 + 2H_2O$$

The other hypophosphites are obtained by double decomposition of Calcium ash. Thus the sodium and potassium hypophosphite are prepared by the reaction of calcium hypophosphite with potassium carbonate or sodium carbonate.

Sodium Carbonate + Calcium Hypophosphate ⟶ Sodium Hypophosphate + Calcium Carbonate

$$Na_2CO_3 + Ca(H_2PO_2)_2 \longrightarrow 2NaH_2PO_2 + CaCO_3$$

The insoluble calcium Carbonate is filtered off and the solution is then evaporated to crystallization or granulation under reduced pressure.

Properties of Hypophosphites

(*i*) When heated; hypophosphites decompose to give off steam and phosphene. A precipitate of the pyrophosphate of the metal being left behind for *e.g.*

Calcium Hypophosphate ⟶ Calcium Pyrophosphate + Phosphine Water

$$2Ca(H_2PO_2)_2 \longrightarrow Ca_2P_2O \quad 2PH_3 + H_2O$$

(*ii*) When a solution of hypophosphite is added to a solution of mercuric chloride, a white precipitate of mercurous chloride is produced; on adding of excess of hypophosphite, the precipitate turns gray owing to the conversion of mercurous salt to metallic mercury.

Calcium Hypophosphate + Mercuric Chloride ⟶ Mercurous Chloride + Phosphoric Acid + Calcium Chloride

$$Ca(H_2PO_2)_2 + 4HgCl_2 \longrightarrow 2Hg_2Cl_2 + 2H_3PO_4 \quad 2CaCl_2$$

Mecurous Chloride + Calcium Hypophosphate ⟶ Mercury + Calcium Chloride + Phosphoric Acid

$$2Hg_2Cl_2 + 2Ca(H_2PO_2)_2 \longrightarrow 4Hg + 2CaCl_2 + 2H_3PO_4$$

(*iii*) When silver nitrate reacts with hypophosphorus acid, white precipitate of silver hypophosphite is produced, which is immediately reduced to metallic silver, sometimes with or sometimes with the evolution of hydrogen.

$$\text{Hypophosphorus Acid} + \text{Silver Nitrate} + \text{Water} \longrightarrow \text{Silver Hypophosphate} + \text{Nitrous Acid}$$
$$2H_3PO_2 + 4AgNO_3 + H_2O \longrightarrow Ag(H_2PO_2)_2 + 4HNO_2$$
$$\text{Silver Hypophosphate} \longrightarrow \text{Silver} \quad \text{Phosphoric Acid}$$
$$2Ag(H_2PO_2)_2 \longrightarrow 2Ag + 2H_3PO_4$$

Phosphorus Acid (H_3PO_3): Also known as 'orthophosphorus acid'. It can be obtained by the reaction of phosphorus trioxide on water that is why phosphorus trioxide is known as the anhydride of phosphorus acid.

$$\text{Phosphorus Trioxide} + \text{Water} \longrightarrow \text{Phosphoric Acid}$$
$$P_2O_3 \quad 3H_2O \longrightarrow 2H_3PO_3$$

The acid is best prepared by the action of phosphorus trichloride on water.

$$\text{Phosphorus Trichloride} + \text{Water} \longrightarrow \text{Phosphorus Acid} + \text{Hydrogen Chloride}$$
$$PCl_3 + 3H_2O \longrightarrow 2H_3PO_3 + 3HCl$$

The solution of the reaction mixture is concentrated by evaporation until the temperature rises to 180°C when hydrogen chloride is driven off. The concentrated solution on cooling yields crystals of phosphorus acid.

Properties: Phosphorus acid forms colourless deliquescent crystalline solids which melts at 73.6°C. When heated it decomposes into phosphene and orthophosphoric acid.

$$\text{Phosphorus Acid} \longrightarrow \text{Phosphoric Acid} + \text{Phosphine}$$
$$4H_3PO_33 \longrightarrow 3H_3PO_44 \quad PH_3$$

Though there are three hydrogen atoms in the molecule H_3PO_3 it is a dibasic acid of medium strength; hence the acid two series of salts (phosphites and hydrogen phosphites) for *e.g.* sodium phosphite, ($NaHPO_3$) and sodium hydrogen phosphite (NaH_2PO_3). Most of the metallic phosphites except those of sodium, potassium and calcium are soluble in water.

On standing, the aqueous solution of phosphorus acid is gradually oxidized by atmospheric oxygen to phosphoric acid.

$$\text{Phosphorus Acid} + \text{Oxygen} \longrightarrow \text{Phosphoric Acid}$$
$$2H_3PO_3 + O_2 \longrightarrow H_3PO_4$$

Phosphorus Acid is also a powerful reducing agent also. Like hypophosphorus acid it reduces gold and silver from their respective salt solutions.

$$\text{Phosphorus Acid} + \text{Gold Chloride} + \text{Water} \longrightarrow \text{Phosphoric Acid} + \text{Hydrogen Chloride} + \text{Gold}$$
$$2H_3PO_3 + 2AuCl_3 + 3H_2OO \longrightarrow 3H_3PO_4 + 6HCl + 2Au$$
$$\text{Silver Nitrate} + \text{Phosphorus Acid} + \text{Water} \longrightarrow \text{Phosphoric Acid} + \text{Nitric Acid} + \text{Silver}$$
$$2AgNO_3 + H_3PO_3 + H_2O \longrightarrow H_3PO_4 + 2HNO_3 + 2Ag$$
$$\text{Phosphorus Acid} + \text{Sulphurous Acid} \longrightarrow \text{Phosphoric Acid} + \text{Water} + \text{Sulphur}$$
$$2H_3PO_3 + H_2SO_3 \longrightarrow 2H_3PO_4 \quad H_2O + S$$

Structure

Phosphorus acid is best represented by the structure:

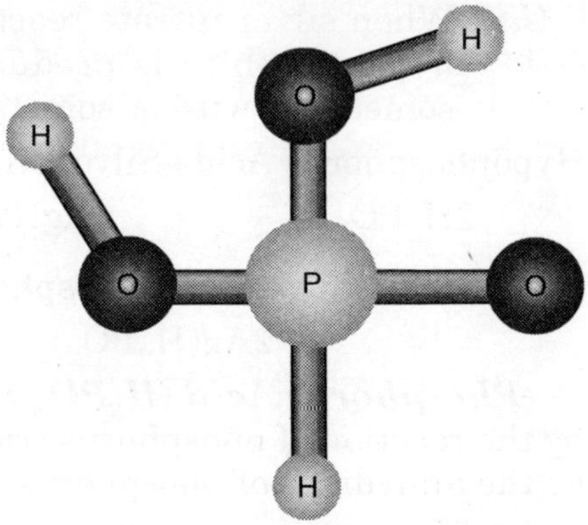

Structure of Phosphorus Acid

This structure explains dibasic character and it's reducing properties. Its reducing properties are due to the hydrogen atom directly linked with phosphorus atom. Only the two hydrogen atoms, which are linked, with oxygen atom ionize to give hydrogen ions. For this reason, each molecule of phosphoric acid when dissolved in water give to hydrogen ions, hence it acts as a dibasic acid.

Metaphosphoric acid [$(HPO_3)_n$]: This acid is obtained on treating phosphorus pentoxide with a small amount of water.

$$\text{Phosphorus Pentoxide} + \text{Water} \longrightarrow \text{Metaphosphoric Acid}$$
$$2P_2O_5 + 2H_2O \longrightarrow 4HPO_3$$

It may be prepared by dehydrating phosphoric or pyrophosphoric acid

$$\text{Phosphoric Acid} \xrightarrow{610°C} \text{Metaphosphoric Acid} + \text{Water}$$
$$2H_3PO_4 \longrightarrow 2HPO_3 + 2HO$$
$$\text{Pyrophosphoric Acid} \longrightarrow \text{Metaphosphoric Acid} \quad \text{Water}$$
$$H_4P_2O_7 \longrightarrow 2HPO_3 \quad 2H_2O$$

Properties

(*i*) It is a transparent glassy solid and is therefore also known as "Glacial Phosphoric acid". It melts at 38°C. It is also soluble in water.

(*ii*) The solution of the acid in water changes into orthophosphoric acid on boiling.

(*iii*) It is monobasic acid, forming salts known as metaphosphate *e.g.* Sodium metaphosphate, $NaPO_3$

Structure: Metaphosphoric acid is not a monomer having the formula HPO_3. Its simplest form known to be a di-polymer $(HPO_3)_n$. Higher polymers of the acid are also known. The acid it therefore, represented as $(HPO_3)_n$ where n is the number of monomer. The simplest form of the metaphosphorus acid is shown in the structure.

As expected it behaves as a dibasic acid, the dissociation constant being KT = 1.6×10^{-2} and $K_2 = 7 \times 10^{-7}$. The corresponding ion as the structure.

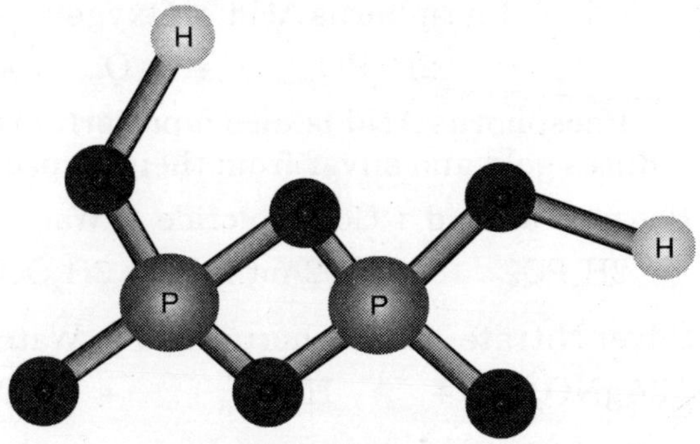

Structure of Metaphosphoric Acid

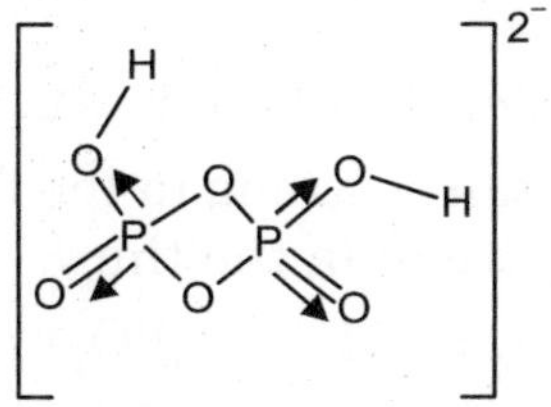

Sodium hexametaphosphate, a complex compound of phosphorus is most important complex of phosphorus having a common formula of $(NaPO_3)_6$ is used as a water softener under the commercial name *Calgon.*

Phosphoric Acid, Orthophosphoric Acid (H_3PO_4): It is the most important acid of phosphorus. This acid may be obtained in the laboratory by anyone of the following methods:

(*i*) *By adding phosphorus pentoxide* in water very cautiously and then boiling the solution so as to convert any metaphosphoric acid, first formed into the ortho acid.

Phosphorus Pentoxide + Water ⟶ Phosphoric Acid

$$P_2O_5 + 3H_2O \longrightarrow 2H_3PO_4$$

The usual laboratory process is boiling the mixture of red lead or violet phosphorus and nitric acid (in the ratio 1:30) in a large size flask. Here iodine acts as a catalyst.

Iodine used as a catalyst in the preparation of phosphoric acid

(*ii*) Rushing brown fumes of nitrogen dioxide are given off and phosphorus is oxidized to phosphoric acid.

Phosphorus + Nitric Acid ⟶ Phosphoric Acid + Water

$$P \quad 5HNO_3 \quad H_3PO_4 + H_2O$$

When the reaction is over the liquid is evaporated until no more acid fumes are evolved and temperature has a desiccator, the phosphoric acid separates out as colourless crystals.

(*iii*) Phosphoric acid can also be prepared by digesting calcium phosphate with a calculated quantity of dilute sulphuric acid for several hours and filtering off the insoluble calcium sulphate.

Calcium Phosphate + Sulphuric Acid ⟶ Calcium Sulphate + Phosphoric Acid

$$Ca(PO_4)_2 + H_2SO_4 \longrightarrow 3CaSO_4 + 2H_3PO_4$$

Physical Properties

Physical state	Crystal
Melting Point	42°C
Boiling Point	130°C
Enthalpy of Fusion	13.48 kJ/mol
Standard Enthalpy of Formation	–1284.4 kJ/mol
Standard Entropy at 22°C	–1124.3 kJ/mol
Heat of Capacity	106.1 J/mol K

Chemical Properties

(*i*) Phosphoric acid is syrupy colourless liquid, which on cooling solidifies into crystals.

(*ii*) Phosphoric acid is tribasic, fairly strong in the first dissociation constant, weak in the second and exceedingly weak in the third.

$$H_3PO_4 \rightleftharpoons H^+ + H_2PO_4)^-$$
$$(H_2PO_4)^- \rightleftharpoons H^+ + (HPO_4)^-$$
$$(HPO_4)^- \rightleftharpoons H^+ + (PH_4)$$

(*iii*) Phosphoric acid possesses no oxidizing properties.

(*iv*) The phosphoric acid produces compounds called *phosphates* or *orthophosphates*. These are divided into three groups:

(*a*) Primary, Sodium Dihydrogen Phosphate (NaH_2PO_4).

(*b*) Secondary, Disodium Hydrogen Phosphate (Na_2HPO_4).

(*c*) Tertiary, Disodium Phosphate, Trisodium Phosphate (Na_3PO_4).

Sodium Dihydrogen Phosphate (NaH_2PO_4): This compound is prepared by the addition of sodium hydroxide solution to phosphoric acid to which a few hydroxides is continued till the solution turns pink. The phenolphthalein changes colour from colourless to pink in the presence of sodium dihydrogen phosphate. The solution is then evaporated to crystallization. Of this compound is slightly acidic in nature because during dissociation it gives out H^+ ions.

Disodium Hydrogen Phosphate (Na_2HPO_4): The secondary phosphate is prepared by dissolving the disodium salt unlike a precipitate does not form, when a drop of barium chloride is added to the solution. On evaporating the resulting mixture white crystals of sodium dihydrogen phosphate separates out. If the disodium hydrogen phosphate is dissolved in water the aqueous solution behaves like a strong alkali because they undergo hydrolysis in water.

Disodium Hydrogen Phosphate +Water ⟶Sodium Orthophosphate+Sodium Hydroxide

$$Na_2HPO_4 + H_2O \longrightarrow NaH_2PO_4 + NaOH$$

Trisodium Phosphate (Na_3PO_4): This tertiary compound of phosphoric acid is prepared by dissolving disodium salt in excess of sodium hydroxide, in tertiary phosphate, ($Na_3PO_4.2H_2O$).

All the primary phosphates are soluble in water, among the secondary and the tertiary phosphates only the sodium and the potassium phosphate are soluble in water.

Graham's Salt: Graham's salt was formerly believed to be the sodium hexametaphosphate, but now it is proved that this is a complex polymer. Graham's salt is prepared by the process of heating of sodium dihydrogen phosphate, until a molten mass is formed. This on cooling yields a glossy material that is Graham's salt. The total reaction equation is shown below:

Sodium Dihydrogen Phosphate $\xrightarrow{230°C}$ Sodium Cyclotriphosphate $\xrightarrow{630°C}$ Graham's Salt

$$NaH_2PO_4 \longrightarrow Na_3(P_2O_2) \longrightarrow (NaPO_3)_n$$

Graham's salt is transparent glossy solid. It is hydroscopic and readily dissolves in water, without limit. Graham's salt is an excellent water softener. It actually removes calcium Cu^{++} and magnesium $Mg+$ ions. That is why it called Calgon (Calcium Gone).

Industrial Preparation of Orthophosphoric Acid by Dorr Strong-Acid Process

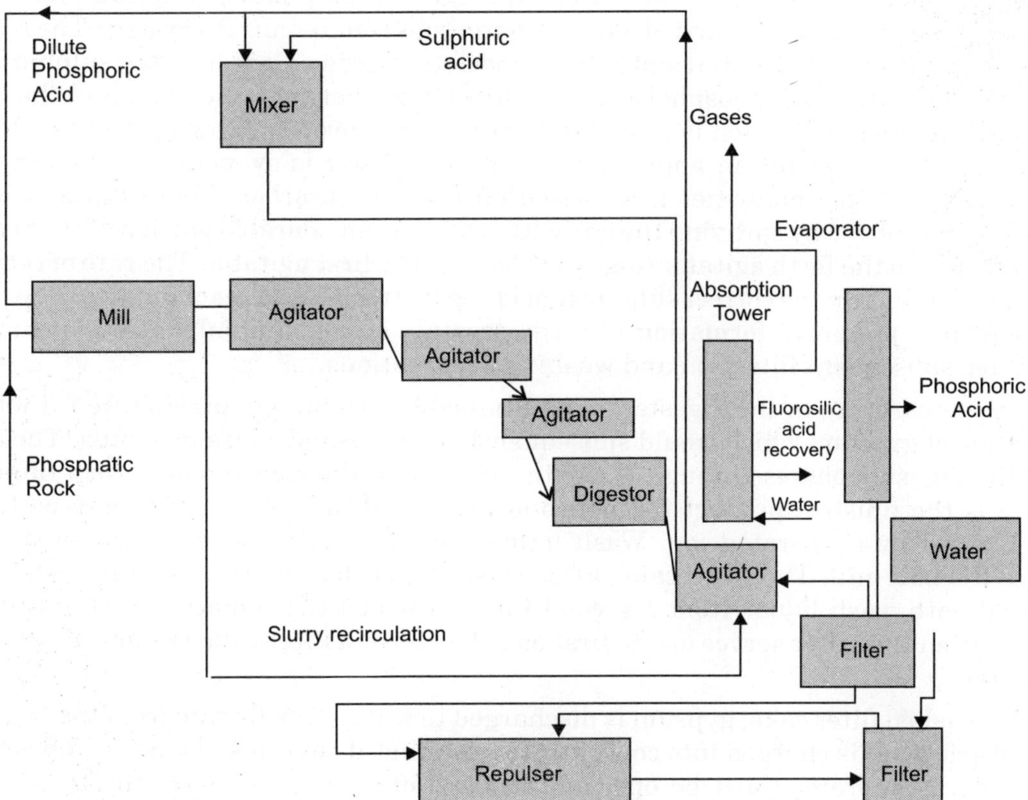

Fig. 13.16. The Industrial Preparation of Phosphoric Acid

Reaction Equation

Phosphate rock + Sulphuric Acid + Water ⟶ Orthophosphoric Acid + Calcium Sulphate

$$Ca_3(PO_4)_2 + 3H_2SO_4 + 6H_2O \longrightarrow 2H_3PO_4 + 3(CaSO_4, 2H_2O)$$

Materials Required

Phosphatic rock	2215.798874 kg
Sulphuric acid	1834.781258 kg

Process: The sulphuric acid produces phosphoric acid when it reacts with phosphate rock, where phosphoric acid is produced with the precipitation of calcium sulphate. Phosphate rock is charged in a special ball mil, where it is ground in a dilute solution of phosphoric acid obtained as a filtrate from a by product of another reaction. The slurry passes through a series of Dorr agitators, where reaction with sulphuric acid takes place.

Separation of the resulting acid from the calcium sulphate precipitate, as well as washing of the precipitate is carried out in a continuous countercurrent system. The reaction system consists of five essential agitators connected in series. Concentrated sulphuric acid (94%) diluted with weak phosphoric acid, is introduced between the third and the fourth agitator. The heat of reaction is considerable and is removed by blowing cold air through the solutions. As a result an appreciable current of water is evaporated. The air, water vapour and gaseous impurities are carried off into an absorber, where flurosilicic acid (H2SiF6) is removed by spraying the gas with water. A considerable portion of the reaction is recycled from the forth agitator (digester) back to the first agitator. The rate of recycle is adjusted to maintain the correct sulphuric acid concentration and reaction time. The proper reaction time promotes formation of large crystals of calcium sulphate (gypsum) and facilitates subsequent filtration and washing preparations.

The temperature in the digester is maintained low enough to precipitate rather than anhydride of gypsum, which would subsequently hydrate and cause pipe plug. The slurry from the digester passes through the fifth agitator to the continuous rotary filter. The filtrate is the finished product the phosphoric acid (about 45%), which may be further concentrated by evaporated ion. Wash liquor used to dilute the sulphuric acid and to charge the ball mill. The filter cake, after washing, is charged into the tank where it is replaced with wash liquor from a second filter in which the replaced slurry is filtered. Filtrate from this filter serves as the first one. The second filter cake is washed with fresh hot water.

The washed filter cake, gypsum is discharged to waste. The filtrate from the first filter (phosphoric acid) is charged into concentrate such that it contains about 35% phosphorus pentoxide. Evaporator must be operated below 150°C because above the temperature orthophosphoric acid is converted slowly into pyrophosphoric acid. By means of the process described, about 98% of the phosphorus pentoxide content of the rock is dissolved and of this 97 to 98% is recovered as the product.

Uses

(*i*) It is used for the preparation of phosphates, which are used for the manufacture of fertilizers.

(*ii*) Used as laboratory reagent.

Pyrophosphoric Acid ($H_4P_2O_7$): This acid is prepared by heating orthophosphoric acid at 255°C.

$$\text{Phosphoric Acid} \longrightarrow \text{Pyrophosphoric Acid} + \text{Water}$$
$$2H_3PO_4 \longrightarrow H_4P_2O_7 + H_2O$$

Properties

(*i*) It is a colourless crystalline solid with a melting point of 60.5°C.

(*ii*) It is recovered into orthophosphoric acid on boiling with water.

(*iii*) It is a tribasic acid, however only two series of salts e.g. sodium pyrophosphate. ($Na_4P_2O_7$) and Sodium Hydro pyrophosphate, ($Na_2H_2P_2O_7$) are readily available.

(*iv*) When strongly heated, it yields metaphosphoric acid.

$$\text{Pyrophosphoric Acid} \longrightarrow \text{Metaphosphoric Acid} + \text{Water}$$
$$H_4P_2O_7 \longrightarrow 2HPO_3 + H_2O$$

Structure: Pyrophosphoric acid has been arranged according to the following structure:

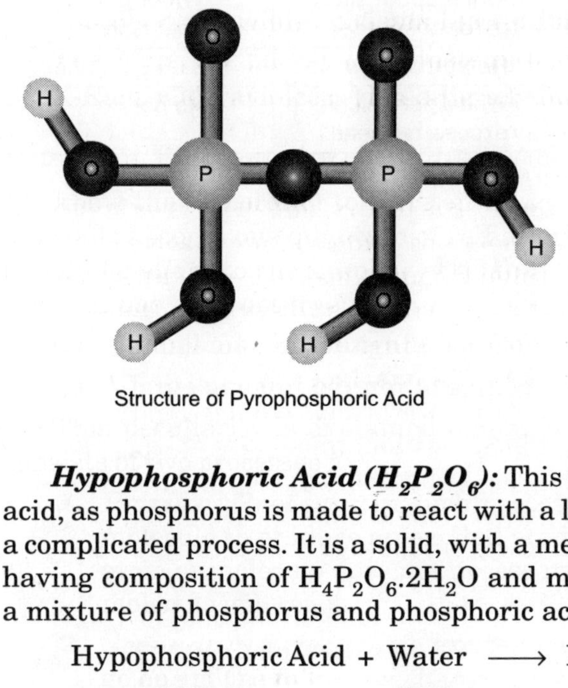

Structure of Pyrophosphoric Acid

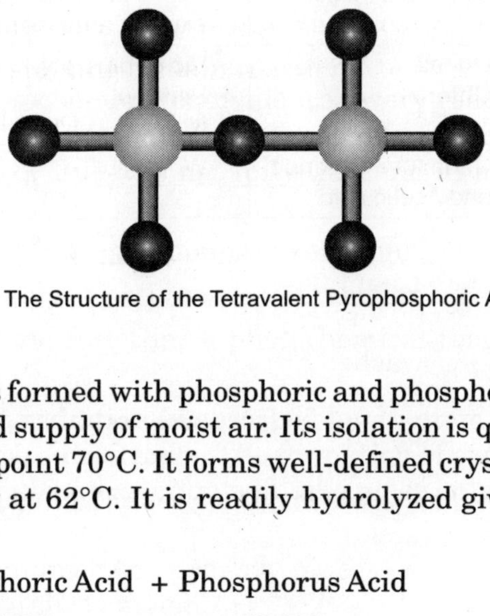

The Structure of the Tetravalent Pyrophosphoric Acid

Hypophosphoric Acid ($H_2P_2O_6$): This acid is formed with phosphoric and phosphorus acid, as phosphorus is made to react with a limited supply of moist air. Its isolation is quite a complicated process. It is a solid, with a melting point 70°C. It forms well-defined crystals having composition of $H_4P_2O_6.2H_2O$ and melting at 62°C. It is readily hydrolyzed giving a mixture of phosphorus and phosphoric acid.

$$\text{Hypophosphoric Acid} + \text{Water} \longrightarrow \text{Phosphoric Acid} + \text{Phosphorus Acid}$$
$$H_4P_2O_6 \quad H_2O \longrightarrow H_3PO_4 \quad H_3PO_3$$

Structure

Hypophosphoric acid is a tetrabasic acid, the defined structured is shown below:

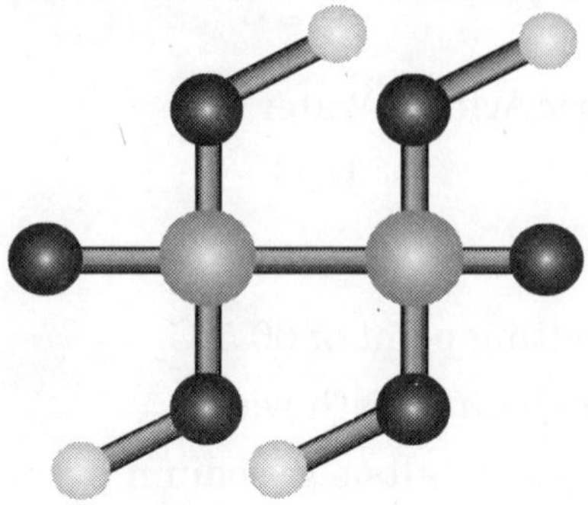

Structure of Hypophosphoric Acid

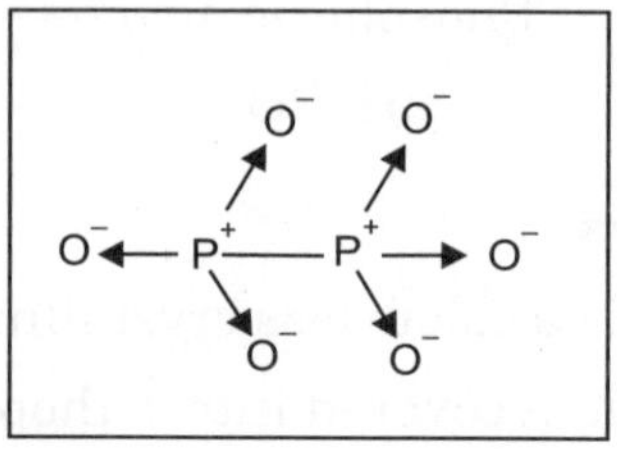

Hypophosphite Ion

Detection of Phosphoric Pyrophosphoric and Met Aphosphoric Acid

Reagents	Orthophosphoric Acid or Orthophosphate	Pyrophosphoric Acid or Prophosphate	Metaphosphoric Acid or Metaphosphate
1. Silver Nitrate	Yellow ppt of silver orthophosphate	White ppt of silver pyrophosphate	White gelatinous ppt of silver
2. Barium Chloride	No ppt	No ppt	Metaphosphate
3. Magnesia Mixture	White ppt insoluble in excess reagent	White ppt soluble in excess reagent	No ppt
4. Cobalt Chloride	Violet blue ppt Pink ppt No ppt		
5. Cadmium Chloride and Acetic acid	White ppt	No ppt	